D1617497

COMPARATIVE MORPHOLOGY OF THE MAMMALIAN OVARY

Comparative morphology of the mammalian ovary

Harland W. Mossman & Kenneth L. Duke

THE UNIVERSITY OF WISCONSIN PRESS

Published 1973
The University of Wisconsin Press
Box 1379, Madison, Wisconsin 53701

The University of Wisconsin Press, Ltd.
70 Great Russell Street, London

First printing

Printed in the United States of America

For LC CIP information see the colophon
ISBN 0-299-05930-8

Publication of this book was supported in part
by NIH grant 5 R01 LM 00348
from the National Library of Medicine

TO

R. J. M.

L. B. D.

Contents

Illustrations

Tables

Preface

The writers of scientific books should be allowed the indulgence of a long preface; it is the one portion they do not need to document, and documentation can be boring even to scientists. They should also be permitted the rather trite ritual of justifying the writing of the book, for almost inevitably they have not only spent many years of their lives in amassing the material for its preparation, but they have also spent large sums of other people's money in research funds, "leave of absence" support, and so on, for which they should feel deep gratitude and also anxious responsibility to return value received.

It may seem that there is no good reason to devote a whole book, even a small one, to such a small part of the body as the ovary. Yet it, like its counterpart in the male, wields an influence in the world out of all proportion to its insignificant size and lowly location. Although the sex glands or gonads are not vital to an individual in the way that the brain and heart are, they certainly give a zestful meaning to life for which the brain or the heart often gets credit, but which neither could provide without them.

The influence and control of the gonads on the individual begins in some still unknown way very early in embryonic development when the sway of the hereditary endowment of sexuality handed down in the germ cells from the parental gonads is gradually concentrated in and taken over by the gonads of the embryo. This power of the gonads over the individual is then continued throughout life. It is especially manifest in four great periods: (1) sex differentiation during embryonic life,

(2) puberty, (3) sexual maturity, and (4) involution (when the gonads demonstrate their former power by withdrawing their support!).

No doubt any morphologist or physiologist who has studied an organ long and diligently has enough information at his command, which, mixed with a little philosophy, would produce a book-sized article suitably dramatic for the supposed tastes of the layman. The gonads would be fit subjects for this sort of thing. One could play up their influence over body and mind of beast and man. One could find much to discuss in order to explain the good physical vigor often seen in castrated animals and the remarkable adjustments as well as physical and mental vigor which have been or are being displayed by some human eunuchs, eunuchoids, castrates, and postclimacterics both male and female. Such a book would bore most people of a scientific turn of mind who wish to learn something concrete about the subject.

This book will bore most people who are not of a scientific turn of mind, for it is designed for the use of scientists. Whether it bores them or not depends on their field of interest, and on the use they are able to make of it. It is our hope that it will serve well a considerable group of teachers and researchers in the general fields of zoology, normal and morbid morphology, embryology, physiology, and endocrinology, and that it will also be of interest to practitioners of medicine, particularly those involved in obstetrics and gynecology.

The book has been written for two major purposes: first, to present in an organized form the mass of basic information that has already been published on the morphology of the ovary of eutherian (placental) mammals; second, to present significant portions of new information, mostly unpublished, which the authors have acquired over a period of about forty years of study of the ovary from the comparative standpoint. It is hoped that the book will serve to reduce the misunderstanding and misinformation regarding the mammalian ovary which one encounters on every hand. Textbooks of zoology, histology, embryology, obstetrics, and gynecology are full of these errors. Consequently one seldom listens to a discussion of the ovary among medical men or medical students, or even among morphologists, physiologists, or endocrinologists in which it does not almost immediately become obvious that there is great confusion because of lack of familiarity with known facts and even because of misunderstanding of the meaning of terms. If you want a concrete illustration of this, all you need to do is to start a discussion with one or two colleagues on such a topic as the question of germ-cell origin, the nature of "interstitial tissue," or just what is meant by "luteinization." You will even find a surprising amount of variation in your colleagues' concepts of the nature, location, and origin of such easily demonstrated entities as corpora hemorrhagica, rete, and follicular epithelium. However, you may be pleased with the unanimity of opinion on such things as the idea that all corpora lutea arise from ovulated follicles, or that all corpora albicantia represent degenerated corpora lutea, but you should not be, for these are highly erroneous.

To make this book as convenient and useful as possible, the first three chapters are designed for general orientation. They attempt to give a clear general account of the gross anatomy, histology, and embryology of the ovary of eutherian mammals. In order to accomplish this the discussion has been kept on a relatively elementary and uncritical level, although far more detail is presented than is ordinarily found in textbook accounts. Inevitably this has led to some repetition, especially of material in later chapters, but perhaps because of our long experience as teachers we believe that some repetition is justifiable.

The fourth chapter is designed to give as complete an exposition as practicable of all the crucial stages through which a typical mammalian ovary passes during the life cycle in a single species. The North American red squirrel (*Tamiasciurus hudsonicus*) was chosen for this because its ovary is more nearly "typical" than that of most other species of mammal with which we are familiar. By typical we mean that it shows clearly and almost diagrammatically, but seldom in an exaggerated or extreme manner, most of the commonly encountered structures and changes characteristic of mammalian ovaries.

The fifth chapter is directed toward those primarily interested in the human ovary. Its arrangement is similar to the description of the "typical" ovary. It is hoped that this will facilitate correlation of knowledge of the human ovary with the comparative material.

From the comparative morphologist's standpoint the most important chapter is the sixth. It is in part a digest and synthesis of the information presented in systematic order in the tables of the appendix. It aims to show the known range of variation in ovarian morphology in the subclass Eutheria and to correlate this information with various other fields, particularly with physiology and phylogeny.

Chapter 7 reviews some of the more recent experimental and descriptive work bearing on such controversial subjects as germ-cell origin, regeneration, cyclic structural changes within the ovary, ovarian tumors, and so forth. It is hoped that this will make apparent a number of promising lines of attack for further investigation of ovarian problems.

The last chapter is an overall look at the problem of comparative morphology of the mammalian ovary. In it we try to point out the inadequacy not only of present knowledge of the subject but also of past efforts to study it. However, from the material at hand we draw certain general conclusions and make certain suppositions and hypotheses. We also suggest methods and directions for further investigations in this field.

An appendix consists of synoptic tables and explanatory notes of comparative data available for the various orders and families of mammals, arranged systematically. It is hoped that careful reading of Chapters 4 and 6 will furnish enough illustrative material to make these data intelligible.

A glossary seemed necessary, not only for the convenience of the younger or more inexperienced reader, but also to help clear up the unfortunate confusion that exists in the field in the use of terms.

In the text, references to the literature are given by author and date, so that the reader need not turn to the bibliography to identify a reference. Full titles are provided in the single alphabetical list at the end of the book. Except for certain classical literature, or publications from which specific information has been obtained, only the most pertinent recent references are cited, since earlier work can always be located from these.

The illustrations are of two very different types, serving two very different purposes. Most of them are halftone reproductions of photomicrographs to show normal gross and microscopic anatomy and to document statements made. A few relatively simple halftone and line sketches or diagrams attempt to reduce to the simplest terms the processes or morphology illustrated and to summarize comparative materials so that its range can be grasped at a glance. The illustrations must be regarded as representative of present knowledge, not as of the total range of variation that may actually exist.

Certain omissions have been intentional. Most of the literature on ovulation and on general ovarian endocrinology and histochemistry has been and will be treated by authors much more competent to do so than we are. Ovarian pathology has been mentioned chiefly in respect to a few generalizations regarding tumors. The cytology and cytogenesis of the ova and their manner of cytolysis in atretic follicles have also been omitted as being a specialized subject having little bearing on the general ovarian morphology and development which we wish to emphasize. For much the same reason, such subjects as blood and nerve supply and lymph drainage are treated rather superficially. We have made use of the relatively meager ultrastructural literature pertaining to the origin of the zona pellucida and the glandular nature of certain cell types, but we have not included that concerned principally with such matters as the detailed cytology of ova, primarily because too little has been done to have much meaning from the comparative standpoint. Unfortunately histochemical or electron-microscopic techniques have been used to study very few of the problems of a comparative nature with which we are concerned. These methods could conceivably facilitate the accurate characterization of the various interstitial gland cell types, or the clear demonstration of whether or not oogonia can arise from cells other than primordial germ cells. Consideration of the ovaries of monotremes and marsupials is largely confined to the synoptic data, since we have little first-hand knowledge of these, and because the biology of reproduction of these two groups is so different from that of eutherians.

We pooled our resources originally in order to broaden the coverage. Mossman concentrated on material from the northern and western states, East Africa, and southern Africa; Duke collected and studied material primarily from the southeastern and mountain states, and from southeastern Asia. The senior author must take responsibility for much of the format, contents, and opinions expressed, although we have found that we are in surprisingly close agreement in most things presented. So far as possible, we have checked with each other and have coordinated our writing throughout.

Harland W. Mossman
Kenneth L. Duke

December 1971

Acknowledgments

Our investigations of the ovary have extended over so many years that it is impossible to recall all who have aided us. Much gross material was given to us and much was purchased at a nominal price from museum personnel, students, and others who were making field collections of mammals. At various times, material or help was provided by the Vilas Park Zoo, Madison, Wisconsin, and by the Wisconsin Department of Natural Resources. We owe special thanks to Bernard Bradle, game manager, Crandon, Wisconsin, for his enthusiastic cooperation. We are also indebted to Abdul Rahman, collector, Kuala Lumpur, Malaysia.

We are grateful for the use of the following anatomical collections: T. G. Lee Collection, Department of Anatomy, University of Minnesota; J. P. Hill Collection, Department of Zoology, University College, London; A. A. W. Hubrecht Collection, Hubrecht Laboratory, Utrecht; C. J. van der Horst Collection, Department of Zoology, University of Witwatersrand, Johannesburg. Material was also provided by the Chicago Natural History Museum and in large amounts by the National Museums of Rhodesia through the courtesy of Reay H. N. Smithers, director.

We are indebted to the National Museum of Natural History, Washington, D.C., especially to Don E. Wilson, for advice on certain problems concerning taxonomic nomenclature.

Many individuals provided us with considerable help. Laboratory specialists were Mrs. Aud Bockman-Pedersen, Mrs. Phyllis Hall, Mrs. Fred R. Jones, Grayson Scott, Mrs. A. F. Zevnik, and William K. Zung.

We take pleasure in acknowledging assistance from the following biological scientists: D. Allred, B. I. Balinsky, F. W. R. Brambell, Robert P. Breitenbach, James W. Brooks, Helmut K. Buechner, C. D. Bunker, Clinton H. Conaway, Walter W. Dalquest, W. Robert Eadie, Allen C. Enders, Robert K. Enders, Theodore V. Fischer, H. S. Gentry, Oliver J. Ginther, Alfred Gropp, W. J. Hamilton (London), Frances Hamerstrom, Frederick Hamerstrom, Carl G. Hartman, C. Lynn Hayward, Frederick L. Hisaw, Roger A. Hoffman, Lloyd B. Keith, Karl W. Kenyon, C. M. Kirkpatrick, Marilyn J. Koering, David A. Langebartel, Winter P. Luckett, John A. Morrison, Archie S. Mossman, Margaret Ward Orsini, Noel Owers, Ben Pansky, Douglas H. Pimlott, M. R. N. Prasad, Richard Shackelford, Akhouri A. Sinha, Fritz Strauss, Enrique Valdivia, Barbara J. Weir, W. I. Welker, William A. Wimsatt, Clinton N. Woolsey, and Philip L. Wright.

Grants from the following institutions have made this work possible: China Medical Board of New York; Duke University Council on Research; Wisconsin Alumni Research Foundation; National Institutes of Health; and the Department of Health, Education, and Welfare.

A note on nomenclature

Anatomical terminology. With few exceptions, we have followed the *Nomina Anatomica* (*NA*), and *Nomina Embryologica* (*NE*). *Nomina Anatomica Veterinaria* (*NAV*) terms have been used when more suitable. However, inconsistencies, errors, and inadequacies occur unavoidably in all of these nomina, so it is proper for specialists in any field to use terminology which in their judgment is the most appropriate. Should this differ from the nomina, it is hoped that it would eventually be adopted by the committees responsible for their revision. After all, these nomina are intended as guides not as mandates.

We have coined several new vernacular terms. In most cases we have thought it neither necessary nor advisable to dignify these with Latin equivalents. In the Glossary we have, where it seemed appropriate, set forth our reasons for the use of these terms. Technical terms not in *NA*, *NE*, or *NAV* are italicized at their first occurrence in the book. In the index these terms are followed by the equivalent terms from the nomina.

English technical terms are used unless their Latin equivalents are in accepted English usage, such as ovum and uterus. In each chapter, at the first occurrence of an English technical term that might not be commonly understood, its Latin equivalent follows in parentheses. If there is an alternative vernacular term, the preferred one precedes, and the alternative is given only once.

The alternative terms occur or have occurred fairly often in the literature. It is not our ambition to relegate all of them to oblivion, although we admit that we think many of them should be; nor do we intend to set up our preference as the only and all-enduring one. Many a properly fertilized egg eventually aborts owing to extrinsic factors neither considered nor under control at the time of conception! We have tried to take into account such matters as previous usage, brevity, freedom from confusion with names of different meaning because of similarity of sound or spelling, descriptive appropriateness, and ease of transposition into other forms of speech.

Mammalian nomenclature. For the classification of major groups we have followed G. G. Simpson, *The Principles of Classification and a Classification of Mammals*, Amer. Mus. Natur. Hist. Bull. 85 (New York, 1945). For the scientific and vernacular names of North American genera and species we have used the nomenclature of E. Raymond Hall and Keith R. Kelson, *The Mammals of North America*, 2 vols. (Ronald Press, New York, 1959). For all others, so far as possible, we have followed Ernest P. Walker, *Mammals of the World*, 3 vols. (Johns Hopkins Press, Baltimore, 1964). Where the scope of Walker's book has precluded the mention of every species within certain genera, we have kept the nomenclature of the various authors being cited.

English taxonomic terms are followed by scientific names in parentheses. The vernacular name is used thereafter, except in those cases where only the scientific name is apt to be more significant or more easily recognizable. The scientific names of common domestic or laboratory animals are omitted in the text, but are given in the figure legends and in the synoptic tables. The reader must also understand that it is often necessary to speak of some group, such as canids, as having certain characteristics, even though not all members of the group have been investigated. If known exceptions occur, they are usually mentioned or reference is made to the synoptic tables where the exception is recorded.

COMPARATIVE MORPHOLOGY OF THE MAMMALIAN OVARY

One

Gross anatomy of the mammalian ovary

The location and gross appearance of the ovary are important not only from the purely comparative standpoint, but also in a practical way if the ovary is to be involved in any surgical or experimental procedures. Is it in an accessible position? Is it hidden within a membranous bursa? How is the infundibulum of the oviduct related to it? Are its attachments loose enough to allow easy manipulation? Does it normally change in appearance with age or reproductive condition? What is the nature and location of its blood and lymph vessels and of its nerves? Are the ripening follicles and corpora lutea easily recognizable in the living animal? This chapter, together with the synoptic tables at the end of the book, aims to answer some of these practical questions about specific groups, as well as to provide comparative data which may have other scientific value.

Location

The location of the mammalian ovary varies considerably among orders. Usually it is in the more caudal portion of the lumbar region, but not in the sacral or true pelvic region as it is in man. This immediately suggests that its cephalocaudal position is correlated either with posture or with the type of uterus, or both. In animals with long, essentially straight uterine horns the ovarian position is relatively far cephalad — i.e., opposite the caudal portion of the kidney as it is in the guinea pig, or in the midlumbar region as in squirrels (Sciuridae), pocket gophers (Geomyidae), the mink (*Mustela vison*), and the dog. Zietschmann,

Ackerknecht, and Grau (1943) showed that the ovaries of the dog and the horse lie opposite the third and fourth lumbar vertebrae, respectively. From our observations most insectivores, lagomorphs, rodents, and carnivores also belong in this category. The most caudal position is found in groups having a simplex uterus or one with very short horns. The best examples of these groups are the primates (excluding the lemuroids, which usually have short, straight bicornuate uteri), certain chiropterans, and the edentates. An intermediate ovarian position occurs in the remaining groups, those having bicornuate uteri with relatively short or coiled horns. In many of these, as in the bovids, the uterine horns are actually relatively long but are helically coiled except in late pregnancy, so that even in nonpregnant females their tubal ends, and therefore the oviducts and ovaries, are often more caudal than much of the main portion of each cornu. This condition is characteristic of the ruminating artiodactyls. Zietschmann, Ackerknecht, and Grau (1943) demonstrated that the ovaries of the domestic cow and sow are opposite the sixth, or last, lumbar vertebra. Although the cornua of the sow are long, they are profusely kinked and the overall shape is a ventrolateral coil much as in the bovids. The cornua of the mare are short and somewhat laterally curved, but are not coiled, and thus the ovaries are relatively cephalic compared with those of artiodactyls. The position of the ovary, therefore, seems to be most directly correlated with the type of uterus. The uterine type may itself be correlated with such factors as posture and method of locomotion, but we know of no attempt to examine this problem.

The position of the mammalian ovary is much less variable among orders than is that of the testis. Even the tremendous changes in uterine size due to pregnancy cycles do not greatly alter, even temporarily, the position of the ovaries (Hibbard, 1961). Although the ovarian ligamentous structures are often relatively long and loose, they still anchor the ovary well and undergo relatively little growth during pregnancy. This is the more surprising because the similar and contiguous ligaments of the uterus are altered greatly by the growth accompanying pregnancy and again by postpartum involution.

Attachment and peritoneal relationships

The ovary is a peritoneal organ. It is attached to the broad ligament (ligamentum latum) or directly to the abdominal or pelvic wall by a thin fold of peritoneum — the ovarian mesentery, or mesovarium (Figs.

1.1 and 1.2). The cephalic portion of the broad ligament is the suspensory ligament of the ovary (ligamentum suspensorium ovarii).

The oviduct and uterus are likewise peritoneal organs attached by a continuous mesentery, the broad ligament, of which the more caudal portion attached to the uterus is the mesometrium and the more cephalic portion attached to the oviduct is the mesosalpinx. A thickened fibromuscular fold in the caudolateral surface of the broad ligament runs from the tubo-uterine junction laterally and caudally to the deep inguinal ring (anulus inguinalis profundus). This cordlike thickening marking the boundary between the mesometrium and the mesosalpinx is the round ligament of the uterus (ligamentum teres uteri). A similar fibromuscular fold extends from the cephalomedial side of the tubo-uterine junction cephalad along the medial surface of the broad ligament to the caudal pole of the ovary. It forms the ventrocaudal free edge of the mesovarium and is called the proper, or round, ligament of the ovary (lig. ovarii proprium). These two round ligaments together, that of the ovary and of the uterus, develop from the gubernacular fold (plica gubernacularis) of the embryonic gonad, and together they correspond to the gubernaculum testis.

Each fetal gubernaculum thus extends as a peritoneal fold from the caudal pole of the gonad almost directly caudad to the deep inguinal ring on the anterior abdominal wall just cephalic and lateral to the pubic symphysis. In so doing, each crosses the female duct (ductus femininus), or Müllerian duct (d. paramesonephricus), where this duct turns medialward just caudal to the mesonephros. This point of crossing marks the future tubo-uterine junction. That part of the female duct cephalic to it becomes the oviduct, and that part which is caudal becomes the uterus. This is true whether the uterus is duplex, bipartite, bicornuate, or simplex. Thus, the proper ligament of the ovary and the round ligament of the uterus at first always attach at opposite sides of the tubo-uterine junction as a result of their development from the gubernaculum and of the latter's relation to the female duct. In some mammals, notably the artiodactyls, the uterine end of the round ligament of the uterus eventually becomes indistinct and disappears at about the center of the mesometrium. Likewise in artiodactyls, the proper ligament of the ovary may disappear in the mesosalpinx before it reaches the tubo-uterine junction (see Fig. 1.21*a*).

In adult mammals the mesovarium is usually a relatively inconspicuous offshoot of the medial surface of the cephalic portion of the mesosalpinx. The mesosalpinx is much more obvious, partly because it is a direct continuation of the mesometrium, and partly because of its attachment to the relatively long and conspicuous oviduct. Actually, however, the mesovarium is developmentally the older structure, having arisen as the mesentery of the gonad while the female ducts were still in a

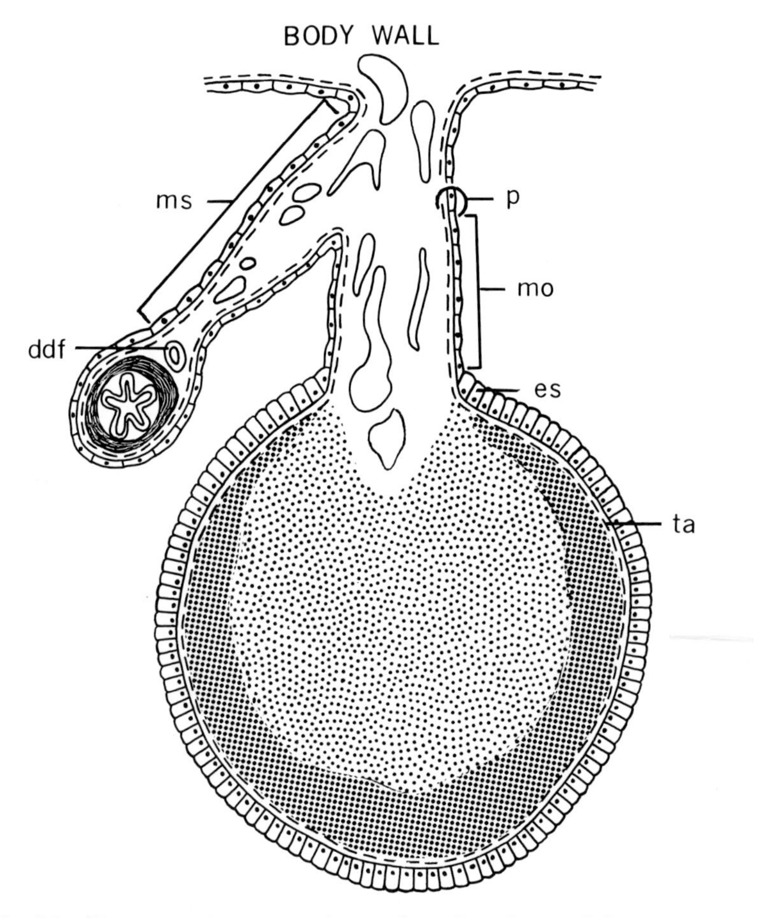

Fig. 1.1 Diagrammatic cross-section to show the relation of the mesenteries of the ovary and oviduct to one another, of the peritoneum to the surface epithelium and tunica albuginea of the ovary, and of the medulla and cortex. *ddf*, ductus deferens femininus; *es*, epithelium superficiale; *mo*, mesovarium; *ms*, mesosalpinx; *p*, peritoneum; *ta*, tunica albuginea; *close stipple*, cortex; *open stipple*, medulla.

completely retroperitoneal position. This more fundamental nature of the mesovarium perhaps accounts for the fact that the ovarian vessels reach the ovary through it and only secondarily send branches from its base into the mesosalpinx and thence to the cephalic portion of the oviduct where they anastomose with the tubal branches of the uterine vessels.

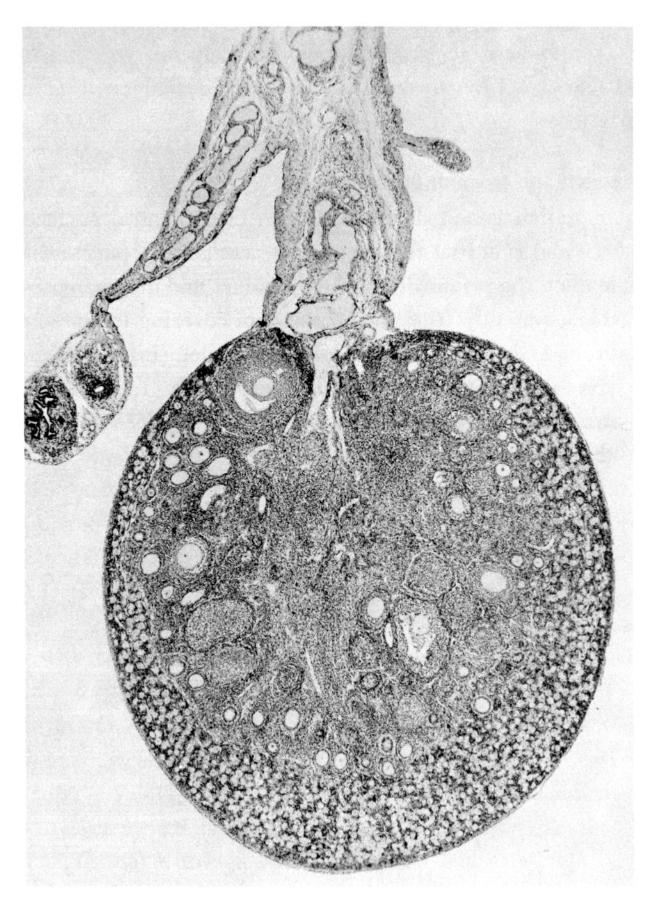

Fig. 1.2 Histological section, comparable with Figure 1.1., of the oviduct and ovary of a rodent, a juvenile sewellel (*Aplodontia rufa*). Note the distinct boundary between the medulla with its primary, secondary, and early vesicular follicles and the still-infantile cortex with its "naked ova" and primordial follicles. × 14.

Ovarian bursae and oviduct patterns

The ovaries of many mammals lie in membranous pouches which partly or even completely isolate them from the general peritoneal cavity. These ovarian pouches (bursae ovaricae) are primarily modifications of the mesosalpinx, hence the oviducts course in their walls. If the bursa is "complete," the internal opening of the oviduct (ostium abdominale tubae uterinae) is within it. Since the definitive anatomy of this area differs so much between groups and often appears so complex, it is worthwhile at this time to consider its basic development (Figs. 1.3–1.12).

Development and types of bursae

The mammalian gonad develops on the medioventral surface of the mesonephros and is at first fully exposed to the general peritoneal cavity. The female duct, the primordium of the oviduct and uterus, arises lateral to the ovary, apparently from the peritoneum covering the mesonephros. Its cephalic end, the future funnel of the oviduct (infundibulum tubae uterinae), is somewhat cephalic to the ovary (Fig. 1.3). As the mesonephros atrophies, its peritoneum becomes part of the mesenteries suspending the ovary and oviduct from the dorsal body wall — the mesovarium and mesosalpinx, respectively. Thus the ovary eventually hangs by its mesovarium from the base of the cephalic portion of the mesosalpinx, which, like the oviduct, extends a short distance cephalic from the ovary. In a few mammals (e.g., apes, man), this embryonic relationship of the ovary to the mesosalpinx, with the oviduct in its free margin, continues throughout life (Fig. 1.5), but in most others modifications of the mesosalpinx, of its position, and the position of the oviduct result in the formation of a small membranous pouch around the ovary (Figs. 1.7–1.12). This pouch may have a wide opening into the general peritoneal cavity and actually enclose only the cephalic portion of the ovary (see Figs. 1.7, 1.10, 1.15, 1.19, and 1.27–1.29); it may enclose all of the ovary and yet have a large aperture (see Figs. 1.8, 1.9, 1.16–1.18, 1.20, and 1.21); or the aperture may range from this condition to a mere pore (see Figs. 1.11, 1.12, 1.22, and 1.23), or even to no opening at all (see Fig. 1.24).

The morphologically caudal edge of the oviduct funnel often becomes attached directly to the cephalic end of the ovary, and the funnel mouth is turned mediocaudally so that it faces the ovary. If the funnel is "fim-

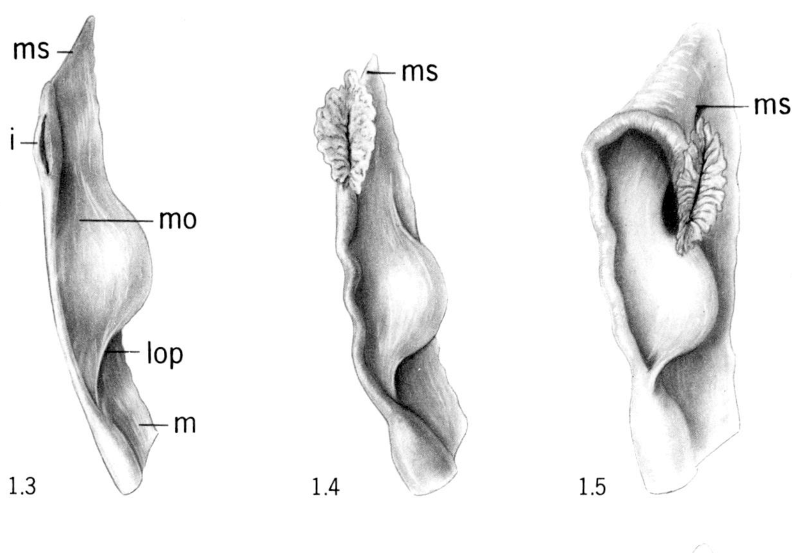

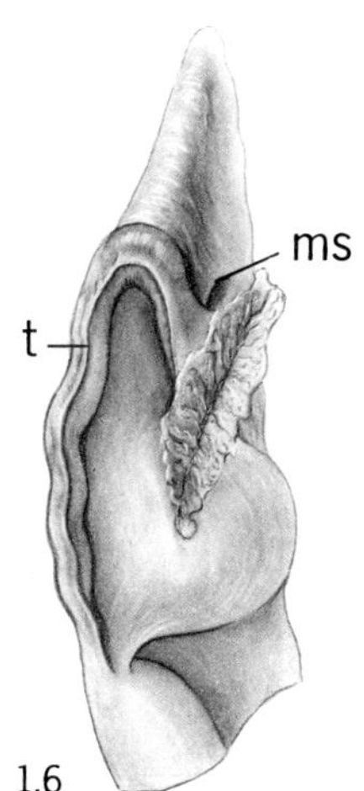

Fig. 1.3 Late fetal condition.

Fig. 1.4 Primitive condition. (Almost entirely confined to infantile and juvenile periods.)

Fig. 1.5 A common condition with no bursa. (Anthropoids and man.)

Fig. 1.6 Rudimentary tubal membrane. (At least one genus of New World monkey, the cacajaos [*Cacajao*] and one anteater, the tamandua [*Tamandua*].)

Figs. 1.3–1.12 Semischematic drawings to illustrate the relations of the ovary, oviduct, and uterus, and of their ligaments and mesenteries to each other, and to show the various degrees of development of the ovarian bursa. All are ventral views of the right side except Figure 1.9, which represents a medial view of Figure 1.8. *ms*, cephalic edge of mesosalpinx; *i*, infundibulum; *m*, mesometrium; *mo*, mesovarium; *lop*, proper ligament of the ovary; *t*, ventral free edge of tubal membrane.

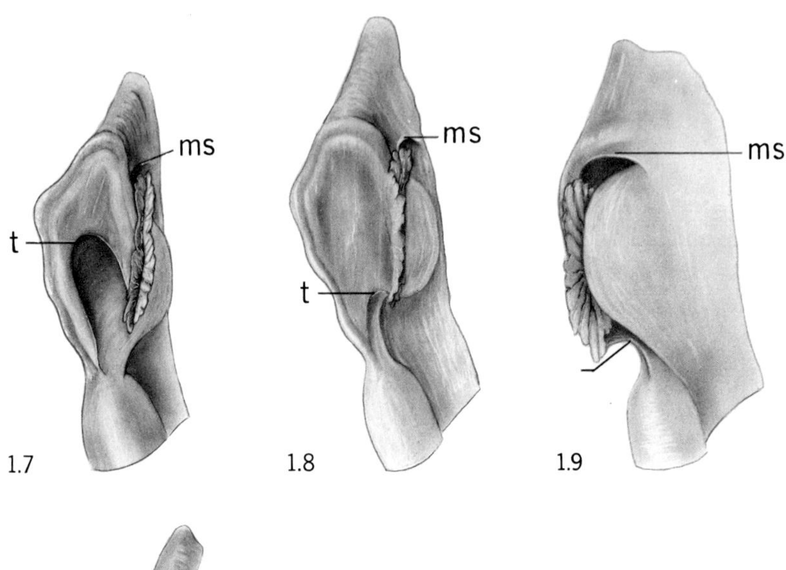

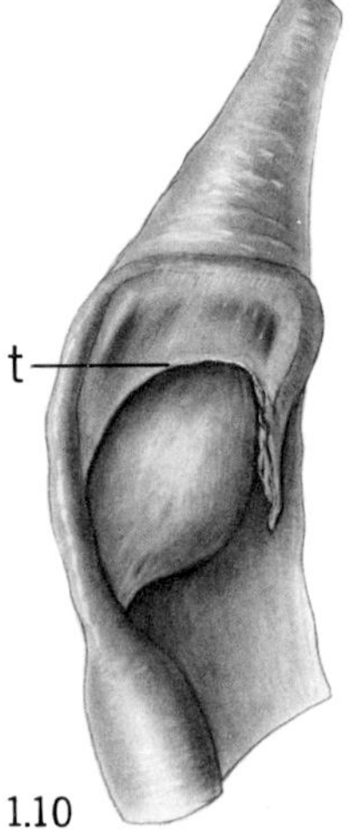

Fig. 1.7 Partial bursa with the oviduct recurving laterally to the ovary. (Aardvark [*Orycteropus afer*] and nine-banded armadillo [*Dasypus novemcinctus*].)

Fig. 1.8 Complete bursa with the oviduct recurving laterally to the ovary, hence a long but often narrow orifice. (Cats [Felidae] and cattle [Bovidae].)

Fig. 1.9 Complete bursa. Medial view of Figure 1.8.

Fig. 1.10 Partial bursa with the oviduct recurving medially to the ovary. (Rabbits and hares [Leporidae] and squirrels [Sciuridae].)

briated," the portion attached to the ovary is known as the ovarian fimbriae (fimbriae ovaricae). Whether directly attached to the ovary or not, a portion of the cephalic end of the oviduct almost always becomes recurved caudad alongside the ovary. Obviously, this recurving of the oviduct with its mesovarium tends to form a peritoneal niche in which the ovary lies (see Fig. 1.5). However, the formation of a definite

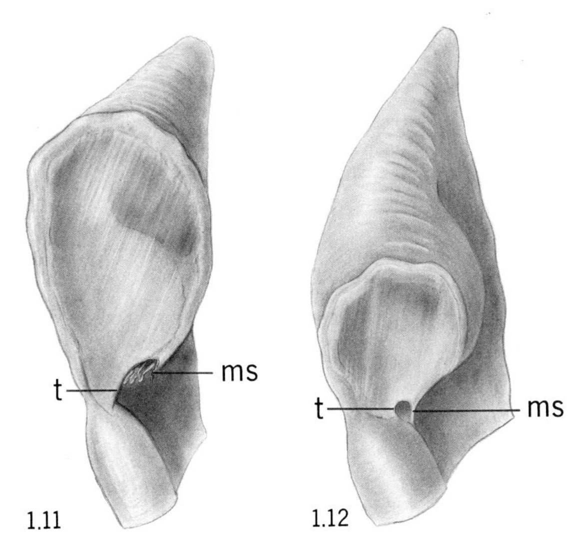

Fig. 1.11 Complete bursa with the oviduct recurving medially to the ovary and with fimbriae extending through the relatively small orifice. (Carnivores, except felids and *Vulpes*; sciurid rodents.)

Fig. 1.12 Complete bursa with the oviduct recurving medially to the ovary and with no fimbriae extending through the porelike orifice. In the muskrat (*Ondatra zibethicus*) and golden hamster (*Mesocricetus auratus*), this orifice is represented by only a small blind pit visible from the peritoneal side, i.e., there is no connection between the ovarian bursa and the general peritoneal cavity. (Myomorpha.)

ovarian bursa depends upon the development of a secondary peritoneal fold along the ventral side of the oviduct. This fold, which we suggest be named the *tubal membrane* (*membrana tubae uterinae*), is analogous to a partial ventral mesosalpinx (see Figs. 1.6–1.12), but probably should not be called that, since there is no known case in higher vertebrates, even developmentally, of a mesentery connecting the oviducts with the ventral body wall. In its simplest form, the tubal membrane is merely a thin narrow antimesosalpingian ridge (Figs. 1.6, 1.13, and 1.14). When somewhat more advanced, it stretches across the hooked portion of the oviduct as a sickle-shaped membrane connecting the morphologically ventral edges of the hooked portion, thus creating a partial ventral floor to the peritoneal cul-de-sac formed by the recurved

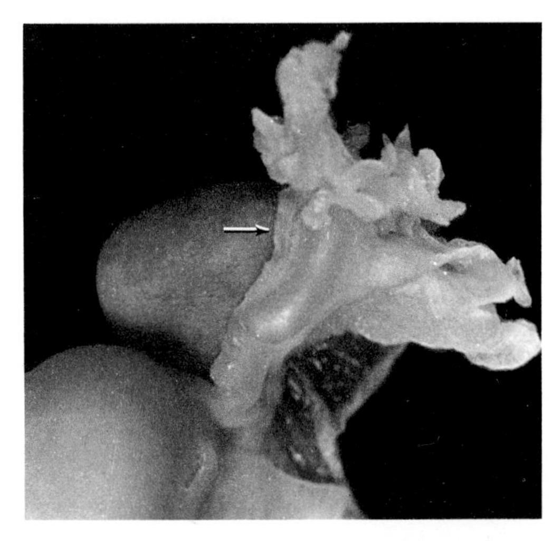

Fig. 1.13 Left oviduct and ovary and the simplex uterus of a genus of New World monkey, the cacajaos (*Cacajao*), showing the rudimentary tubal membrane, a thin low antimesosalpingian peritoneal "fold" (*arrow*) along the oviduct. × 3.

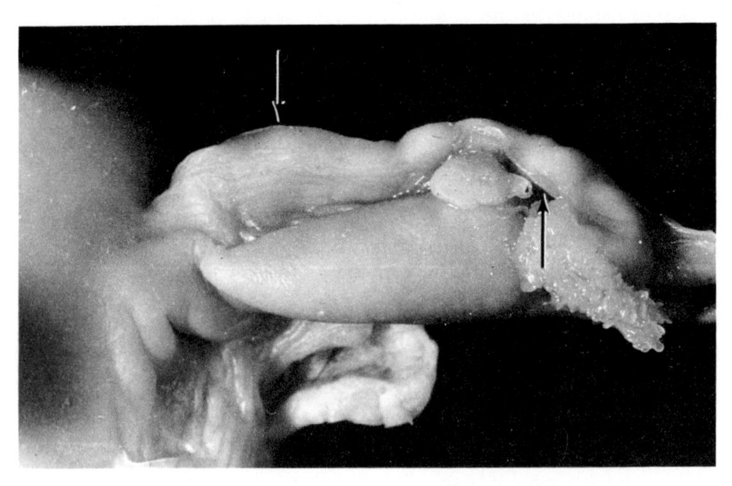

Fig. 1.14 Right oviduct and ovary and right half of the simplex uterus of an anteater, the tamandua (*Tamandua*), showing the rudimentary tubal membrane (*arrows*), a thin low antimesosalpingian peritoneal fold along the oviduct. × 3.

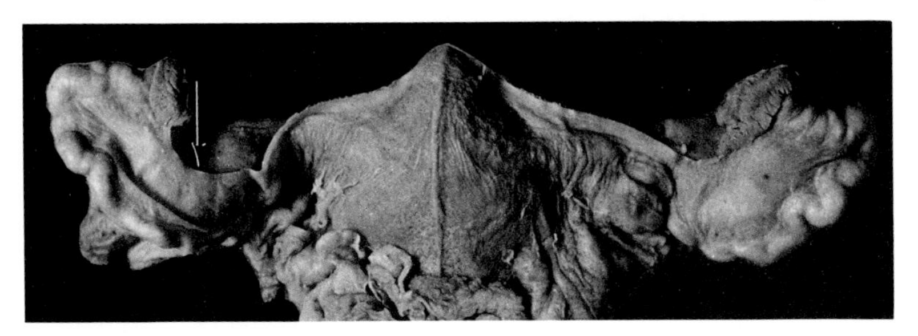

Fig. 1.15 Simplex uterus, oviducts, and ovaries of the nine-banded armadillo (*Dasypus novemcinctus*), showing how the tubal membrane (*arrow*) forms a partial bursa and extends almost to the midline of the cephalic end of the uterus. × 3.

portion of the mesosalpinx (Figs. 1.7 and 1.15). Such a partial bursa usually encloses at least the cephalic pole of the ovary.

The relationship of the tubal membrane to the ovary depends not only on the extent of its development but also on whether the recurved portion of the oviduct lies medial to the ovary or more ventrolateral to it. When the recurved portion lies medial to the ovary (see Fig. 1.10), it is easy to see how the oviduct and its mesovarium can completely surround the ovary (see Fig. 1.11), and how the neck of the funnel is brought very close to the tubo-uterine junction. In this case the tubal membrane, extending as a broad diaphragm across the floor of the bursa, is attached to the entire length of the oviduct, with only a very short free edge connecting the funnel with the tubo-uterine junction. A complete bursa such as this with a small porelike opening into the general peritoneal cavity near the tubo-uterine junction is typical of many mammals, e.g., muroids, canids, and mustelids (see Figs. 1.11, 1.12, 1.22, and 1.23). Complete fusion of the short free edge of the membrane also occurs, providing a completely closed bursa, e.g., golden hamster (Clewe, 1965), and muskrat (*Ondatra zibethicus*) (see Figs. 1.24 and 1.25). If the recurved portion of the oviduct does not encircle the ovary but lies ventrolateral to it (see Figs. 1.8 and 1.9), then even though the funnel is very close to the tubo-uterine junction and the tubal membrane fills the entire oviduct loop, a large part of the medial and even ventral surface of the ovary may be exposed to the general peritoneal cavity through a long slitlike opening, e.g., guinea pig and other hystricomorph rodents, cats (Felidae), perissodactyls, bovines (see Figs. 1.8, 1.9, 1.20, 1.21, 1.28, and 1.29).

Fig. 1.16 Ventral view of right partial ovarian bursa of a gray squirrel (*Sciurus carolinensis*), comparable with the type shown in Figure 1.11, but with a strongly looped oviduct. × 2.5.

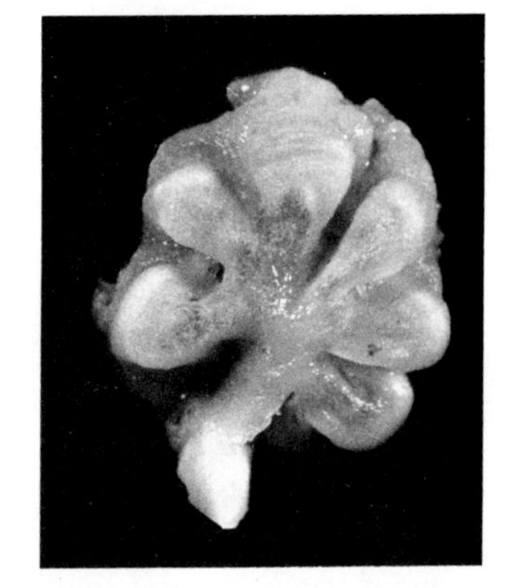

Fig. 1.17 Ventral view of the bursa of a gray squirrel (*Sciurus carolinensis*) in which the oviduct is more uniformly looped than usual. × 4.

A few mammals whose bursae we have examined are somewhat unusual and somewhat difficult to fit into the patterns represented by Figures 1.4–1.12. The toothed whales (Odontoceti), hyraxes (Hyracoidea), elephants (Elephantidae), and pigs (Suidae) have infundibula which are relatively very capacious and which constitute a large portion of the pouch covering the ovary (see Figs. 1.26 and 1.27). Basically, however, these infundibular relationships seem to be comparable with those of felids and bovids (see Figs. 1.8 and 1.9). A somewhat peculiar situation occurs in the horse because the only portion of the adult ovary covered by surface germinal epithelium is the relatively small area of the ovulation pit and its rim. The rest of the ovary is, morphologically speaking, embedded in the mesovarium (see Figs. 1.28–1.30). The infundibulum

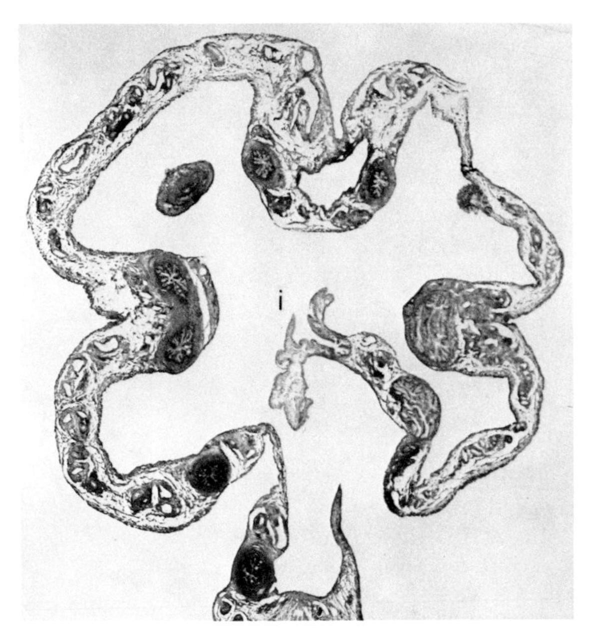

Fig. 1.18 Histological section through the bursa and oviduct pictured in Figure 1.17, showing the uniform looping of the oviduct and the inwardly directed infundibulum (*i*). × 11.5.

faces into the cephalic portion of the pit, but the basic arrangement is again most like that in Figures 1.8 and 1.9.

Gross patterns of oviducts

While a casual examination of the relation of the oviduct to the ovary in most mammals gives the impression of a somewhat haphazard arrangement, closer scrutiny reveals considerable uniformity within any single taxonomic group. In some cases the gross pattern of the oviduct is remarkably regular. The oviduct of the gray squirrel (*Sciurus carolinensis*) is commonly symmetrically looped (see Figs. 1.16–1.18). In most of the carnivores, it is sharply kinked, but has an overall circular pattern over the ventrocaudal portion of the bursa (see Figs. 1.22 and 1.23). The mink oviduct often forms a strikingly regular filigreelike setting for the ovary (see Fig. 1.23). Other mustelids closely resemble the mink in this character. In the geomyoids (pocket gophers, kangaroo rats, pocket mice, etc.), the oviduct is tightly and uniformly kinked

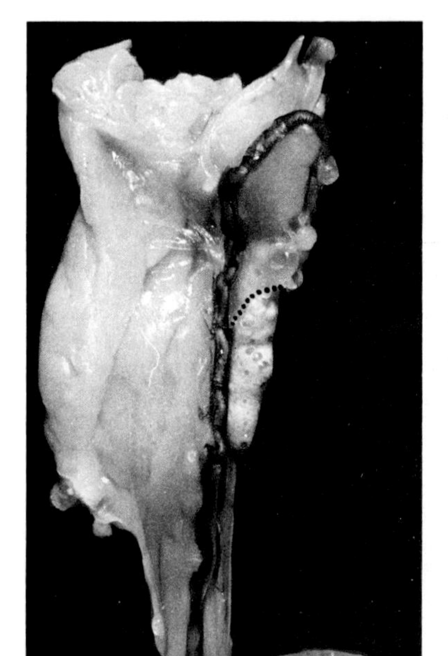

Fig. 1.19 Ventromedial view of the right oviduct, ovary, and cornu of a domestic rabbit (*Oryctolagus*). Because the tubal membrane of rabbits is extremely delicate and transparent, its ventral edge has been indicated by a dotted line. This bursa is comparable with the type shown in Figure 1.10. × 1.

and gradually enlarges as it extends from the uterus toward the ovary. Each kink or coil is adherent to the next and the whole is ensheathed by an extension of the longitudinal musculature of the uterus so that a slender spindle-shaped structure results, except at the recurved cephalic end, which is not enclosed in the muscular sheath (Fig. 1.31).

Just why these specific patterns develop in certain mammalian groups is, of course, unknown. For experimental study of the oviduct, such patterns as those of pocket gophers and of mink would be ideal if it were important to repeat procedures or to interpret results in terms of specific portions of the oviduct.

Of course, man and the few laboratory-adapted species of other primates have oviducts which are simple enough or large enough to make possible the identification of comparable areas. However, the very tight and apparently irregular "knotting" of this organ in most laboratory

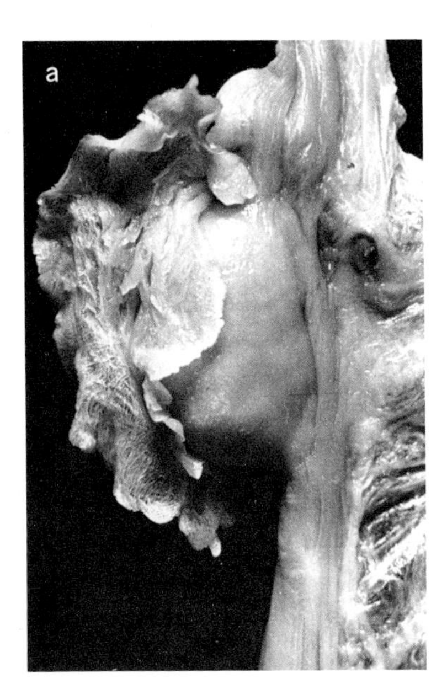

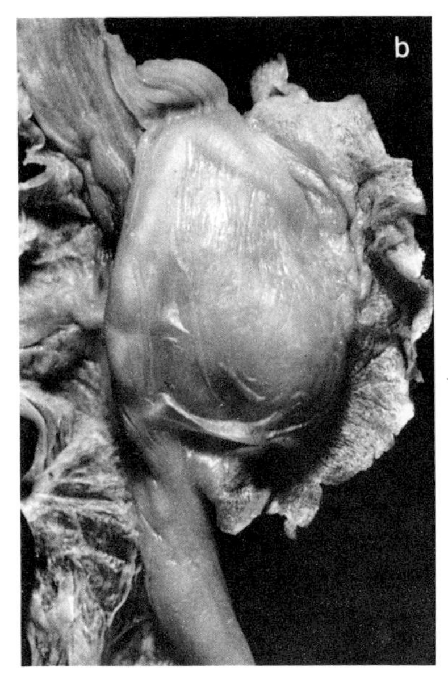

Fig. 1.20*a* Medial view of the right ovary and funnel of the oviduct of a bobcat (*Lynx rufus*). This is comparable with the type shown in Figure 1.8, except that in order to show the whole funnel its dorsal edge was pulled out of the bursa. × 2.5.

Fig. 1.20*b* Lateral view of the specimen shown in Figure 1.20*a*. × 2.5.

rodents makes accurate identification of specific regions difficult. Fortunately, the oviduct of the domestic rabbit is one of the simplest and most accessible surgically (see Fig. 1.19).

Classification

Several questions concerning the presence or absence and the nature of the ovarian bursa pertain to the physiology and pathology of reproduction: the possibility of migration of ova across the peritoneal cavity from one ovary to the opposite oviduct; the chances for an abdominal pregnancy; the possible entrance into the general peritoneal cavity of sperm, blood from ovarian hemorrhage, and other contaminative material; the drainage of peritoneal fluid, either normal or pathological, into the genital tract; and the assurance of normal entrance of ova into the oviduct. Then too, from a purely practical standpoint, one wants to know whether, upon opening the peritoneal cavity, the ovary can be

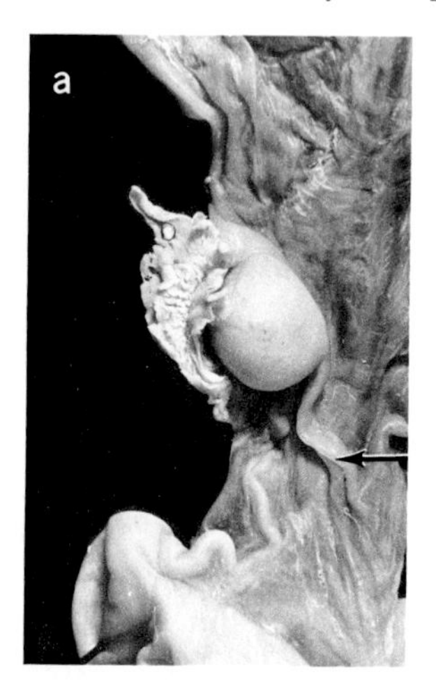

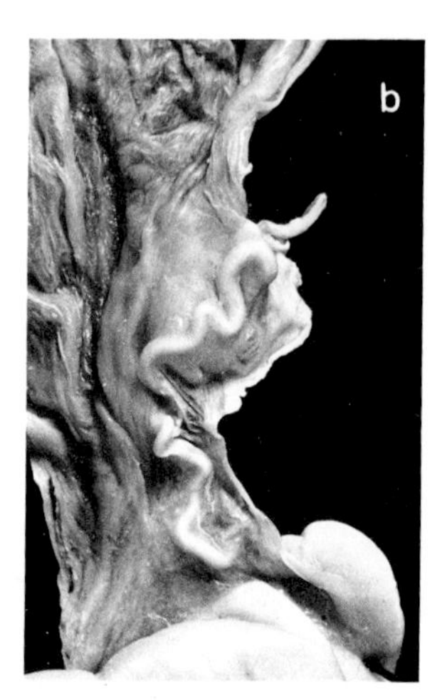

Fig. 1.21*a* Medial view of the right ovary and oviduct of an antelope, the gray duiker (*Sylvicapra grimmia*). The proper ligament (*arrow*) of the ovary does not reach the tubo-uterine junction. Comparable with the type shown in Figure 1.8. × 1.3.

Fig. 1.21*b* Lateral view of the specimen shown in Figure 1.21*a*. × 1.3.

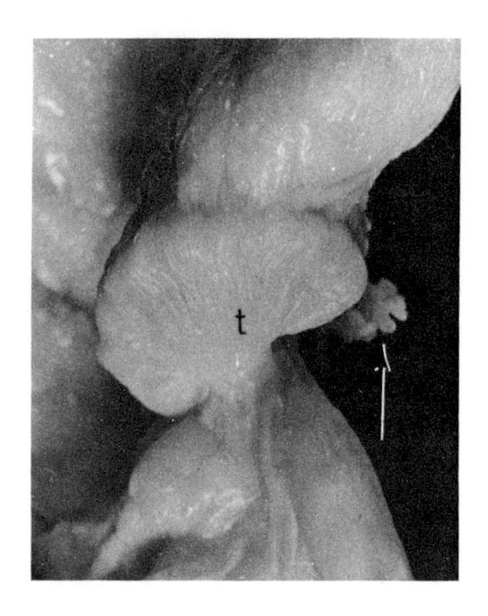

Fig. 1.22 Medial view of the complete right bursa of a black bear (*Ursus americanus*), showing the uniformly looped and encircling oviduct and fimbriae (*arrow*) extending out of the orifice. *t*, tubal membrane. Comparable with the type shown in Figure 1.11. × 2.

seen without further dissection. With these things in mind we have formulated the following classification of ovarian bursae.

1) No bursa. (No tubal membrane or only a rudimentary one; cf. Figs. 1.5 and 1.6.)
2) Partial bursa. (A fairly well developed tubal membrane, but which

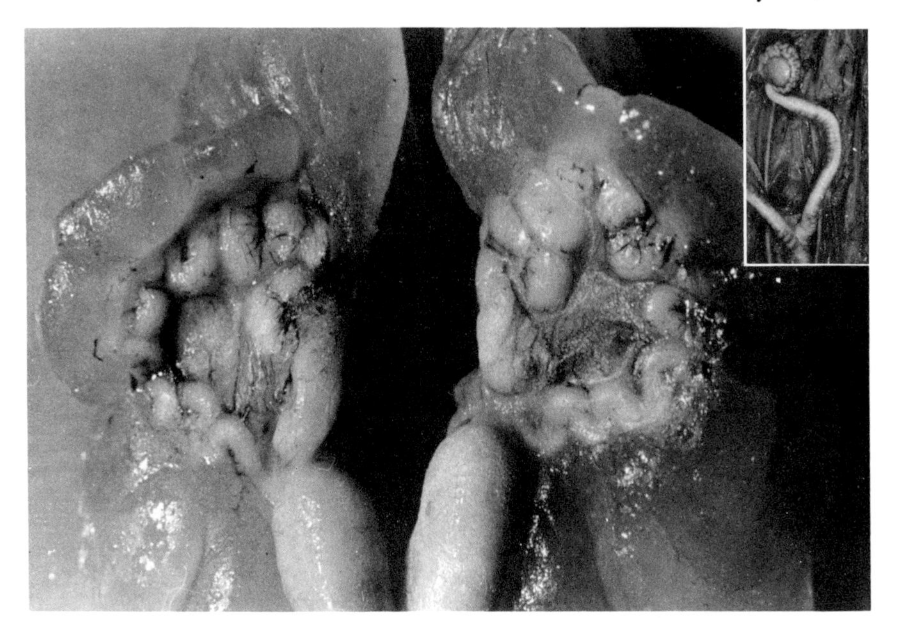

Fig. 1.23 Ventral views of the oviducts and complete bursae of the mink (*Mustela vison*), showing the complete encirclement of the bursal wall by the uniformly kinked oviduct. Comparable with the type shown in Figure 1.11, but with very slight or no protrusion of fimbriae. From a very fat animal; and from a very lean one (*inset*). × 5; *inset* × 1.

allows examination of the whole free surface of the ovary without artificial enlargement of the opening into the general peritoneal cavity; cf. Figs. 1.7 and 1.10.)

3) Complete bursa. (An opening too small to allow examination of the whole free surface of the ovary.)
   a) Cleftlike orifice. (Orifice a long slit usually bordered ventrally by the infundibulum; cf. Figs. 1.8 and 1.9.)
   b) Large to small porelike orifice. (A few fimbriae may extend out through the orifice, and there are sometimes one or two additional small pores; cf. Figs. 1.11 and 1.12.)
   c) No orifice. (In very small mammals, careful injection technique or study of serial sections may be necessary to determine whether or not the bursa is actually completely closed.)

Since classification is no concern of Nature, only of men, this one like all others breaks down in the cases of intergrading types, and so

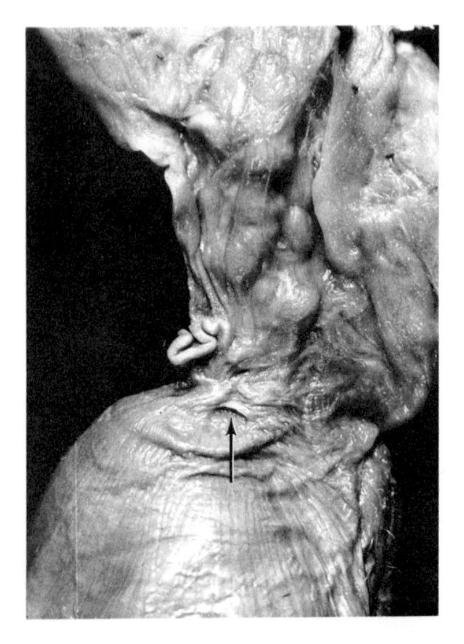

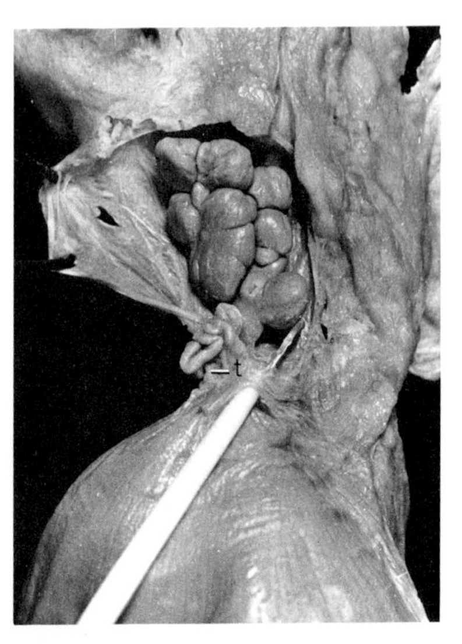

Fig. 1.24 (*left*) Medial view of the right bursa and tubo-uterine junction of a muskrat (*Ondatra zibethicus*) in late pregnancy, showing the relatively short, tightly looped oviduct confined to the narrow uterine end of the bursa and the porelike, probably blind, orifice (*arrow*). × 2.5.

Fig. 1.25 (*right*) The muskrat (*Ondatra zibethicus*) specimen shown in Figure 1.24, with the ventromedial wall of the bursa reflected to show the lobulate ovary with numerous corpora lutea. The reflected portion is dorsal mesosalpinx, and the tubal membrane (*t*) lies between the loops of oviduct and the uterus and is therefore very small, forming only a minor portion of the bursa. The probe was forced through an apparently completely closed deep segment of the orifice to show the relation of the blind orifice to the bursal cavity. As shown in Figure 1.12, the sickle-shaped margin of the orifice is formed by the morphologically cephalic free edge of the dorsal mesosalpinx. × 2.5.

will be subject to different interpretations by different people. Several other attempts to classify ovarian bursae have been made, and the reader may prefer one of them (Zuckerkandl, 1897; Gerhardt, 1905; Strauss, 1966); however, the one above has been used in the synoptic data and elsewhere in this book.

Literature

The best comparative studies of the morphology of ovarian bursae were made by Zuckerkandl (1897) and Gerhardt (1905). Zuckerkandl

Fig. 1.26 Partial bursa and right ovary of a juvenile Asiatic elephant (*Elephas maximus*). The hoodlike bursa is a combination of the tubal membrane and the voluminous infundibulum. × 1.

studied its development in the dog, and described and figured the definitive oviduct and bursa, if one was present, in monotremes, marsupials, and several orders of eutherians. Gerhardt (1905) made comparative studies of the same extent, but did not mention Zuckerkandl's work. In a thorough study in which he used reconstructions, Agduhr (1927) worked with the development of the oviduct and bursa in the mouse. While his descriptions and data are no doubt the most detailed and accurate available, he has contributed to confusion about the bursa in two ways. First, he refers to the mesosalpinx as the "mesotubarium inferius," in spite of the fact that it is clearly a dorsal "mesentery." He also speaks of the ventrally located tubal membrane as the "mesotubarium superius." Second, he emphatically states that the tubal membrane is unnecessary to the formation of a bursa and that, when present, it is of little significance. He was, of course, influenced in this direction by the smallness of the tubal membrane in the mouse. We believe that Agduhr was completely wrong in both of these matters, and we agree with

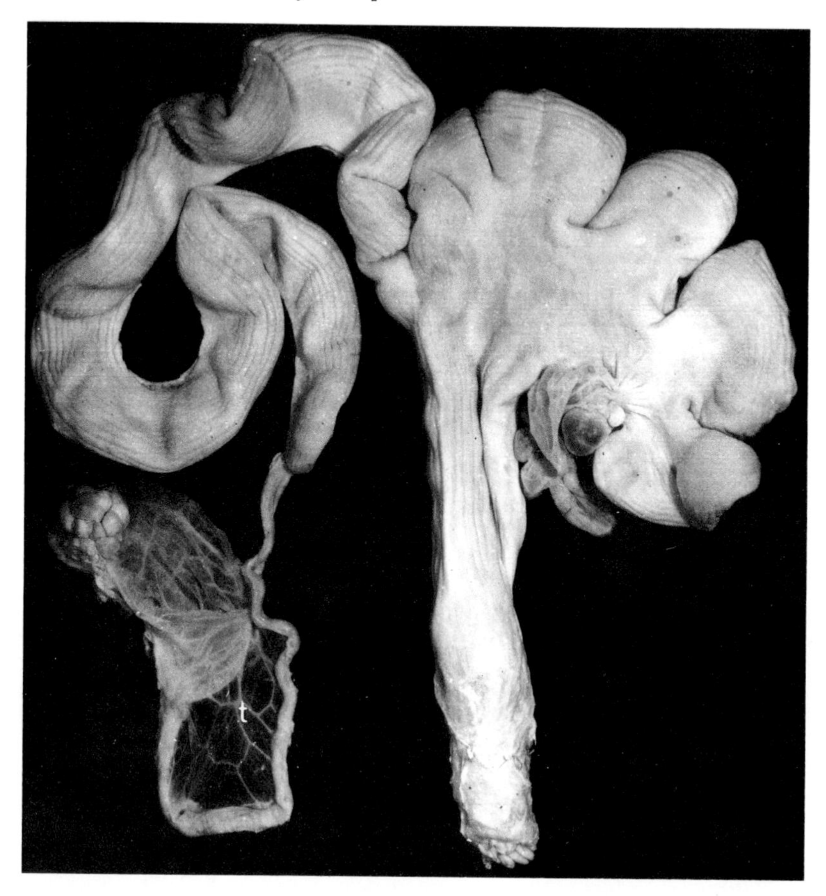

Fig. 1.27 Ventral view of the reproductive tract of a parous sow (*Sus scrofa*). The right mesometrium and dorsal mesosalpinx have been removed to show more clearly the extensive tubal membrane (*t*) and the voluminous infundibulum. The partial bursa is much like that of the elephant, with the funnel making up a large part of it. × 0.2.

Zuckerkandl that the tubal membrane is necessary to the formation of an ovarian bursa as it occurs in mammals. Wislocki (1932), apparently following Agduhr's terminology, also equated the mesosalpinx with the "inferior mesosalpinx" in his description of the primate ovarian bursa.

The development and anatomy of the ovarian bursa in the rat have been described by Kellogg (1941). Alden (1942) and Wimsatt and Waldo (1945) have shown clearly that in the adult rat a small orifice connects the cavity of the bursa with the general peritoneal cavity.

Fig. 1.28 Ventral view of the left ovary and oviduct of a Burchell's zebra (*Equus burchelli*), showing at the upper left from above downward the kinked ampullary end of the oviduct, the funnel, the ovulation pit of the ovary, and the proper ligament of the ovary. On the right is the ruffled free edge of the tubal membrane (*arrow*). The partial bursa is similar to that shown in Figures 1.8 and 1.9. × 1.5.

Clewe (1965) demonstrated the lack of a peritoneal connection in the golden hamster.

Relatively little experimental work has ever been done to illustrate the function of the bursa. Kelly (1939) opened the bursae of 18 rats on one side, placed the animals with males, and examined them 14 days later. The average number of fetuses on the operated side was 4.28 compared with 1.0 on the control side. These data mean little, partly because complete regeneration of the bursae had occurred in 56% of the animals. Butcher (1947) excised the bursa in young rats and found some retardation of ovarian growth. In the hamster, ligation of the oviduct near the uterus results in fluid accumulation and "ballooning" of the bursa. We are ignorant of the implications of this phenomenon with respect to normal function. The usual notion that a bursa assures entrance of ova into the oviduct, while probably true, seems of little importance from the comparative standpoint since many species which ovulate only one egg have no bursa, and numerous polyovular species have complete bursae. That such a specialized structure would have no physiological purpose seems unlikely, but practically nothing is known about the function of ovarian bursae.

Blood vessels

In every mammal so far described in the literature or examined by the authors, the ovary has a double blood supply (Fig. 1.32). The main and primary one, developmentally, is from the ovarian artery and vein which are direct or almost direct branches of the dorsal aorta and in-

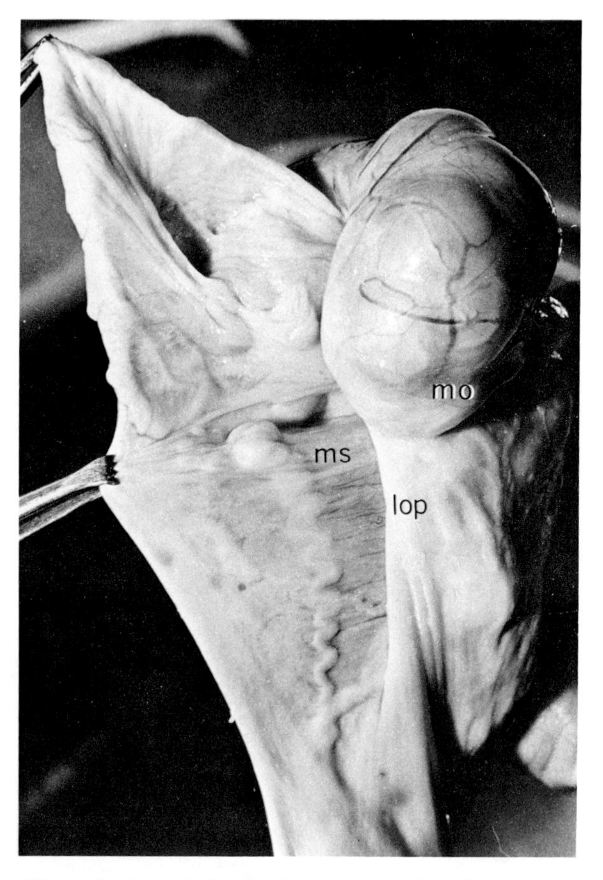

Fig. 1.29 Ventral view of the fresh unpreserved right ovary, oviduct, mesosalpinx, and tubal membrane of a domestic mare (*Equus caballus*). At the upper left, the funnel is held open by forceps. The lower forceps holds the tubal membrane, which is separated from the mesosalpinx (*ms*) by the oviduct. The ovulation pit is hidden by the caudal pole of the ovary. The ovary is covered by mesovarial tissue (see Fig. 1.30), hence the network of surface blood vessels. *lop*, proper ligament of the ovary; *mo*, mesovarium. × 0.8.

ferior vena cava, respectively, in the upper lumbar region at about the same level as the renal vessels. In fact, the left ovarian vein often drains into the left renal vein close to the vena cava. Normally, the ovarian vessels' plane of origin from the aorta and vena cava is more ventral than that of the lumbar or renal vessels. If the ovaries occupy a pelvic position, then most of the cephalic portion of each artery and vein lies in a retroperitoneal position a short distance lateral to the aorta and almost parallel with it. If the ovaries are lumbar in position, then the vessels run laterally from the aorta at a less acute angle and are retroperitoneal for only a short distance. In either case, they leave the body wall by entering the mesovarium through its cephalic attachment and travel within it to the hilus of the ovary. In species such as rabbits and hares (Leporidae) and man where the ovary has essentially retained its embryonic shape and has a relatively long mesovarial edge, the major vessels enter it near the cephalic portion of this edge, although some branches usually extend caudad along the mesovarium to turn toward the ovary and enter it midway or even well toward the caudal end. If the organ is more globular and its mesovarium is more pediclelike, as it is in the case of the rat, then the vessels appear to enter centrally, although actually the same fundamental relation still exists. In either case, some branch or branches of fair size extend caudad from the main ovarian artery and vein, or from their more caudal branches, often from both, to anastomose with the ovarian branches of the uterine vessels. These anastomotic vessels usually lie well proximal at the base of the mesovarium or even more proximally in the mesosalpinx. The

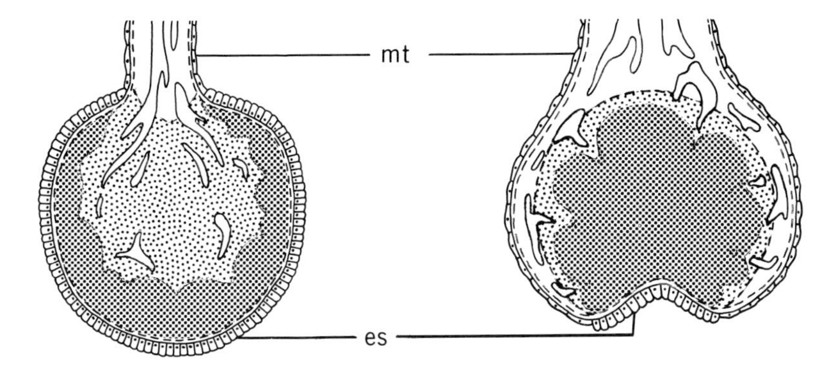

Fig. 1.30 Diagrammatic cross-sections of a typical mammalian ovary (*left*) and that of a horse (*right*) to show the peculiar relations of the mesovarium. The surface germinal epithelium of the horse is confined to the ovulation pit and its rim, and the thin medulla surrounds most of the cortex. A similar condition occurs in the nine-banded armadillo (*Dasypus novemcinctus*). *es*, epithelium superficiale; *mt*, mesothelium; *close stipple*, cortex; *open stipple*, medulla.

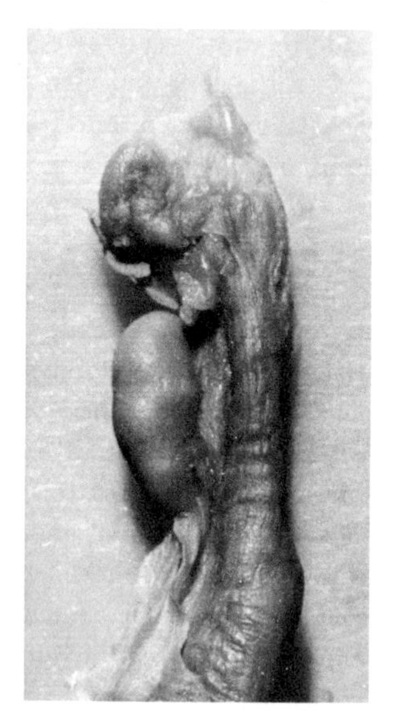

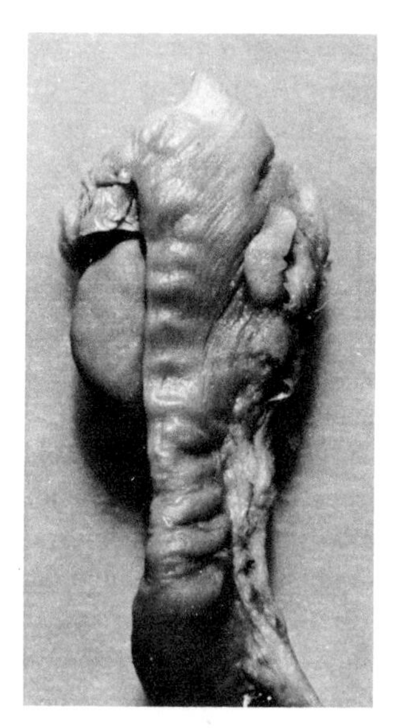

Fig. 1.31 Left oviduct and ovary of two plains pocket gophers (*Geomys bursarius*), showing the extension of the longitudinal musculature of the uterus as a sheath enclosing the tightly kinked major portion of the oviduct exclusive of the funnel and distal portion of the ampulla. × 4.

ovario-uterine anastomosis therefore makes a pattern comparable with the vascular arcades of the intestine.

Similar but usually longer and finer caliber branches of the uterine vessels extend along the mesosalpinx and supply the oviduct. The cephalic ends of these reach the ampulla and funnel of the oviduct and there have minor anastomoses with the ovarian vessels, usually with the finer branches of the cephalic end of the ovary, but sometimes directly with intra-ovarian vessels at the regions where the funnel or the fimbriae are fused to the ovarian surface. Usually, relatively very fine caliber anastomotic vessels course in the proper ligament of the ovary (see Fig. 1.32).

Because of this good anastomotic blood supply to the ovaries, the main ovarian vessels of most mammals could undoubtedly be tied without serious interference with ovarian nutrition. This has been shown to be true in the porcupine (*Erethizon dorsatum*) (Mossman, unpublished). Conversely, the ovarian vessels help supply the uterus with blood, and are known to enlarge tremendously during pregnancy. This hypertrophy

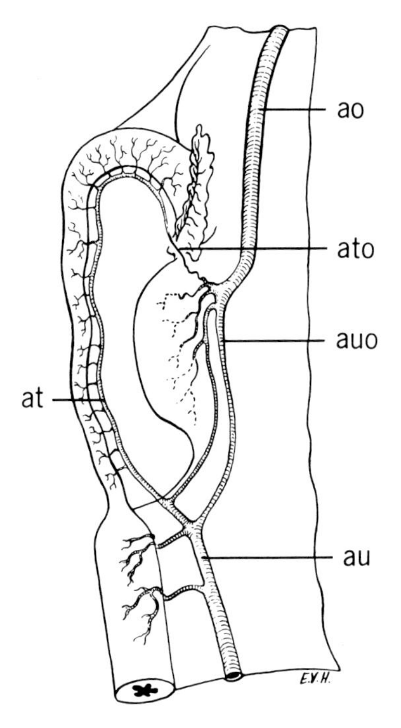

Fig. 1.32 Diagram of the gross arterial blood supply of a typical mammalian ovary. Minor branches to the mesosalpinx, etc., are not shown. *ao*, arteria ovarica; *at*, a. tubaria; *ato*, anastomosis tubo-ovarica; *au*, arteria uterina; *auo*, anastomosis utero-ovarica (parallel to this is the small anastomotic ramus of the ligamentum ovarii proprium).

mainly involves that portion up to and including the uterine anastomoses, but even the proper ovarian branches (rami ovarici) are often considerably enlarged. Whether or not there is a physiological reason for such an increased supply to the ovary during gestation is unknown, although a small increase in vascular need could result from the presence of corpora lutea. Perhaps the enlargement of the branches to the ovary is only in response to the same unknown direct stimulus which causes hypertrophy of the uterine vessels. Obstruction or removal of the anastomotic channels before pregnancy and observation of the effect on both ovarian vessels and ovaries during gestation might help to answer this question.

In some species, particularly in man, the ovarian veins break up into a coarse plexus near the ovary. This is homologous to the pampiniform plexus of the testis. Both are derived from the venous plexus of the mesonephros from which the immediate ovarian branches arise in the embryo. Presumably, animals such as the pig, which have a highly developed mesonephros, would have a large pampiniform plexus while those with a small mesonephros, such as the cat, would have a small one. This has not been verified, however, and such other considerations as animal size and stance may be modifying factors. In male mammals with "descended" testes, the pampiniform plexus acts as a heat exchanger, but such a function seems remote for abdominal testes or for ovaries.

Lymph drainage

The ovary is permeated by a rich lymphatic capillary plexus which drains through the medulla by way of numerous large channels, valved in man and other large mammals, to vessels paralleling the blood vascular supply (Polano, 1903; Poirier and Charpy, 1912; Andersen, 1926; Rouvière, 1938; Wislocki and Dempsey, 1939; Bachmann, 1949; Eichner and Bove, 1954; and Morris and Sass, 1966). The main drainage is along the ovarian vessels to the middle lumbar nodes. In man, a secondary drainage by way of the uterine anastomoses to the sacral nodes has been reported (Rouvière, 1938). In the red squirrel (*Tamiasciurus hudsonicus*) and one of the bush babies (*Galago demidovii*), a lymph node in the mesovarium (Figs. 1.33 and 6.84) probably receives most of the lymph from the ovary.

Nerve supply

The classical description of the nerve supply to the human ovary (Kuntz, 1929), and that given in many textbooks of human anatomy, chiefly involves the plexuses along the ovarian vessels. These plexuses contain afferents from the tenth thoracic nerve, as well as efferents presumably composed of both sympathetic and parasympathetic fibers, i.e., thoracolumbar and vagal fibers. According to Kuntz (1919), the dog ovary is similarly innervated. R. T. Hill (1949), using rabbits, mice, rats, guinea pigs, and cats, found no effect on the ovarian cycle and function when he cut the uterine vessel nerve plexuses, but a marked effect when he cut those of the ovarian vessels or performed a thoracic vagotomy. He believed he had sufficient control data to indicate that these effects were not caused by direct interference with blood supply or by the serious general gastrointestinal consequences of vagotomy.

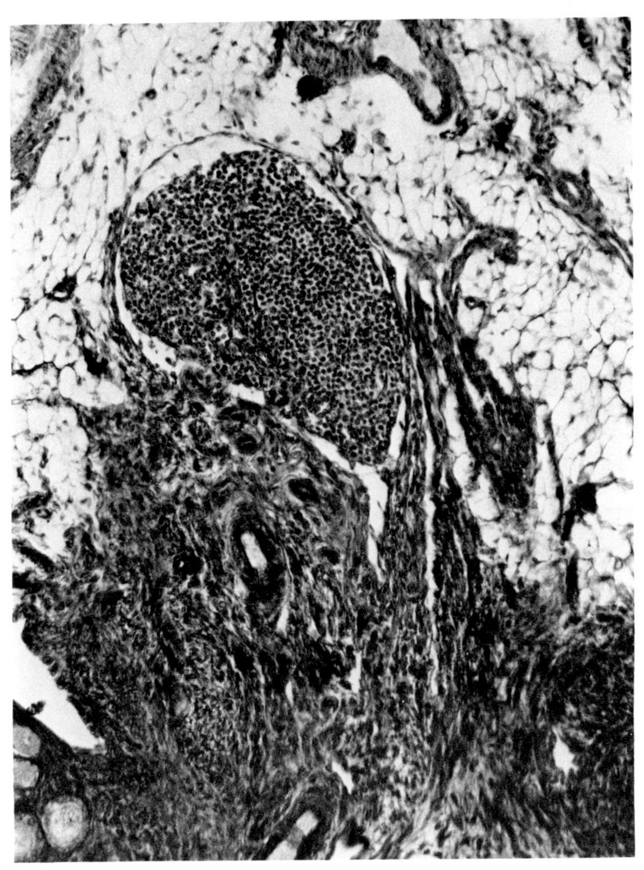

Fig. 1.33 Section showing a regional ovarian lymph "node" in the mesovarium of a red squirrel (*Tamiasciurus hudsonicus*) within 1 mm of the hilus of the ovary. Small lymph "nodes" of small mammals such as this one do not have the complex structure of true lymph nodes of larger mammals, yet they apparently perform the same functions. See also Fig. 6.84. *T. hudsonicus*, 3. × 130.

Carlson and DeFeo (1965) showed that cutting of the pelvic nerves of the rat severed the afferents responsible for the induction of pseudopregnancy, but that such cutting had no direct effect on ovarian function. They also found that extensive abdominal sympathectomy had no effect on ovarian or uterine function.

Pines and Schapiro (1930), working with cats, rabbits, cows, pigs, and mice, demonstrated the presence of nerve endings in the theca interna but not in the granulosa, in the interstitial gland masses, in corpora lutea but not in corpora albicantia, and in the tunica albuginea and connective tissue of the hilus. The latter two they considered to be sensory nerve endings. Jacobowitz and Wallach (1967), using a method involving fluorescence, saw catecholamine-containing fibers in the stroma, especially in relation to blood vessels in cats, women, and rhesus monkeys (*Macaca mulatta*). Owman and Sjöberg (1966), using rabbits, found adrenergic fibers which ended in the cortical parenchyma and were contiguous to follicles but which had no obvious relation to blood vessels. Neilson et al. (1970) reviewed the literature on the innervation of human ovaries, but found no clear evidence of innervation of follicles. Owman, Rosengren, and Sjöberg (1967) observed numerous adrenergic fibers ending in the parenchyma and on blood vessels in human ovaries. Fink and Schofield (1971), using electron microscopy, demonstrated nerve fibers ending at the basement membrane of primary follicles of cats. With light microscopy, they saw them entering the granulosa of maturing and mature follicles. Also they demonstrated experimentally that all were either afferent or postganglionic, and that they either arose in, or traversed, the aortico-adrenal ganglia.

Wallart (1936), using the Bielschowsky technique, demonstrated a rich sympathetic innervation of the epithelium of the epoophoron. He had previously shown a similar innervation of the rete epithelium.

Apparently, then, the ovary receives all of its innervation by way of the perivascular plexuses along the ovarian vessels. The afferents are from the lower thoracic nerves. Both vagal fibers and adrenergic fibers from the thoracolumbar sympathetics are probably present. The adrenergic fibers may be partly secretory as well as vasomotor, since some endings have been found to be unrelated to the blood vessels.

Embryonic vestiges associated with the ovary

The so-called embryonic vestiges associated with the ovary are homologs of structures associated with the testis. In the male most are clearly functional; some of those of the female may have minor functions. The nomenclature of these structures, particularly those of the female, has been complicated by lack of understanding of the homologies between the two sexes. We feel that homologous structures in the two sexes should be designated by similar names, and have therefore not always followed the *Nomina Anatomica* (*NA*) and *Nomina Embryologica* (*NE*), but we have indicated these divergences.

Beginning at the gonadal area, these structures are: the rete (rete ovarii), the efferent ductules of the ovary (*NE*, ductuli efferentes ovarii; *NA*, ductuli transversi), and the *female ductus deferens*, or canal of Gartner (*ductus deferens femininus*; *NE*, ductus deferens vestigialis). The ovarian end of the female ductus deferens receives the efferent ductules and is specifically named the duct of the epoophoron (*NE*, ductus epoophorontis; *NA*, ductus epoophori longitudinalis). This portion, together with the efferent ductules, is the epoophoron, the homolog of the epididymis. The major portions of the efferent ductules are derived from the tubules of the middle segments of the mesonephros. Some of the more rostral and more caudal mesonephric tubules remain as isolated vestigial blind tubules, and are known in both sexes as the rostral aberrant ductules (ductuli aberrantes rostrales) and the caudal aberrant ductules (ductuli aberrantes caudales). The latter are collectively called the paroophoron in the female and the paradidymis in the male.

All of these are seldom thought of as gross structures, yet in larger mammals they are often visible with the naked eye. There are two reasons why they are not ordinarily seen even in large mammals: one is the difficulty in distinguishing their elements from surrounding tissue, especially in preserved specimens; the other is simply ignorance as to just where to look for them (Fig. 1.34).

The rete ovarii, an irregular network of spaces lined by simple low columnar epithelium and homologous to the rete testis, is located at the cephalic end of the hilar region of the ovary and often extends both deep into the ovarian medulla and well out into the mesovarium. It is quite large and compact in many species and is often partially cystic, as in the guinea pig. Many human ovarian cysts are also believed to be derived from the rete, in spite of the fact that it is relatively small in man. In species where it is fairly well developed, it has about one-sixth the linear dimensions of the whole ovary, but it may be larger, as in the tarsier (*Tarsius*) and the pangolins (*Manis*).

The efferent ductules of the ovary extend from the rete through the mesovarium to its region of junction with the mesosalpinx; there they join the more cephalic portions of the rudimentary female ductus deferens to collectively make up the epoophoron. In fresh specimens one can often see the ductules and duct by examining the mesovarium and broad ligament by transmitted light.

From our experience in examining serial sections of the ovary and

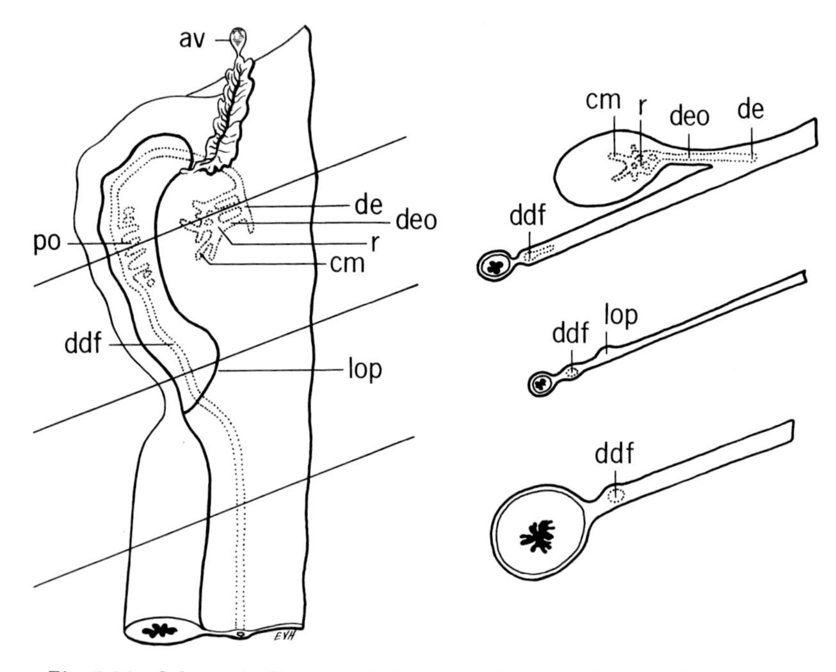

Fig. 1.34 Schematic diagram of the upper right female genital tract of a mammal to show the usual position of the major vestigial structures. Sections at the three levels indicated are on the right. *av*, appendix vesicularis of the oviduct; *cm*, chordae medullares; *ddf*, ductus deferens femininus; *de*, ductus epoophorontis; *deo*, ductuli efferentes ovarii; *lop*, ligamentum ovarii proprium; *po*, paroophoron; *r*, rete ovarii.

oviduct with their adjacent mesenteries in small mammals, the paroophoron is much less constant than is the epoophoron. This is also true of the mesometrial portion of the female ductus deferens. When present, the paroophoron usually forms a small compact mass of contorted tubules in the mesosalpinx about opposite the midportion of the ovary. In larger animals such as the pig, the female ductus deferens is often easily seen in the broad ligament close to and paralleling the oviduct and uterine horn.

Summary

The ovaries in eutherian mammals are variously located from a position opposite the caudal poles of the kidneys in the lumbar region to one well within the pelvis. They are free peritoneal organs, but may have a relatively thick mesovarium, hence a broad bare area. In one genus (*Tragulus*), they have a very short mesovarium which disappears during pregnancy, so that they are then attached directly to the uterine wall.

In most mammals, the ovary is enclosed in a bursa formed from the mesosalpinx and a ventral mesenterylike tubal membrane. This bursa partly, and in a few groups completely, isolates the ovary from the general peritoneal cavity.

The major ovarian blood and lymph vessels and nerves reach the ovary through the more cephalic part of the mesovarium. The ovarian vessels are always anastomotic at the base of the mesovarium or in the mesosalpinx with one or more branches of the uterine vessels, so that the ovary, oviduct, and uterus have an efficient anastomotic circulation with one another.

In larger mammals, many of the vestigial remnants of adjacent embryonic structures may be grossly visible. These include the epoophoron and the main portion of the ductus deferens femininus.

Two

General microscopic structure of the mammalian ovary

It is somewhat paradoxical to describe the general microscopic structure of a mammalian ovary in a book in which the main theme is description of the changes in this organ during the reproductive cycles of the individual and of the differences which distinguish the ovaries of the various mammalian groups. Yet, for the sake of those who may wish to find a foundation of facts here rather than to depend upon their memory or to turn to a textbook of histology, we felt it advisable to present a brief general microscopic description of the ovary, based on the usual conditions found in the mature organ. Figure 2.1 is designed to show diagrammatically the principal ovarian structures and their normal changes and fate during a pregnancy cycle.

Surface epithelium and capsule

As was pointed out in Chapter 1, the ovary is a free, or peritoneal, organ suspended more or less directly from the dorsal body wall by a mesentery, the mesovarium. It is therefore covered on all sides, except at the line, or area, of attachment to the mesovarium, by a modified serous membrane homologous to the peritoneum of other visceral organs (see Fig. 1.1). Like the rest of the peritoneum, this serous membrane typically consists of a simple surface epithelium supported by a delicate fibrous layer continuous with the underlying connective tissue of the organ. However, in the case of the ovary, the relation of the surface epithelium to the underlying tissue may be much more intimate than with most other viscera. It may rest directly on the ovarian stroma with no distinct intervening fibrous layer corresponding to the fibrous layer of the peritoneum. This condition is common in very small species such as the shrews (Soricidae). Larger species rather commonly have a distinct thick fibrous tunica albuginea ovarii which, however, is never as thick or collagenic as the tunica albuginea testis. The majority of mammals exhibit a condition intermediate between these two extremes, i.e., there is a thin delicate subepithelial fibrous layer which is often overlooked unless the section is stained with a differential collagen stain. Always in the embryo and commonly in the adult, the tunica albuginea, regardless of its thickness, is indented or pierced by numerous crypts, cords, or tubules of surface epithelium. Unlike the peritoneal mesothelium, which is usually simple squamous in type, the ovarian surface epithelium is usually simple low columnar but may be squamous, especially on bulging areas such as those caused by large follicles and corpora lutea.

Cortex and medulla

The main ovarian mass is usually divided for purposes of description into two portions: an outer zone, the cortex; and an inner core, the medulla. In the adult, the cortex contains most of the characteristically gonadal structures such as follicles, ova, corpora lutea, and other glandular elements. The medulla is made up mainly of the larger blood and lymph vessels with their supporting connective tissue, but also contains such more or less rudimentary epithelial structures as the rete ovarii and medullary cords (chordae medullares) (Fig. 1.34), and often much interstitial gland tissue. The connective tissue, or stroma, of the medulla is of the ordinary loose adult type; that of the cortex is "embryonic" in type — that is, compact, cellular, and relatively lacking in fibers. Actually, there is usually no clear line or even limited zone of demarcation between cortex and medulla. In almost every specimen one examines, some cortical elements extend into what would ordinarily be considered medulla, and some medullary tissues may be found well out in the cortex. Hence the terms "cortex" and "medulla" are generally thought of as merely convenient expressions to denote rough topographical levels rather than definitive anatomical entities, but more is said of this below (pp. 59 and 125–27). We may now consider the definitive structures found either in the cortex or medulla, or both.

Ova

The most important structures in the ovary are, of course, the ova. Yet, strictly speaking, there are no mature ova in most mammalian

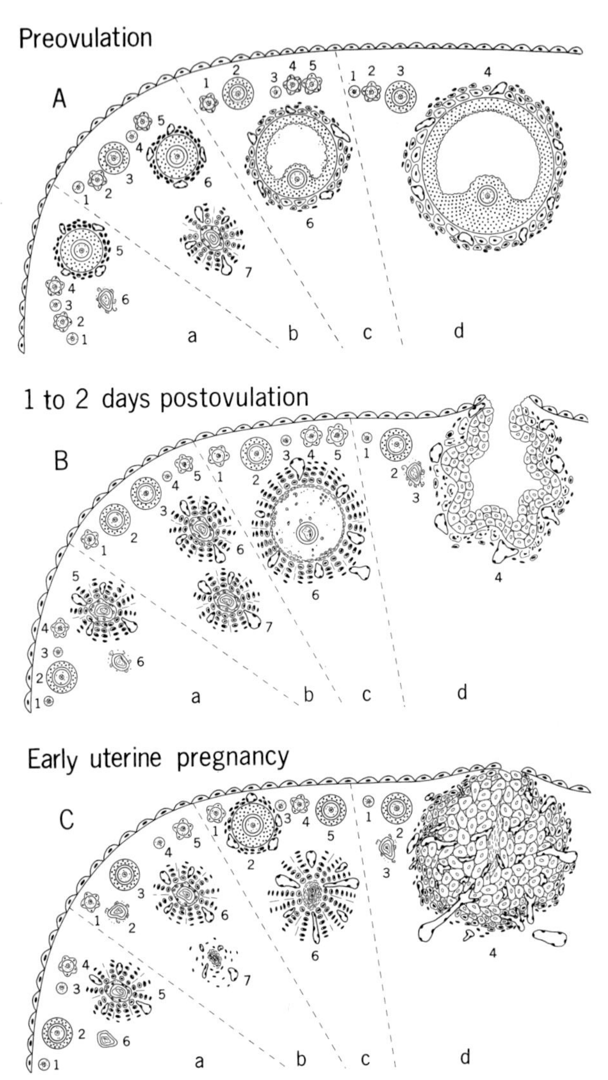

Fig. 2.1 Theoretical diagrams of sections of the same portion of a mammalian ovary at six different phases of the reproductive cycle to show the types of follicles and follicle derivatives characteristic of each phase, as well as the history of specific individual follicular structures during the cycle. Each sketch has been arbitrarily divided into sectors *a*, *b*, *c*, and *d*, and the structures within each sector numbered. For example, *Ad4* shows a maturing follicle which can be traced through its rupture and the degeneration of its thecal gland—*Bd4*; the transformation of its granulosa cells into luteal cells and the differentiation of additional luteal (paraluteal) cells from the surrounding stroma—*Cd4*; to a mature corpus luteum—*Dd4*; a degenerating corpus luteum—*Ed4*; and a corpus albicans—*Fd4*.

Midpregnancy

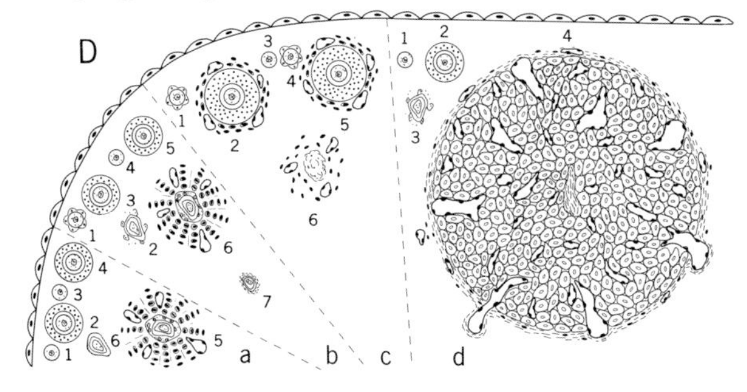

Late pregnancy to lactation

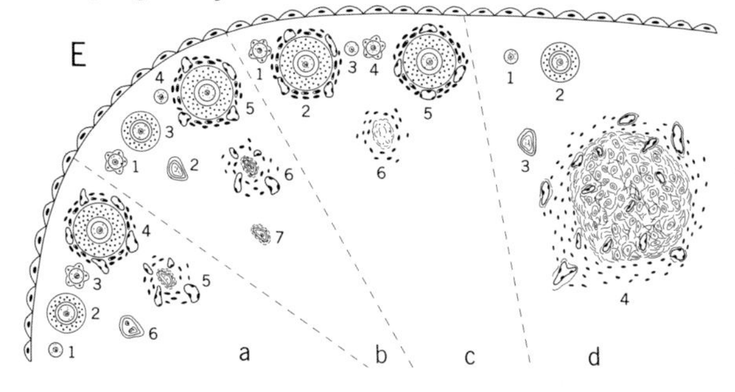

Late lactation

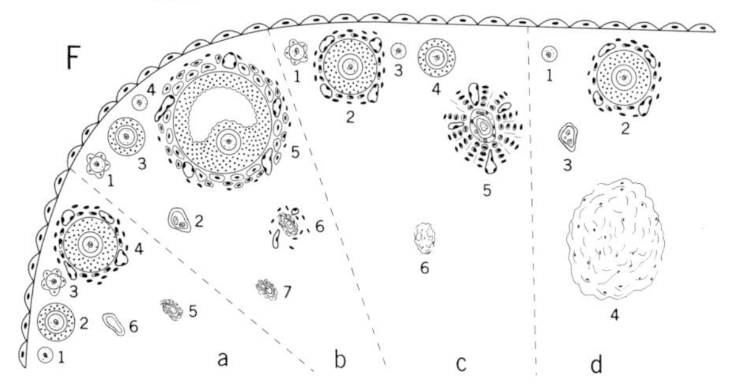

Likewise, the atresia of follicles and formation of interstitial gland masses from their thecae internae can be followed—*Ac6* to *Bc6*; *Aa5* to *Ba5*; etc. *Fc6* shows that some large atretic follicles result in small corpora albicantia indistinguishable except by size from those derived from corpora lutea.

While this diagram is primarily intended to represent the events of a pregnancy cycle, essentially the same events occur in an estrous cycle, except that the corpora lutea and interstitial glands do not become as well differentiated or as voluminous. As represented here in *B* and *C*, interstitial gland tissue is usually best developed during estrus and early pregnancy, in other words, at the time of greatest follicular atresia.

ovaries; instead, there are stages in the maturation of female germ cells, namely, oogonia, primary oocytes, and secondary oocytes. With certain exceptions (pp. 43–44 and 143), the ova are shed from the ovary as secondary oocytes. After entering the oviduct, they are penetrated by sperm and then give off the second polar cell (polocytus secundus), thus becoming mature ova already in the process of fertilization. Upon fusion of the sperm pronucleus and the egg pronucleus the zygote stage is reached. Hence, in mammals there is a remarkable overlapping of the processes of fertilization with the last maturation division of the female germ cell.

It is, nevertheless, convenient to speak of the female germ cells in the ovary as ova, although technically they are almost all primary oocytes. In ordinary histological sections oogonia are difficult to distinguish from the germinal epithelium of the surface and of cortical cords (see Chap. 7, pp. 248–57). Without considering the details of nuclear structure, it can be said that as soon as oogonia reach a diameter of 10–20 μ, depending upon the species, they are primary oocytes (oocyti primarii). These are soon enclosed in follicular epithelium, which somewhat later produces the zona pellucida (pp. 44–46). They remain as primary oocytes throughout the growth of the follicle until just before ovulation. In most mammals, within a few hours of that event, they undergo the first maturation, or reduction division, by giving off the first polar cell (polocytus primus) and thereby becoming secondary oocytes (oocyti secundarii). They are then ovulated and fertilized as described above. The reader of this or other literature on the mammalian ovary must keep in mind that the term "ovum" is commonly used in the broad sense, and that the so-called ova found in follicles are usually in the form of primary oocytes, or else in some phase of degeneration, if the follicle is atretic. (For a detailed account of the biology of the mammalian egg, see Austin, 1961.)

Normal ovarian follicles

Types

Most microscopic sections of adult ovaries contain several ovarian follicles in various phases of development or atresia (see Fig. 2.1). The genesis of follicles is discussed in Chapters 3 and 4, but it may be said here that the four stages of follicles usually recognized are: primordial, primary, secondary, and vesicular (Figs. 2.2, 2.3, 6.1, and 6.4). Further subdivisions of some of these categories have been made and are useful

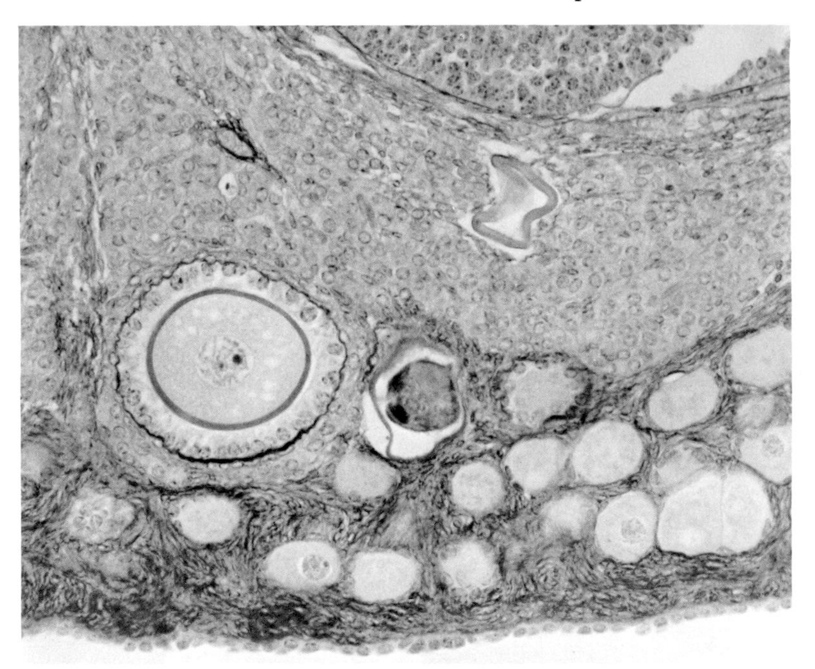

Fig. 2.2 Superficial cortical area of the ovary of a fox squirrel (*Sciurus niger*) in proestrus in June following a spring pregnancy. Note the surface epithelium, the indefinite tunica albuginea, and the primordial and primary follicles. The largest primary follicle shows the zona pellucida, the basement membrane of the follicular epithelium, and the theca interna. Adjacent to it on the right is an atretic primary or secondary follicle. Two corpora aretica with their zones of thecal type interstitial gland tissue lie between the large primary follicle and the small sector of the wall of a large vesicular follicle at the upper right. One of these still retains the zona pellucida. *S. niger*, 50. × 250.

at times. The final stages of a vesicular follicle just before rupture are quite obvious (see Figs. 6.23 and 6.24), thus the term "ripe" or "preovulatory" follicle is often used. The term "ovulated" or "ruptured" follicle is also useful, but the moment ovulation is completed is usually thought of as marking the end of the follicle and the beginning of the corpus luteum. Degeneration is a normal and very frequent occurrence at all stages of development, except in the primordial and the fully ripe stages, where it does occur, but less often. Such degenerating follicles are usually called atretic follicles (folliculi atretici), and if the changes are not too advanced may often be recognized as having reached one

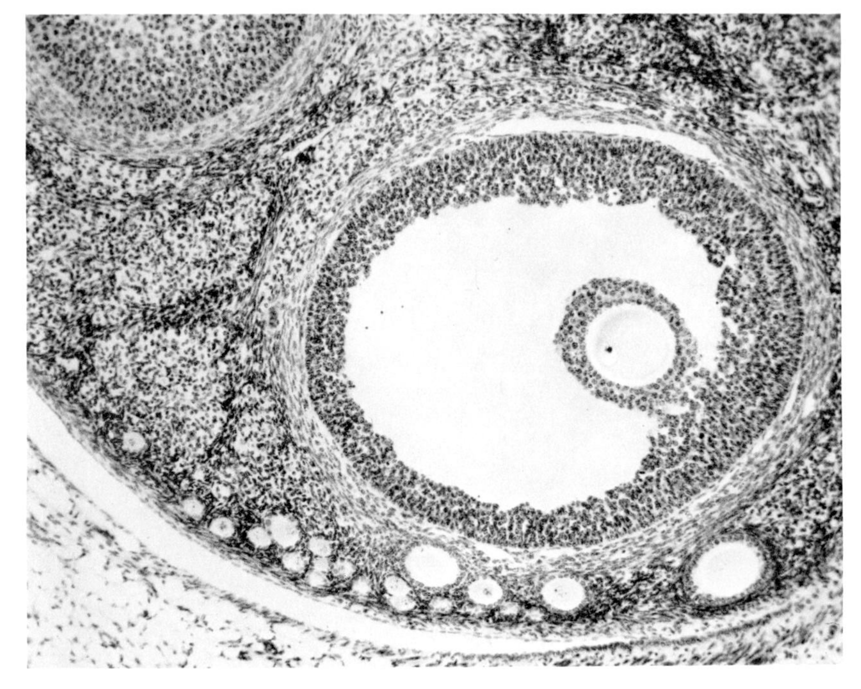

Fig. 2.3 Large vesicular follicle of a summer-born red squirrel (*Tamiasciurus hudsonicus*) in her first proestrus the following April. This follicle shows the first signs of atresia—a relatively thin cumulus oophorus and a few pyknotic nuclei and cell fragments bordering the lumen or floating in the liquor. Masses of thecal type interstitial gland tissue are numerous. *T. hudsonicus*, 62. × 100.

of the four normal stages, even though degeneration has begun (Figs. 2.2–2.4 and 2.6). Most ovaries contain several times as many atretic secondary and vesicular follicles as normal follicles in the same stages.

An egg is considered to be in a follicle as soon as a definite layer of cells becomes arranged around it as a uniform capsule of follicular epithelium. In most mammals the best evidence indicates that the follicular epithelial cells are derived from embryonic cells (i.e., undifferentiated pluripotential cells) of the stroma. Such follicles with a simple squamous epithelium around the egg are called primordial follicles (folliculi primordiales) (see Fig. 2.2).

The simple squamous cells eventually enlarge to form a simple low columnar epithelium, and in this way the primordial follicles are converted into primary follicles (folliculi primarii) (see Fig. 2.2).

The simple columnar cells of primary follicles multiply, first forming a double, and finally a stratified cuboidal follicular epithelium, or granulosa. These are secondary follicles (folliculi secundarii) (see Figs. 6.4 and 6.36). Although subdivisions of this stage have been made by certain authors to suit their particular purposes (Pincus and Enzmann, 1937), they are omitted here because there is no general agreement as to their delimitations and because such minute partitioning is rarely useful.

Multilayered secondary follicles transform into vesicular, or Graafian, follicles (folliculi vesiculosi) by the progressive formation of large intercellular fluid-filled spaces about halfway between the oocyte and the outer surface of the follicle (see Figs. 4.23, 6.32, 6.33, and 6.63). In the majority of mammals these spaces enlarge until they coalesce into a single large cavity, the follicular antrum (antrum folliculare), which partially surrounds the oocyte ensheathed by several layers of epithelial cells. The oocyte with these surrounding cells once occupied the center of the secondary follicle, but now forms a hillock, the cumulus oophorus, attached at one side to the follicular wall and projecting into the otherwise spherical antrum (see Figs. 4.5, 6.9, 6.23, 6.24, and 6.38). In many small mammals (shrews, moles [Talpidae], bats [Microchiroptera]), the antral space remains thin and the oocyte with its cumulus is only slightly displaced from its original central position (see Figs. 6.31–6.35). In other small mammals (mice and rats [Muroidea], squirrels [Sciuridae]) and in most larger ones, the follicular fluid accumulates rapidly, greatly enlarging the antrum and leaving the oocyte in its cumulus in a very eccentric position, so that, in species the size of man, fully grown follicles may have an antrum diameter of 20 mm with a cumulus of only about 0.5 mm fastened inconspicuously at some point inside the follicular wall. In a few species (pikas [Ochotonidae], rabbits and hares [Leporidae]), the follicle remains small, and, although the antrum is relatively much larger than that of shrews or bats, the cumulus remains in a nearly central position and is attached more or less all around by several radiating trabeculae of follicular epithelium (see Fig. 6.25). In the large Madagascar "hedgehog" (*Setifer setosus*) (Strauss, 1938), and probably in some related genera, no antrum forms and the ripe follicle is structurally like a multilayered secondary follicle (see p. 143 and Fig. 6.26).

Since changes in the cumulus are intimately involved in the process of ovulation, its detailed structure must be considered. The oocyte is surrounded by a very thin perivitelline space enclosed by the zona

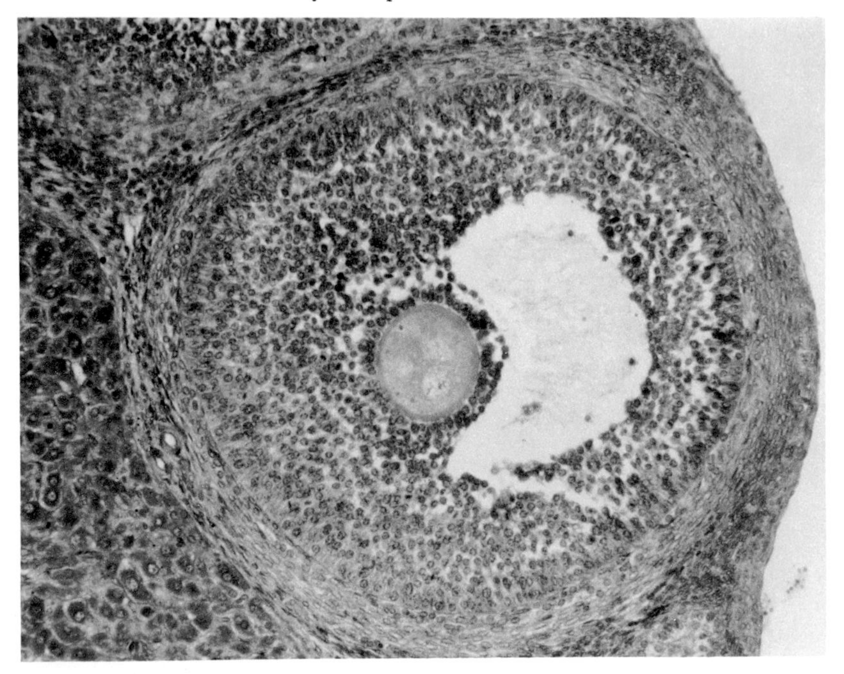

Fig. 2.4 Medium-sized vesicular follicle of a parous red squirrel (*Tamiasciurus hudsonicus*) in late July early in her second pregnancy of the year. A portion of a corpus luteum of the current pregnancy is at the lower left. *T. hudsonicus*, 125. × 150.

pellucida (Figs. 2.5, 6.24, and 6.25). A layer of columnar or pseudostratified columnar cells, the corona radiata, immediately surrounds the zona pellucida. Outside of these is a layer of ordinary stratified cuboidal epithelium most of which borders on the antrum but part of which is continuous with the inner surface of the follicular epithelium of the follicular wall.

Growth and structure

Those vesicular follicles destined to ripen grow rapidly during proestrus. Increase in diameter of the liquor-filled antrum is the most obvious feature of this growth, although there is, of course, an accompanying multiplication of granulosa cells and an increase in vascularity of the theca interna and in size and number of the glandular cells of this layer. There appears to be some stretching of the follicular wall from increased

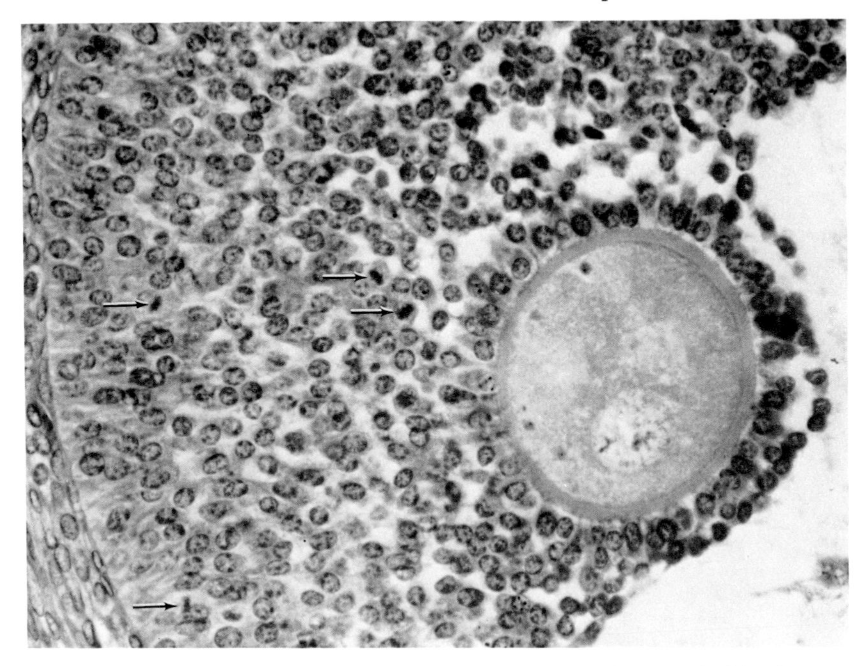

Fig. 2.5 Higher magnification of the follicle of the red squirrel (*Tamiasciurus hudsonicus*) shown in Figure 2.4. Mitotic figures (*arrows*) are apparent in the outer granulosa in spite of early atresia. The zona pellucida is well developed and at this magnification appears to be in direct contact with the oocyte. *T. hudsonicus*, 125. × 300.

pressure of the liquor, yet there is little, if any, demonstrable increase in intrafollicular pressure during this period (Asdell, 1960; Blandau and Rumery, 1963; Espey and Lipner, 1963; Rondell, 1964*a*, 1964*b*). The final preovulatory hours are characterized by a swelling of the inner granulosa cells, particularly those of the cumulus oophorus. In the latter area, the cells become widely separated by liquor-filled spaces which then make up part of the general antrum; this process thus frees, or almost frees, the egg with its corona cells from the follicular wall (see Figs. 4.6, 4.7, 6.23–6.25). Such a follicle with loosened cumulus is properly called a ripe follicle (folliculus maturus). When rupture of the follicle occurs, both the egg with its corona cells and the loosened cumulus cells normally float out into the peritoneal cavity or cavity of the bursa ovarica along with the liquor.

In at least one species, *Setifer setosus*, no antrum forms (Strauss,

1938). In these animals, the ova mature in follicles still having the structure of the secondary stage (see Fig. 6.26). Even more curiously, the sperms penetrate these follicles and fertilize the eggs in situ (see Figs. 6.27 and 6.28). The fertilized egg in the pronuclear stage is then extruded from the ovary by a coordinated process of rupture and contraction of the sheath of the follicle and by a shift in position of the granulosa cells to expose the zygote on the ovarian surface (see Figs. 6.29 and 6.30).

Thecae

A description of the sheaths, or thecae, of normal follicles should precede a discussion of the fate of ruptured and atretic follicles and of corpora lutea. In all mammals we have studied, there is a more or less complete layer, or zone, of glandular cells mingled with a rich capillary plexus immediately adjacent to the granulosa of the ripe follicles. This is undoubtedly an endocrine gland and is called the *thecal gland* (Mossman, 1937; Stafford, Collins, and Mossman, 1942). It appears in an apparently functional state around the larger follicles at the beginning of proestrus (see Figs. 6.37–6.44) and degenerates after follicular rupture, usually during the period that the cleaving eggs are still in the oviduct (see Fig. 6.45). In secondary and early vesicular follicles, it is in an embryonic condition and consists of a zone of rounded cells with many mitotic figures and a prominent developing capillary bed (see Fig. 6.36). In this period it may be called an embryonic thecal gland, or simply the theca interna, which was the term applied to it before its glandular nature in the mature follicle was emphasized.

In many mammals, especially those of medium size (domestic cat) to large size, a thin enveloping layer of spindle-shaped cells forms on the outside of the theca interna somewhat after the latter first appears. This is the theca externa (see Fig. 6.41). Its cells may possibly be contractile, but probably are never typical smooth muscle cells (Rocereto, Jacobowitz, and Wallach, 1969). Upon follicular rupture, it thickens, apparently by contraction, but becomes dispersed and soon after loses its identity. The fibrous capsule which surrounds older corpora lutea in some species (see Figs. 4.14 and 4.15) is not derived directly from it.

Zona pellucida

The zona pellucida is a normal component of ovarian follicles. It encases the oocyte, separating it from the neighboring follicular cells. In fresh preparations, it has a jellylike consistency (Sotelo and Porter, 1959), but in fixed material the appearance varies: it may resemble concentric lamellae, matted fibrils, radial striations, or vacuolations.

The origin of the zona pellucida has long been a controversial subject. The oocyte and the follicular cells have both been implicated, individually and in concert. Based on their observations of polyovular follicles, Beneden (1880) and Hartman (1926) concluded that the zona pellucida was a product of the oocyte alone, since mutually opposing surfaces of adjacent oocytes were covered by a zona pellucida and no follicular cells were seen in the area. However, Duke has occasionally observed in the ovaries of the dog and an oriental civet (*Viverra tangalunga*) growing follicles in which a zona pellucida surrounded a central core of granulosa cells and no oocyte was seen (unpublished).

Recent studies, in which the techniques of histochemistry and electron microscopy were utilized, strongly indicate that follicular cells are the chief source of the zona pellucida (Chiquoine, 1960). Chemically, the material appears to consist of mucopolysaccharides (Stegner and Wartenberg, 1961; Stegner, 1967) with a sialic acid component (Soupart and Noyes, 1964), which result from the synthetic activity of the rough endoplasmic reticulum (Merker, 1961). This organelle is much more elaborately developed in the follicular epithelial cells than in the oocyte. As visualized by both light and electron microscopy, the first signs of zona pellucida formation are isolated wedges of material between adjacent primary follicular cells rather than a uniform deposit around the surface of the oocyte (Aykroyd, 1938; Odor, 1960). Odor and Blandau (1969*b*) have identified the zona pellucida in primordial follicles of early postnatal mice; this is the earliest such identification in ontogeny and in the follicular cycle. The thickness of the zona pellucida gradually increases until the layer measures up to 10, occasionally 15, microns, but surprisingly few studies have included precise measurements. Stegner (1967) gives an excellent review of the ultrastructure from the oogonium to early cleavage, including follicular cells, the zona pellucida, and fertilization.

Just why an accumulation of such material should develop between the oocyte and its surrounding follicular cells has been a puzzle to many students of reproduction, especially when this occurs at the time of most rapid oocyte growth. Heape (1886) described the radial striations in the zona pellucida of the common European mole (*Talpa europaea*) oocyte and suggested that protoplasmic processes from the follicular cells transferred nutriment to the oocyte via the radial canals. Utilization of this pathway for the transfer of fatty materials was shown experimentally by Wotton and Village (1951) in the oocytes of kittens. Practically every electron-microscopic study of the ovarian oocyte and the follicular cell relationship has a description of long protoplasmic processes from the follicular cells and of microvilli from the oocyte lying within or piercing the zona pellucida. Cellular exchange of metabolites has been the functional activity usually attributed to them.

The zona pellucida of some species (rabbit, rat, and hamster) is permeable to all known essential nutrients, to the majority of pharmacologically active compounds, and to natural steroid hormones. There is also some evidence to suggest that it is involved in the sperm-block mechanism to prevent polyspermy (Austin, 1961).

The zona pellucida is thus not just an inert, amorphous barrier, but is a functional part of the ovarian follicle — one that invites further study.

Luteal glands

Usually, examination of the ovaries of an animal that has recently been in estrus, or is pregnant, reveals one or more stages of corpora lutea. These masses of glandular tissue, formed from ovulated follicles, could preferably be called *luteal glands.* Fully developed ones are usually composed of a solid spheroidal mass of large polyhedral gland cells richly supplied with sinusoidal blood and lymph capillaries. They occupy the sites of ruptured follicles and are formed mainly by enlargement and multiplication of the follicular epithelium, often supplemented by apparently identical gland cells added by metaplasia of the adjacent stromal cells. Ordinarily, they grow even larger in diameter than the ripe follicles from which they are derived. These glands usually begin to degenerate late in pregnancy, or during lactation, sooner if pregnancy does not occur. They are gradually reduced to small masses of connective tissue often containing numerous macrophages with a deep yellow, red, or brown pigment. These connective tissue bodies, known as corpora albicantia, tend to sink deep into the ovary and are last seen near or deep to the corticomedullary boundary. Their pigment usually fades rapidly in preserving fluids. Because the corpora albicantia contain relatively few collagenic fibers in most mammals, they soon blend with the general ovarian stroma and disappear completely. In a few, especially bovids, their pigmentation persists for months and becomes more intense as they shrink in size. A cow ovary may contain dozens of these bodies, ranging in size from about one-third the diameter of an active corpus luteum (15–25 mm) to small specks, and in color from pale yellow to a bright deep orange red. In human ovaries, they are true to their name "albicans," for they contain little pigment and few cells, and masses of very fine collagenic fiber bundles account for their distinctly white appearance in gross specimens (see Fig. 5.5).

Atretic follicles and interstitial gland tissue

Manner of degeneration

Since atresia of follicles is a normal circumstance, and since it often results in the formation of an interstitial gland mass at the former site of the follicle, no description of an ovary is complete without an account of the beginning and course of follicular atresia. Degeneration of primary and primordial follicles (see Fig. 2.2), and even of naked ova and oogonia undoubtedly occurs, but these are such small and scattered structures that the changes do not significantly modify the microscopic appearance of the organ as a whole. However, atresia of secondary and vesicular follicles is conspicuous. Such follicles present a variety of appearances depending (1) on the stage of the follicle at the time degeneration begins; (2) on the exact nature of the process, which varies considerably in different species and probably even in the same species; and (3) on the extent of the retrogressive changes at the time (Figs. 2.3–2.6).

The questions are still unsettled as to what determines that certain follicles must degenerate, and as to whether atresia starts with changes in the egg or in the follicular epithelium (Ingram, 1962). The first easily recognizable changes are in the epithelial cells immediately adjacent to the lumen of vesicular follicles and in the layers just external to the corona radiata of secondary follicles. The first signs of the process in these regions are pyknosis and karyorrhexis and an absence of mitoses. In vesicular follicles, these nuclear fragments and pyknotic cells tend to float centralward into the liquor where they are easily seen. This soon results in a noticeable shrinkage of the cumulus oophorus and smoothing of its antral surface (Figs. 2.4–2.6), a process which apparently proceeds quite rapidly and allows the ovum, covered only by its zona pellucida, to enter the follicular liquor. The egg, with its zona, then either lies bare against the inside of the degenerating granulosa or floats freely in the now-disappearing liquor folliculi. In most species, the

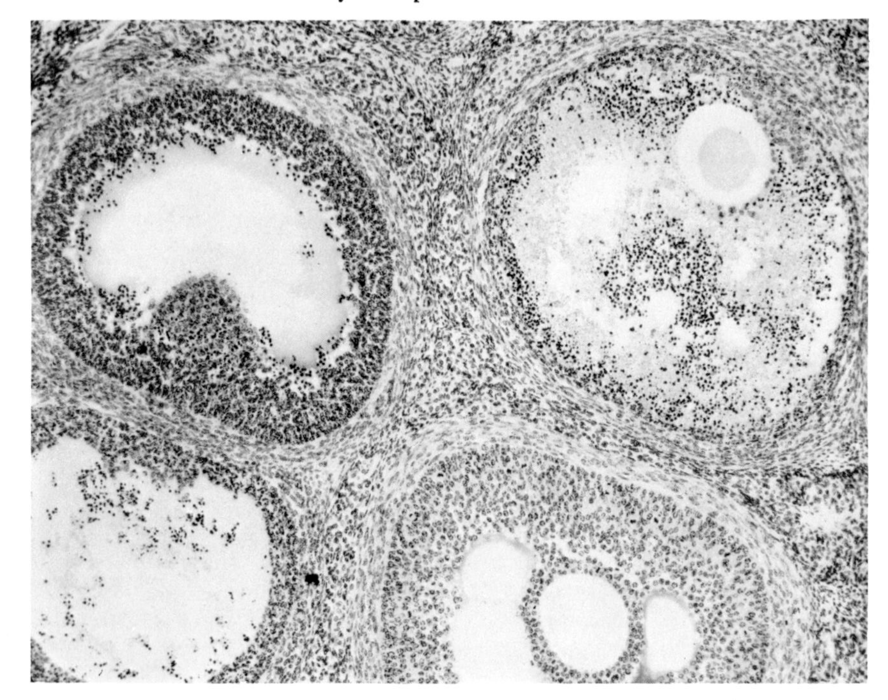

Fig. 2.6 Parts of four medium-sized vesicular follicles in various stages of early atresia, from the ovary of a summer-born red squirrel (*Tamiasciurus hudsonicus*) in early April late in her first proestrus. *T. hudsonicus*, 62. × 100.

granulosa disappears completely, and fibroblasts and various types of leukocytes invade and replace the residual liquor in the cavity. The egg eventually fragments and disappears, often with the aid of polymorphonuclear leukocytes which penetrate it through its apparently still-intact zona pellucida. In fact, in some species (guinea pig and porcupine [*Erethizon dorsatum*]), the last remnant of an atretic follicle is a shriveled zona pellucida lying in a space in the stroma just big enough to contain it. Usually, however, the zona pellucida undergoes dissolution soon after the egg does. In other species, notably in man, atresia of medium to large vesicular follicles eventually results in the formation of small corpora albicantia identical in structure to those formed from degenerated corpora lutea (see Fig. 5.6).

Degeneration of secondary follicles appears to take about the same course. It is difficult to prove, but it seems fairly certain that granulosa degeneration usually causes the formation of cavities in the secondary follicles, since vesicular follicles of very small diameter are almost always in early atresia.

In all species, follicular atresia eventually results in disappearance of the granulosa, but in some, the mink (*Mustela vison*), for instance, there is a transient and partial transformation of the granulosa into luteal cells (see Fig. 6.87). In others, like the porcupine, mountain viscacha (*Lagidium peruanum*), and agouti (*Dasyprocta aguti*), complete lutealization of many atretic follicles occurs during pregnancy. In the porcupine (Mossman and Judas, 1949) and the viscacha (O. P. Pearson, 1949), these accessory corpora lutea form in large numbers from atretic follicles, reach the same cytological condition as the primary corpus luteum, and persist just as long (see Figs. 6.100 and 6.101). Thus, in certain species, some atretic follicles may transform into small but otherwise typical and undoubtedly functional corpora lutea, which then have a history parallel in every way to the primary corpus luteum of the species.

Thecal type interstitial gland tissue

A prominent morphological and important physiological feature of many atretic follicles in most species is the formation from their theca interna of interstitial gland tissue. A zone of these glandular cells surrounds many atretic secondary and vesicular follicles (see Figs. 2.1, 6.5, 6.48, 6.56, and 6.59). These interstitial gland masses usually persist long after all remnants of the inner follicular parts have vanished (see Figs. 6.60–6.62). In many species, they finally break up into small groups of cells scattered in the medulla. The fate of these cells is often uncertain (Stafford and Mossman, 1945), but in most species they eventually just dedifferentiate in situ.

The presence of well-differentiated interstitial gland tissue is an obvious feature of most ovaries during proestrus, estrus, and the first half or two-thirds of the gestation period. (See Chapter 6 for a discussion of the occurrence of interstitial gland tissue in fetal, infantile, and juvenile mammals, and after the menopause in women.) It will be recalled that the thecal glands which develop from the theca interna of ripening follicles seem to be in a functional state only during proestrus and estrus. It is remarkable that the process of atresia leads to the metaplasia of the embryonic thecal gland cells (theca interna cells) into interstitial gland cells.

Other types of interstitial gland tissue occur in certain species. One of these types seems to be derived directly from the stroma without any special relation to atretic follicles or other structures. Another type results from glandular hypertrophy of medullary cord epithelium. Some species (artiodactyls) have relatively inconspicuous interstitial gland tissue at any time in the cycle.

Medullary cords, rete, and efferent ductules of the ovary

Medullary cords or tubules of epithelial tissue are usually inconspicuous, but with careful study can be demonstrated in at least the infantile ovaries of most species (see Figs. 4.1, 4.19, 6.16, and 7.2). They are commonly composed of low cuboidal cells with little cytoplasm relative to nuclear size, an indication that they are in an embryonic or undifferentiated condition. These cords and tubes usually consist of interrupted short segments scattered in the medulla and to some extent in the cortex (see Figs. 6.17, 6.80, 6.81, and 6.83). Occasionally, they form long continuous strands connected at one end to the rete ovarii (see Fig. 6.88).

The rete ovarii is probably always present and is usually rather conspicuous in sections of the area of the hilus near the cephalic end of the ovary. Its epithelium is commonly low cuboidal and its network of lumina is sometimes locally dilated or even grossly cystic. Serial sections almost always show well-preserved ductuli efferentes ovarii connecting the rete to the epoophoron portion of the vestigial female ductus deferens in the mesovarium (Fig. 2.7).

Summary

The surface epithelium of the ovary is continuous with, and comparable to, the peritoneal mesothelium, but unlike the general peritoneum it is more commonly simple columnar than squamous. The fibrous layer of the peritoneum may be absent on the ovary, or it may exist in varying degrees of differentiation as the ovarian tunica albuginea, but it is always relatively thinner and less fibrous than the tunica albuginea of the testis.

In the adult ovary, the outer zone, or cortex, contains the follicles and their derivatives; the inner core, or medulla, contains the major intraovarian blood and lymph vessels, remnants of the medullary cords, and some or all of the rete ovarii. Interstitial gland tissue is usually present in both. Adult type irregular fibrous connective tissue is typical of the

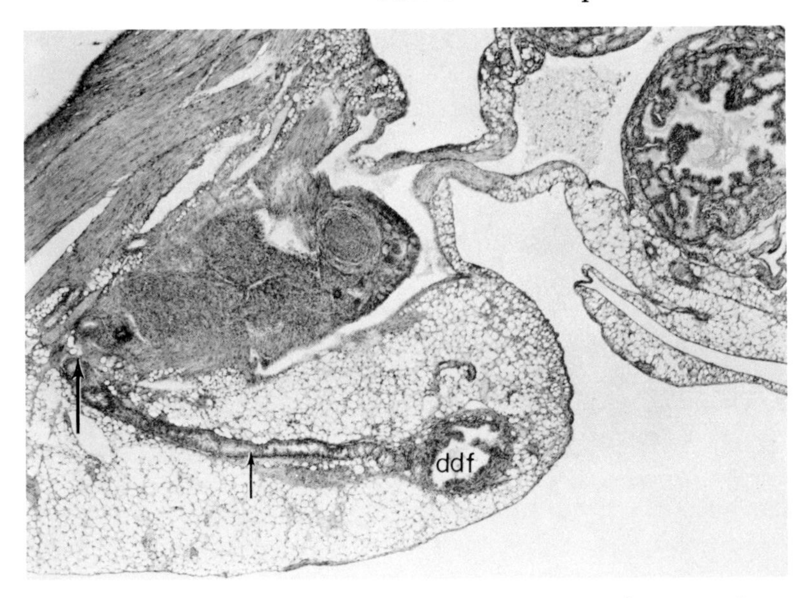

Fig. 2.7 Mesosalpinx, mesovarium, and hilar region of the ovary of a star-nosed mole (*Condylura cristata*), showing the ductus deferens femininus (*ddf*), connected by a ductulus efferentis ovarii (*small arrow*) to the rete ovarii (*large arrow*). Massive smooth muscle occurs in the mesovarium, and a broad band of the muscle enters the ovary. *C. cristata*, 13. × 40.

medulla, while that of the cortex is of the embryonic, that is, cellular type. Nevertheless the boundary between the cortex and medulla is usually very irregular and indistinct.

Almost all of the so-called ova, or eggs, seen in adult mammalian ovaries are primary oocytes. Technically, they are not ova until they have completed the second meiotic division, and in mammals that usually does not occur until these ova have been shed and penetrated by a sperm cell.

As soon as an oocyte is surrounded by a definite single layer of epithelioid cells it is considered to be in a follicle. The usually recognized stages of follicular development are primordial, primary, secondary, vesicular, and ripe or preovulatory. In at least one species, follicles ripen without developing an antrum.

During the late primary follicle stage, a mucopolysaccharide capsule, the zona pellucida, is formed immediately around the oocyte. It is believed to be secreted by the inner layer of follicular cells.

From the secondary stage onward, all mammalian follicles are closely surrounded by a capillary net mingled with potential gland cells. This theca interna becomes a differentiated thecal gland as the follicle ripens. In some species, especially larger ones, a sheath of cellular connective tissue, the theca externa, forms just outside the theca interna. Upon follicular rupture at ovulation, the thecal gland cells rapidly degenerate, but the follicular epithelium transforms into luteal gland tissue, thus forming a corpus luteum. These bodies may grow to some extent by accretion of luteal cells differentiated from the immediately surrounding stroma. Corpora lutea eventually degenerate. In larger animals fibrous and often pigmented "scars," corpora albicantia, mark their sites, sometimes for many months, but in most smaller mammals their degeneration is rapid and no trace remains.

Roughly 95% of ovarian follicles undergo atresia (degeneration). Follicles at any stage of development may become atretic, but in adult ovaries the process is most active during proestrus, estrus, and early pregnancy. Commonly, the undifferentiated gland cells of the thecae internae of degenerating follicles transform into interstitial gland cells.

Medullary cords or tubules are usually inconspicuous or absent in ovaries of adult mammals, but the rete ovarii and epoophoron always persist.

Three

Development of the mammalian ovary

The organs and organ systems of placental mammals generally originate and become almost completely differentiated structurally and often functionally within the first half of the gestation period. Beyond this period, minor differentiation usually continues, but the more obvious activity is growth, which is usually very rapid from the standpoint of absolute increment, although not so rapid on the basis of proportional increase as it was during the earlier period. However, there are a few cases where growth and differentiation are practically inseparable, as for instance in teeth and bones, where the growth of blastemal (membranous) or cartilaginous elements and their differentiation into the definitive tissue of dentine or bone are contemporary processes. Thus, in the skeleton, differentiation is a major process up to the end of the growth period.

The reproductive organs show an even closer and more complex interrelation between growth and differentiation. In the male, both processes become complete at the end of puberty, unless one considers spermatogenesis as a continuation of differentiation, but in the female neither growth nor differentiation of the reproductive organs is complete at puberty.

When pregnancy occurs, the ovary and the uterus undergo still further enlargement and morphological change, as manifested by the corpus luteum and interstitial gland tissue of the ovary, and by the decidua and the great growth of the myometrium and vascular elements of the uterus. Likewise, the mammary gland hypertrophies and undergoes marked differentiation during the first pregnancy. Furthermore, throughout the active reproductive period the internal genitals and the mammae undergo recurring periods of growth and differentiation followed by involution. In other words, throughout the reproductive period of life, the female mammal's major generative organs retain their embryonic nature, being capable of both growth and differentiation (Mossman, 1952).

Perhaps of all these organs, the ovary is the most strikingly "embryonic." It merits this distinction for two reasons: first and more important, it produces germ cells; second, with each reproductive cycle it produces new vesicular follicles and new luteal and interstitial gland tissue. This is not just a reactivation of structures already present, as in the mammary gland, although redifferentiation does account for much interstitial gland tissue. Obviously, both the stroma and epithelium of the ovary have unusual powers of regeneration and differentiation, so far as ovarian tissues and functional units are concerned. It is probably safe to say that both stromal and epithelial ovarian cells are pluripotential until senescence begins. (See discussions of the embryonic nature of ovarian tissue in Chapter 7, pp. 257–59 and 267–68.)

Because the primordium of the gonad occurs in embryos still possessing gill slits and with the first rudiments of limb buds, the period of development and differentiation of the ovary is longer than that of most organs. Since much of this development takes place during puberty and is discussed below (Chapters 4, 5, and 6), we shall deal here mainly with the embryonic and fetal periods.

As with any other organ, the morphology of the ovary cannot be thoroughly understood without knowledge of its development. But unfortunately, in spite of years of study of gonad embryology by descriptive anatomists and experimentalists, there are several fundamental points of much practical and theoretical interest about which there is still considerable disagreement and uncertainty. In fact, these are so numerous that an attempt to describe gonad development in a critical fashion involves so many qualifications that it is almost impossible to present the subject with reasonable continuity and clarity. We shall therefore first present a relatively uncritical account of the development of the ovary, based on the facts as we know them from the literature and our own observations and interpreted in the light of what seem to us the most reasonable theoretical considerations. With this as a basis for discussion, we shall then critically examine the details and shall try to point out how much is demonstrable fact and how much theory. We shall also attempt to show which theories seem to be the more tenable.

Since it is awkward to cite authorities at every point, we here acknowledge that much of the information is taken from the following more recent studies of gonad development: Bookhout, 1945; Fischel, 1930; Gillman, 1948; Gruenwald, 1942; Hargitt, 1925, 1926, 1930*a*, 1930*b*, 1930*c*; Kingsbury, 1913; Sauramo, 1954; Wagenen and Simpson, 1965; and Witschi, 1948, 1963. Two accounts concern the development of the cat ovary: the one by Kingsbury is a most detailed and careful study and gives an adequate review of most of the earlier literature; the other (Sainmont, 1906) is an excellent and well-illustrated description extending well into the postnatal period. Hargitt was probably the first investigator to base a study of the development of the mammalian ovary on an extensive chronological series of accurately dated embryos and neonates. He concluded: "The entire ovary of the rat, including its germ cells, is produced by a proliferation of the peritoneal epithelium; germ cells are not a special type of cell which migrate into the gonad" (1930*a*: 316). Bookhout has done the obviously necessary but still infrequently attempted thing of studying a complete series through the embryonic, fetal, and immature stages to maturity. Gillman and Witschi have provided beautifully illustrated and detailed descriptions of human material. Gruenwald has contributed greatly to an understanding of the fundamentals of ovarian development by his concepts of the close relationship between the mesothelium and mesenchyme of the gonadal ridge and what one might call the equipotentiality of the superficially different histological elements of the gonads. His comparative viewpoint and his open-minded interpretation of demonstrable facts have made his account perhaps more scientifically acceptable, although it is less complete, than many of the others. Wagenen and Simpson have provided a handsomely illustrated and complete atlas of the development of the ovary in man and the rhesus monkey.

Gross development up to sex differentiation

The embryonic gonad, or gonadal ridge (jugum gonadale), is first recognizable at about the time limb buds appear. This is in the late gill slit stage when, in the case of the human embryo, the total length is about 5 mm crown–rump (CR) and the age about 32 days. These gonadal ridges are longitudinal thickenings of the cuboidal coelomic epithelium along the ventromedial aspect of each mesonephros close to the root of the dorsal mesentery. They rapidly become more prominent, changing from narrow ridges to relatively much shorter, thicker, rounded ones, and finally to discrete, roughly cylindrical bodies usually with rounded ends and only about three or four times as long as they are thick. They reach this condition by the time the embryo has passed through the limb bud stage and is in the harelip or very early fetal stage, around 20 mm CR and 40 days in the human embryo.

Gross relationships

The change in shape of the gonad from a narrow longitudinal ridge with ill-defined ends to a discrete ellipsoidal body results in the formation of a dorsal mesentery suspending it from the mesonephros. As the mesonephros degenerates and shrinks in size, the most cephalic portion of the gonadal ligament becomes the suspensory ligament of the gonad (ligamentum suspensorium gonadis), and the most caudal portion becomes the gubernacular fold. These, together with the usually thicker dorsal intermediate portion, compose the mesentery of the gonad, the mesorchium of the male or the mesovarium of the female.

In both sexes, the degenerating mesonephros, together with the blood vessels supplying it, becomes included in the more dorsal and cephalic portions of this mesentery; hence, in the female, the remnants of the mesonephros, the epoophoron, are located in the mesovarium near the cephalic end of the ovary (see Fig. 1.34), and the ovarian artery and vein course through this portion (see Fig. 1.32). As the female duct extends caudally from its place of origin lateral to the gonad, it crosses the gubernacular fold just caudal to the gonad and in a retroperitoneal relation to it, thus dividing the fold into two portions. The part remaining between the female duct and the caudal pole of the gonad in the female will eventually become the proper ligament of the ovary, that from the duct to the inguinal region, the round ligament of the uterus.

Further gross changes in the ovary lead toward (1) the attainment of the particular shape characteristic of the species; (2) the definitive relationship of the ligaments and oviducts to it, including the formation in some species of a more or less complete ovarian bursa; and (3) the descent toward or into the pelvis. This retreat caudalward is apparently correlated with the length and shape of the uterine horns. Without exception, so far as we know, mammals with long, relatively straight uterine cornua have abdominal ovaries, while those with either very short or coiled cornua or a simplex uterus have more caudal or completely pelvic ovaries. Indifferent stage gonads are shaped much like adult testes, but as they differentiate into ovaries they become more or less compressed laterally, hence somewhat bean-shaped. Many, however, eventually become almost as wide as long, and appear as pedunculated, somewhat fungiform masses. Others are highly lobulate and hence very irregular in shape. A few, such as those of the porcupine (*Erethizon dorsatum*) and man, retain the flattened elongate shape, their ends tapering rather gradually into the suspensory ligament cephalically and into the proper ovarian ligament caudally.

Histogenesis before sex differentiation

The histogenesis of the gonads is far more complicated and obscure than the gross development. The gonadal ridge is differentiated from both the coelomic mesothelium and the mesenchyme beneath it. Embedded in both these layers from the beginning are primordial germ cells (cellulae germinales primordiales). These migrate into the gonadal ridge, probably by way of the dorsal mesentery of the gut, from their origin in the endoderm of the yolk sac and rudimentary hindgut at the region of the posterior intestinal portal. (For more details on the germ cell problem in mammals, see Chapter 7, pp. 248–57.)

Gruenwald (1942) pointed out that the first indication of a gonadal ridge is an increase in height of the more or less cuboidal cells of the coelomic lining of the area, and that this is quickly followed by multiplication and concentration of the underlying mesenchymal cells. He also showed that soon after these processes start the basement membrane of the coelomic lining of the ridge disappears, that for some time afterwards there is no clear line of demarcation between the surface epithelium and the underlying condensed epithelioid mass, and that undoubtedly the latter is derived from both the mesenchyme and the epithelium. Mitosis is frequent in the whole mass at this stage, and the primitive germ cells also increase in number by mitotic division.

By the early limb bud stage, around ovulation age (OA) 35 days and 7 mm CR in man, the cells of the gonadal mass begin to arrange themselves into indistinct gonadal, or sex cords (chordae gonadales), which are oriented more or less radially, that is, perpendicular to the gonad surface. It is often assumed that this arrangement is caused by the ingrowth of these cords from the surface epithelium, but Gruenwald had rather convincing evidence that they form in situ, differentiating from the center of the mass outward. Each cord is usually continuous at its outer portion with the epithelioid surface zone, often with the surface epithelium itself.

The latest stage at which a gonad can be called indifferent sexually (about OA 40 days and 17–20 mm CR in man) is composed of a central reticulate mass of irregularly radiating epithelioid gonadal cords narrowly and incompletely separated by mesenchymal cells. Both the cords and the mesenchymatous tissue merge rather abruptly into the still-undifferentiated epithelioid outer zone, the thickness of which varies considerably among species. The surface cells of this zone are flattened on their coelomic surfaces but are otherwise indistinguishable from those beneath, since at this time they have no obvious basement membrane. The central or deep ends of the gonadal cords also end in an obscure manner, appearing to be more or less continuous with the mesenchyme of the hilar region, as if differentiating centripetally from it. Some believe the rete forms in this manner, although this has never been clearly shown.

Histogenesis of sex differentiation

At the end of the indifferent period, characters of the definitive sex first appear in the gonads of male embryos. The earliest distinctly ovarian characters become obvious much later, when primordial follicles begin to form. This is at about menstruation age (MA) 21 weeks and 160 mm CR in the human fetus. Since definite testicular characters appear at the earliest fetal stage and are clear-cut in fetuses which are much smaller than 160 mm, all fetuses of this period are considered females if they do not show testicular characters. In fact, in man the sex can be distinguished from external genitalia alone at about OA 10 weeks and 30 mm CR.

Sex differentiation in the female gonad begins as a continuation of the indifferent stage. It is easier to understand the features of this somewhat obscure early differentiation of the ovary if one first considers the rather definite changes that occur in the testis, keeping in mind that most of these are the things which do *not* happen in the ovary. Briefly, in the testis the following changes occur quickly at sex differentiation: (a) the gonadal cords become sharply delimited by basement membranes and are anastomotic with one another, especially distally and centrally; (b) the mesenchyme between the cords increases considerably in amount, generally separating the cords by as much or more than the cord width; (c) the outer cellular zone is rapidly converted into the peripheral portions of the cords, and mesenchymatous cells in the area form a distinct tunica albuginea; (d) the surface epithelium is reconstituted with a distinct basement membrane.

In the ovary, however, sex differentiation takes place as follows: (a) the gonadal cords remain as a fine-caliber, irregular, and ill-defined three-dimensional net, but soon become much more numerous than in the testis; (b) because the intercord mesenchyme does not increase as rapidly, the gonadal cords continue to be separated by relatively thin, irregular septa; (c) the outer cellular zone persists to at least the seventh fetal month and no clear-cut continuous tunica albuginea forms; (d) the surface epithelium is more or less reconstituted as a continuous layer but without a complete basement membrane and in places is often a double layer, from the inner one of which secondary, or cortical gonadal cords (chordae corticales) continue to invade the outer cellular zone. This condition occurs in the cat at around 30 mm CR (Sainmont, 1906), but in forms such as man there is no distinct difference between the secondary and primary gonadal cords at this time (about OA 40 days and 20 mm CR), for they are more or less connected to each other as well as to the surface epithelium.

Cords thus continue to be organized from the cellular zone while that zone is still proliferating both within itself and from the surface. These proliferations from the surface correspond to secondary cords, but may be indistinctly organized, particularly if no primary ovarian tunica albuginea develops. Therefore the often-expressed concept is not always applicable that the ovarian cortex develops from secondary cord proliferation of the surface epithelium outside of a primary ovarian tunica albuginea corresponding to that of the testis. A more generally applicable concept is that only a barely perceptible primary tunica albuginea or none at all forms in the ovary at this time, and that gonadal cords continue to differentiate from the outer cellular zone of the cortex, which may be covered either by a distinct, often double, layer of surface cells, surface epithelium, or by cells still indistinguishable from those of the outer zone itself. In the presence of a primary tunica albuginea, fairly definite secondary gonadal cords may sometimes be seen growing inward from the deeper of the two epithelial layers, or later even from the outer one. In the absence of a tunica, primary cords can be traced outward into the cellular zone and often through it, where they can be seen to connect with the cells at the surface; yet cords of cells which are not connected to the primary cords also seem to differentiate from the surface zone.

The latter are the closest to the classical concept of secondary gonadal cords that one sees. They probably occur in many species, including man (Gruenwald, 1942).

Gillman (1948) called this period in the human ovary, the time from the beginning of sex differentiation at about OA 37 days and 17 mm CR to that of the first primordial follicle formation in the medulla at about OA 18 weeks and 150 mm CR, the "stage of sexual differentiation and growth." He emphasized growth as the predominating feature, for the ovary at the end of this period is not markedly different histologically from its condition at the beginning. In Gillman's words, "The young ovary therefore consists of a covering layer of coelomic epithelium which merges imperceptibly with the underlying cellular layer. The sex cords are much thicker because of a general enlargement of the constituent cells, but remain in continuity with the rete cords. The primitive sex cells, a prominent feature of previous stages, are still easily seen in the massive sex cords. The innermost segment of the sex cord, where it joins the rete cord, represents the medullary cord of later stages. . . . Capillaries and connective tissue cells are scant except in the rete zone. . . . there is still no structural resemblance to the ovary of older fetuses" (pp. 97–98).

The remainder of the fetal period, OA 18 weeks, 150 mm CR to birth (300 mm CR) in the human species, Gillman called the "stage of formation of fetal stroma and primary follicles." It is better to call this simply the stage of follicle formation, for the primordial follicles, the first ones to form, usually have a simple squamous epithelium in contrast to the later primary follicle stage characterized by a simple cuboidal epithelium. Furthermore, as Gillman mentions, even vesicular follicles may develop before the end of this period.

Primordial germ cells and oogonia are numerous throughout the gonadal cords during the growth period. As the period of follicle formation approaches, the medullary portions of the cords begin to break up into primary medullary follicles (folliculi medullares), their epithelioid cells forming a single layer around the germ cells (now primary oocytes). (Winiwarter and Sainmont, 1908–1909, appear to have been the first to use the term "medullary follicle.") Somewhat later the same thing occurs in the cortical zone, but because of the greater number of cords and germ cells in this area and the lesser amount of intervening stroma, the cortical follicles (folliculi corticales) are usually much more closely packed. Still another difference may exist between the cortical follicles and those of the medullary region in certain species such as the cat (Kingsbury, 1913) and the squirrels (Sciuridae). In these, the medullary cords often do not break up completely but may remain, thus connecting a number of follicles which obviously develop linearly within the cords. Such cordlike extensions of both nonvesicular and vesicular follicles of this region sometimes occur in postnatal animals. (See the discussion of medullary follicles in Chapter 4, pp. 88–91.) During this period of follicle formation the stroma becomes more obvious and loses its mesenchymal character. Gillman (1948) explained this change as an "invasion" of fetal stroma, and said it is "lethal" to all unencapsulated sex cells. It seems far more likely that stromal cells do little invading, but rather that they differentiate from the embryonic stroma or mesenchymal cells already present.

With the formation of the first primordial follicles the gonad has clearly become an ovary. Of course, differentiation is by no means complete, yet this point is usually considered to mark the end of sex differentiation in the female gonad, in the sense that the gonad can now be unmistakably distinguished as an ovary. Insufficient data are available to characterize accurately the stage of development at which this point in differentiation of the female gonad is reached in mammals in general; but it can be roughly estimated to be well into the fetal stage in contrast to the embryonic or very early fetal stage at which clear-cut testicular characteristics appear in the male gonad.

The late fetal and early postnatal ovary

In most species, the ovary remains in this follicle-forming stage until well into postnatal life (Fig. 3.1). In some squirrels, it is known that the first follicles to mature and rupture are those formed from the medullary cords (see Chapter 4, pp. 88–91). It is also almost certain that the large follicles characteristic of the late fetal, newborn, and infant human ovary are of medullary origin (see Fig. 5.1) (see Chapter 5, pp. 100–104).

The human fetal testis (at about 100–140 mm CR and OA 14–18 weeks) (Gillman, 1948) has well-developed interstitial gland tissue apparently derived from the mesenchyme between the gonadal cords. Fetal type interstitial gland tissue also occurs in human ovaries at 190 mm CR and MA 5½ months (Wagenen and Simpson, 1965). Gillman describes thecal cell hypertrophy around atretic follicles of late human fetuses and suggests that this may be comparable to the male interstitial

Fig. 3.1 Cortex and outer medulla of a 2½-year-old child. This is typical of the ovaries of infant large mammals. Compare with the ovary of juvenile rodents, Figures 1.2 and 4.3. Human, 143L. × 80.

gland. Such thecal interstitial gland tissue certainly does not commonly occur in fetal stages of altricial mammals, although there obviously is need for much more extensive observations on the possibility of gland cell formation in both prenatal and immature stages, especially in precocial species. Gerall and Dunlap (1971) showed significant compensatory hypertrophy of the remaining ovary after hemispaying in 10- and 15-day-old rats. This indicates that a gonad-pituitary feedback mechanism was operative in these newborn animals.

Origin and migration of primordial germ cells

It has long been noted that large cells resembling in many ways definitive spermatogonia and oogonia occur in the early gonadal primordia of mammals and other vertebrates. It has also long been known that in birds and mammals there are cells which are identical in appearance and which can be found at this time and even earlier in embryos having only a few somites, in the mesothelium and mesenchyme of the dorsal mesentery of the gut, and in the endoderm and splanchnic mesoderm of the yolk sac and hindgut at and near the region of the posterior intestinal portal. On the basis of a considerable amount of morphological evidence, but of a very little direct experimental evidence, the theory has grown up that these are the first distinct germ cell progenitors; that in mammals they migrate, probably by amoeboid activity, within the dorsal mesentery and coelomic wall to the gonadal ridge; and that they are the ancestors of all future germ cells of the individual. Modifications of this theory are numerous; some workers maintain that all these primordial germ cells eventually degenerate, and that the definitive cells are derived later from the germinal epithelium of the embryonic gonad or, in the case of the female, from the germinal epithelium of the fetal, immature, or even adult ovary.

It has been extremely difficult to obtain evidence which gives clear-cut proof of many of the points on which this theory or its variations have been based. Among the many reviews of the subject that have been written, those by Gillman (1948) and Witschi (1948, 1963) should both be consulted for their evidence and views. Everett (1945) considered very adequately the germ cell problem in the vertebrates as a whole. A more detailed discussion of this problem appears in Chapter 7, pp. 248–57, but for the purpose of general orientation at this point the evidence is reasonably sound that primordial germ cells do arise and migrate as indicated above. However, there is no proof that they are necessarily the ancestors of all future germ cells of the individual.

Summary

The reproductive organs of mammals, particularly those of females, tend to remain potentially "embryonic" throughout life, that is, they undergo recurring periods of growth and differentiation alternating with periods of regression and inactivity. Both ovarian epithelium and stroma behave this way, and both tend to be pluripotential, often apparently giving rise to the same differentiated cell types.

Mammalian gonads first appear as ventromedial ridges on the mesonephros in the late gill slit period of development, and become discrete bodies just before the beginning of the fetal period. At this time the rudiments of their ligaments also appear.

Histogenetically, gonads are derived from the primordial germ cells and the gonadal ridge mesothelium and mesenchyme. Recognizable testicular characters first appear at about OA 40 days in man; but distinctive ovarian features, i.e., primordial follicles and easily distinguished cortex and medulla, first occur at about MA 21 weeks in man. The degree of separate development and distinctness of these two portions of the ovary varies greatly with the species. In a few mammals, the primary medullary portion is clearly separated from the secondary cortical portion by mesenchyme and even in the adult by somewhat fibrous stroma, but in most there is never a clear separation.

Oocytes and follicles appear first in the medullary portion where they are derivatives of the primary gonadal cords. Usually this occurs during the latter half of fetal life, but follicular activity in the medulla often persists into the juvenile period, and in some squirrels even functional ovulatory follicles may develop here at the first estrus. Usually, many oocytes are present in the secondary, or cortical, zone before birth and secondary follicles continue to develop during infancy. Vesicular follicles seldom appear in the cortex until shortly before the first estrus.

There is good experimental evidence in a few laboratory animals and man that all primordial germ cells originate in the more caudal yolk sac endoderm and migrate in the dorsal mesentery and body wall to the gonadal ridge. However, evidence is insufficient to prove that all functional germ cells of mammals are derived from these primordial germ cells.

Four

Morphology and cyclic changes of a representative mammalian ovary

The ovary of the red squirrel, or chickaree (*Tamiasciurus hudsonicus*), shows clearly the basic morphology and cyclic changes generally characteristic of this organ in mammals. There are two probable reasons why the ovary of this species is so representative. First, the squirrel family (Sciuridae) is one of the most primitive of the order Rodentia. Second, in southern Wisconsin the red squirrel has only two reproductive seasons each year with anestrous periods several weeks in length interposed, hence the structural changes of one breeding period are not heavily overlapped by those of the next — a cause of much confusion in interpretation of the ovaries of laboratory rodents.

Red squirrel ovaries are small enough to permit thorough study of serial sections. The cells of their ovarian glandular tissues are relatively easy to distinguish from one another; quite different from those of the rat, for example, whose luteal and interstitial gland cells often look almost alike by light microscopy. Furthermore, the ovary's features are expressed at a median; no important element is exceptionally large or small, plentiful or scarce, or unusual in position.

The guinea pig is the only common laboratory mammal whose ovarian morphology approaches in clarity that of the red squirrel, but it is less suitable in some ways. Squirrels are among the most available of wild species, and several have ovaries closely resembling those of the red squirrel. Some form of tree squirrel, chipmunk, ground squirrel, or marmot is easily available in almost any habitable part of the world, except the Australian region.

Our material consists of over 130 female red squirrels collected by Mossman throughout the year, mainly in south-central Wisconsin, especially in Sauk County. One ovary and oviduct were serially sectioned and mounted, although sometimes series of only every tenth section were made. Sample sections of uteri and vaginae were also taken. Ovaries were not weighed because of possible damage to the material and loss of relationships between the ovary and oviduct. Measurements of the diameter of follicles were made only in the two dimensions of the plane of section and were an average of the greatest diameter and of the least taken at right angles to the greatest. Measurements extended from base to base of the follicular epithelium. Corpora lutea were measured in the same fashion between the outer limits of luteal tissue. In other words, these measurements excluded the thecal gland of the follicle and the fibrous capsule of the corpus luteum.

Reproductive seasons of the red squirrel

In southern Wisconsin estrus occurs in midwinter (February and early March) and in midsummer (July and early August). As is usual among wild species, an occasional aberrant mating occurs: one of our parous animals was in estrus on 6 May. Summer-born young mate during the following February and March, and spring-born females mate during the immediately following July and early August while they are essentially juveniles. We have no evidence of much deviation from this schedule, although it is entirely likely that very late spring-born young might fail to mate until the following winter. (See Layne, 1954.)

Our data are obviously inadequate to establish the length of any phase of the reproductive cycle, yet the fact that signs of proestrus commonly occurred in animals collected as much as a month before any estrous or tubally pregnant specimens were common indicates that proestrus probably lasts two to four weeks. Estrus presumably lasts only one or two days, for very few estrous animals were collected compared with the numerous proestrous and tubally pregnant specimens. The gestation period is believed to be about 40 days (Layne, 1954). It is unknown whether ovulation is spontaneous or induced. It is also unknown whether these animals are monestrous or polyestrous. If mating is prevented at the first estrus of the reproductive season, 13-lined ground squirrels (*Spermophilus tridecemlineatus*) will return to estrus at least once, but this pattern may not be universal among sciurids.

There are two interesting questions with respect to the breeding seasons of this animal. What is the mechanism by which estrus is triggered in the coldest part of the year on increasing daylight, and again in the hottest part of the year on decreasing daylight? Why are essentially juvenile females capable of mating in July along with parous adults while juvenile males are not? It is known that in the same locality spring-born juvenile female eastern chipmunks (*Tamias striatus*) also breed during their first summer, but that their male litter-mates do not ordinarily do so. This pattern increases the chance that new-generation females will mate with older-generation males. This is also true of harem-forming species. In both cases, sires are assured on which natural selection has acted for a longer time than would be the case if new-generation males matured at the same time, or if mating were promiscuous instead of being restricted to the harem male.

Gross anatomy

The red squirrel ovary is globose, nonlobate, and relatively smooth: large follicles and corpora lutea project just enough to be readily seen. The short, relatively thick mesovarium is attached to the mesosalpinx.

There is a complete ovarian bursa with an opening too small to expose the whole ovary. It is like that of the other tree squirrels (see Figs. 1.11, 1.16–1.18). The oviduct is coarsely kinked and encircles the bursa. The uterine horns are moderately long and straight. Since each has its own cervix, the uterus is of the duplex type. The vagina is unusually long and much coiled (Mossman, 1940), the most atypical thing about the female reproductive tract of this species.

The fetal ovary

Only a few embryonic and fetal ovaries were examined. Late fetuses (32 mm CR) had indistinctly separated medullary and cortical zones, many primordial germ cells or oogonia, no primordial follicles, and no obvious fetal interstitial gland cells. These are the features one would expect in an altricial species.

Ovaries of summer-born females

Juveniles before the first proestrus

No nest young were available. Layne (1954) had evidence that red squirrels first leave the nest at six to seven weeks and are weaned at about eight weeks. The smallest summer-born female in our collection had a body length of 170 mm compared with 193 mm for an average adult.

Sagittal sections of the ovaries of these small summer-born, like those of spring-born juveniles before their midsummer proestrus, show a very thin, weakly developed tunica albuginea and a thick cortical zone packed with oocytes (Fig. 4.1). This zone abruptly becomes very thin and disappears completely about 0.5 mm from the attachment of the mesovarium.

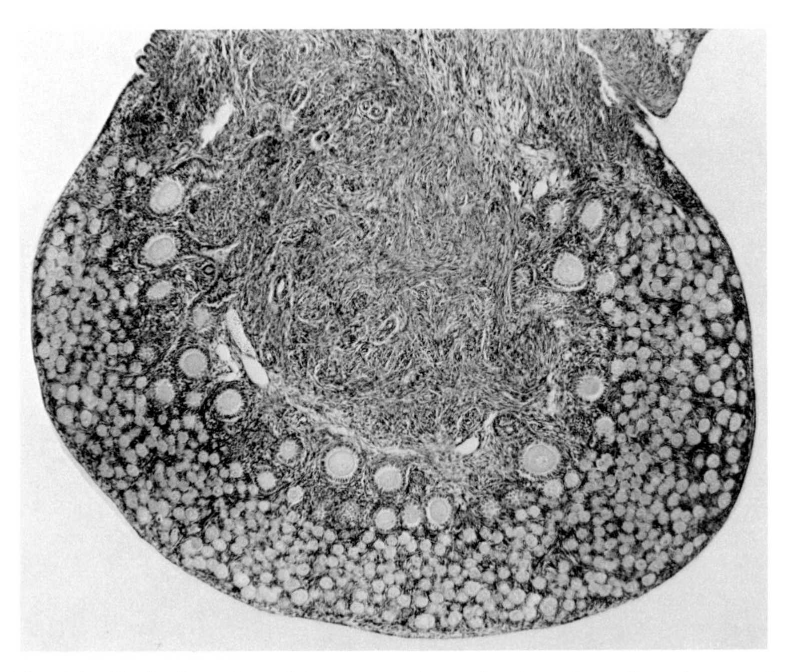

Fig. 4.1 Sagittal section of the ovary of a spring-born juvenile red squirrel (*Tamiasciurus hudsonicus*) in the following July, showing the thin, weakly developed tunica albuginea, the infantile cortex packed with oocytes, and the medullary cords containing eggs. Many medullary cords appear to have already broken up into segments, i.e., medullary follicles, each containing an egg. *T. hudsonicus*, 86. × 62.

Each oocyte of this zone is enclosed in a very inconspicuous simple squamous epithelium. Primordial follicles of this type are characteristic of most mammals, and indicate the probable direct origin of their follicular epithelium from stromal cells. Yet according to Evans and Swezy (1931), the follicular epithelium of the rat, guinea pig, dog, cat, rhesus monkey, and man is more or less columnar from the first and appears to arise directly from the epithelium of cortical crypts and cords. We believe there is some evidence for this interpretation of the origin of the epithelium of cortical follicles in the dog and cat, but not in the others.

Just deep to this zone of primordial follicles is a narrower zone of irregularly scattered large primary follicles (see Fig. 4.1). At first glance, these seem to form the inner layer of the cortex, but careful study of serial sections reveals that in the youngest females most of them are actually in the distal (i.e., cortical) ends of medullary cords, and so are definitely in the medulla. (See the discussion of cortex and medulla in Chapter 6, pp. 125–27 and 230). That the earliest mammalian follicles to pass beyond the primordial stage are often formed in medullary cords is not generally recognized; yet there is clear evidence of this in many rodents and in some other mammals, including man (Chapter 5, pp. 100–103) and the cat (Kingsbury, 1913).

The medullary epithelial cords are much wider and more easily seen in the red squirrel and chipmunk than in other squirrels that we have studied. Tracing by means of serial sections shows that most are isolated, discontinuous segments of cords, yet they often branch and connect with as many as three primary follicles. Toward the hilus they become narrower and commonly unite with one another. It is difficult to determine their relation to the rete, but it is fairly certain that at this stage few are ever directly continuous with it, although in many species, especially in infantile or juvenile stages, such connections are common. It is also hard to define the exact limits of the rete, but more of it lies in the mesovarium than in the ovary. It is clearly continuous with the efferent ductules of the ovary, which in turn lead into the larger, straighter female ductus deferens. The epithelium of the latter structures is simple columnar and heavily ciliated, that of the rete simple low columnar to almost squamous and nonciliated.

Table 4.1 gives vaginal, uterine, and follicular diameter data on 12 females collected between 5 September and 3 February. There is little change in the ovaries during this juvenile sexually inactive period, except for a trend toward larger vesicular follicles and that medullary cords were conspicuous only in the youngest (nos. 108, 111, and 15) taken in September and October, and rare or absent in the others (Fig. 4.2). Atretic follicles in the medulla were more numerous at the later dates, but it was increasingly difficult to be certain whether they really belonged to the medullary cord group or to those derived from cortical follicles. Correlated with greater follicular development and atresia in

Table 4.1 Measurements (in mm) of summer-born juvenile red squirrels before proestrus

Animal number	Date	Greatest diameter		Diameter of largest follicle
		Vagina	Uterus	
108	5 Sept.	1.6	1.1	0.26
110	5 Sept.	0.8	1.0	0.25
111	18 Sept.	1.2	1.0	0.25
113	2 Oct.	2.0	0.9	0.50
15	18 Oct.	1.8	0.9	0.38
17	5 Nov.	1.3	0.7	0.25
28	14 Dec.	—	0.6	0.18
37	10 Jan.	1.9	0.8	0.50
36	10 Jan.	1.8	0.8	0.52
38	10 Jan.	1.0	0.5	0.58
45	3 Feb.	2.0	0.5	0.55
46	3 Feb.	2.0	0.8	0.60

Table 4.2 Measurements (in mm) of summer-born red squirrels during their first proestrus and estrus

Animal number	Date	Greatest diameter		Diameter of largest follicle
		Vagina	Uterus	
		Proestrus		
25	3 Dec.	3.0	1.0	0.64
4	4 Jan.	—	—	0.64
31	8 Jan.	3.0	1.2	0.60
42	3 Feb.	2.8	0.8	0.28
43	3 Feb.	3.5	1.0	0.63
18	21 Feb.	8.0	2.0	0.56
19	21 Feb.	—	—	0.63
81	4 Mar.	2.5	1.4	0.70
82	4 Mar.	7.5	1.8	0.53
62	8 Apr.	7.0	1.5	0.64
		Estrus		
98	11 Mar.	9.0	2.5	1.05
76	17 Mar.	10.0	2.5	1.01

the older animals was a general increase of thecal type interstitial gland tissue, especially in the medulla. However, most of this tissue consisted of undifferentiated cells, except in no. 46, a February animal, which was on the borderline of proestrus. Some of the younger females had scattered differentiated fetal type interstitial gland cells, especially in the

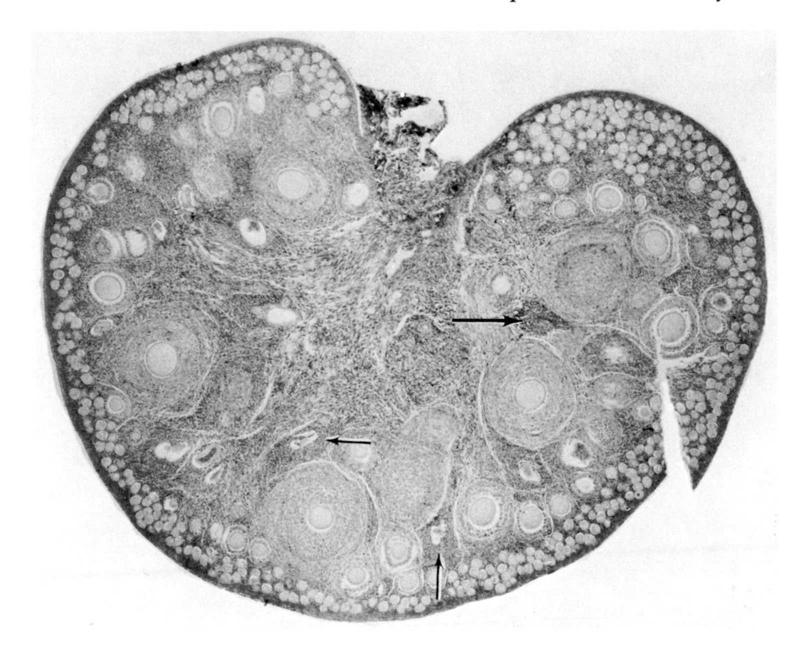

Fig. 4.2 Ovary of a summer-born juvenile in September, showing the juvenile cortex and numerous medullary follicles. Many of these were already corpora atretica with relatively undifferentiated thecal type interstitial gland tissue (*small arrows*). An occasional medullary cord was still present (*large arrow*). *T. hudsonicus.* 108. × 42.

medulla. Spring-born juveniles in July also had this type of interstitial gland cells (Fig. 4.3). These fetal ovarian interstitial gland cells are believed to be homologous to the fetal testicular interstitial gland cells which are often conspicuous in late fetal life.

As would be expected, the vagina tends to increase in diameter during this period. However, the opposite is true of the uterus. There is no evidence that this discrepancy in size of the uterus was caused by different preservation methods, so it may well be the result of either a normal variation in the diameter of this organ, or a variation caused by different degrees of contraction of its musculature at the time of fixation.

During the first proestrus

Proestrous females of this class were taken between 3 December and 8 April (Table 4.2). Enlargement of the vagina is the first grossly visible sign of the beginning of proestrus. Maximum vaginal diameters between

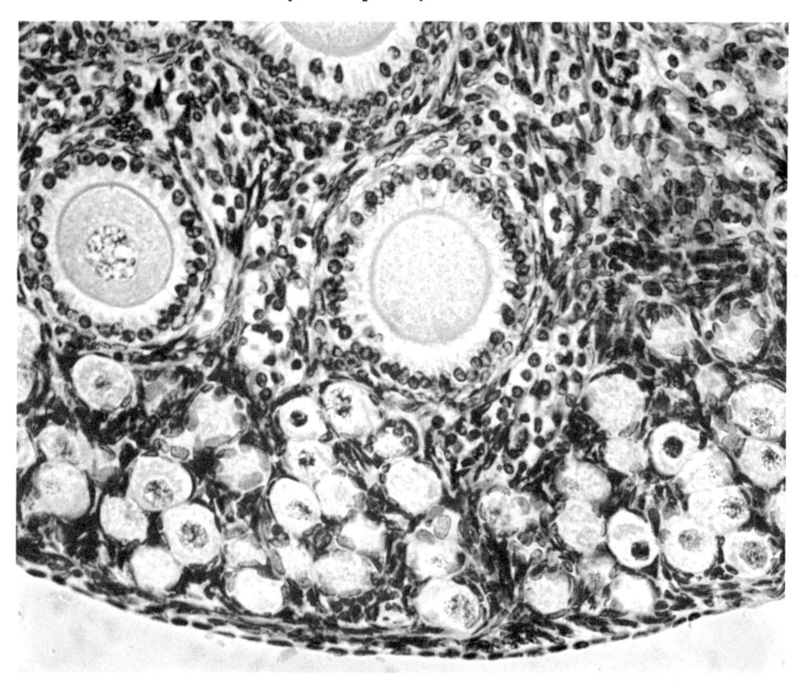

Fig. 4.3 Cortex and corticomedullary junction of a spring-born juvenile in mid-July, showing fetal type interstitial gland cells. *T. hudsonicus*, 124. × 200.

2.1 and 8.0 mm were arbitrarily selected as the limits for proestrus. A diameter of over 8.0 mm was an indication of estrus unless there were ruptured follicles or corpora lutea. The uterine diameter has a slight tendency to increase late in proestrus.

It is possible that some of these females were born the previous spring but had not bred during the summer. However, there is no evidence that this ever occurs: all females collected in late fall and early winter were obviously either parous or summer-born juveniles.

In any species, the ovary at the first proestrus differs from that of a mature animal in that it contains none of the products or vestiges of products of previous ovulatory periods. Some follicular growth and atresia do go on during the prepuberal period, as they do during anestrus of a mature female, hence there are atretic follicles and varying amounts of interstitial gland tissue derived from the thecae internae of these follicles. The ovaries of these proestrous puberal red squirrels conformed to this rule (Figs. 4.4 and 4.5). They also appeared to have a slowly decreasing number of primordial and primary follicles in the cortex, but we could not determine whether this was actually true or merely relative owing to increasing ovarian stroma. The impression after examining serial sections of these ovaries was that there is much individual variation with respect to most of these features. For instance, the diameters of the largest follicles did not correlate directly with the vaginal diameters (Table 4.2).

Each individual of this group had several large follicles, most or all of which were in atresia. This is a common observation among proestrous mammals with anestrous periods several weeks in length, and probably indicates that, during proestrus, groups of follicles are consecutively stimulated to grow, but that most of these undergo atresia. This brings up the interesting possible existence of "lesser anovulatory cycles" within the longer ovulatory estrous cycles.

During the first estrus

Two females of this age taken on 11 and 17 March (Table 4.2) had vaginal diameters of 10.0 mm and 9.0 mm, respectively, and uterine diameters of 2.5 mm, and so were considered to be in estrus. The di-

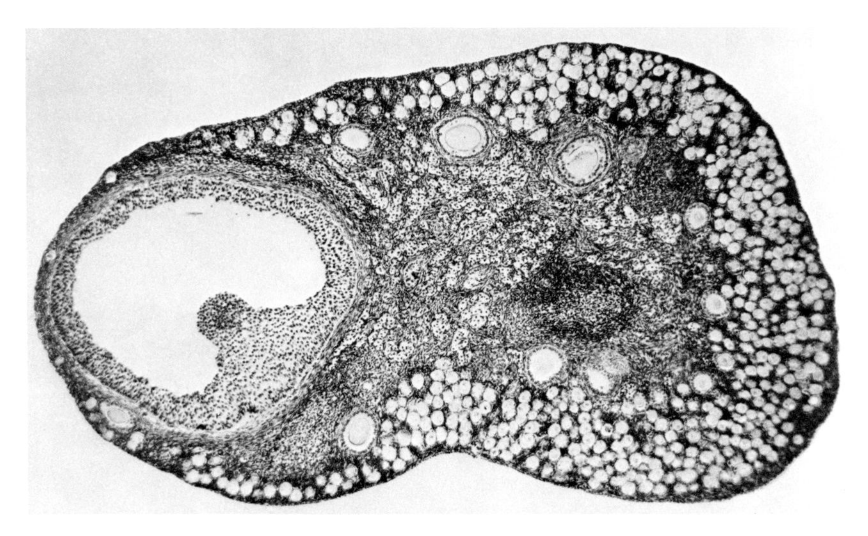

Fig. 4.4 Ovary of a summer-born red squirrel during her first proestrus in early February, showing a large follicle and scattered interstitial gland cells, probably mainly of the thecal type. *T. hudsonicus*, 43. × 55.

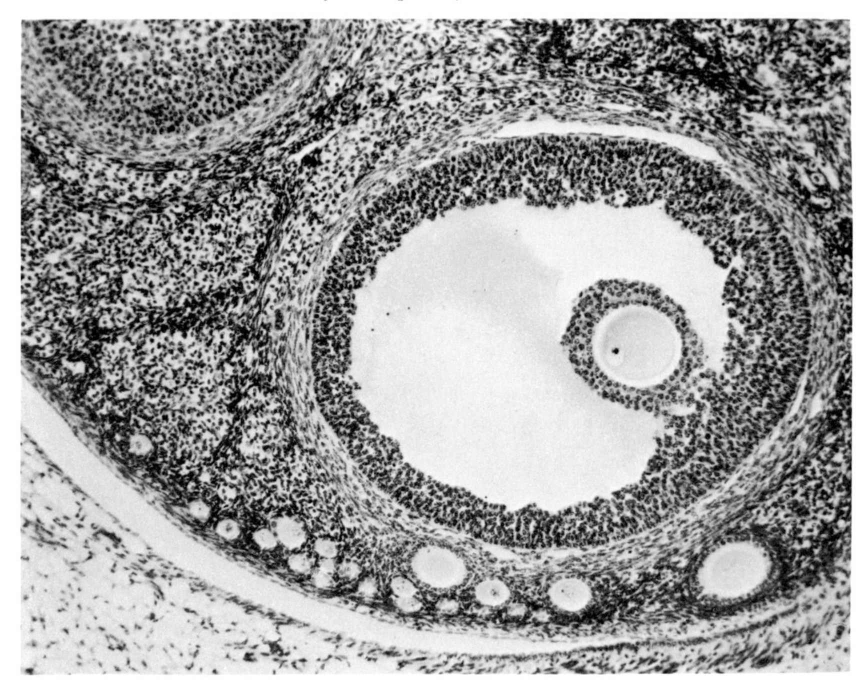

Fig. 4.5 Ovary of a summer-born red squirrel during her first proestrus in early April. The cumulus is relatively thin, an indication that atresia of this follicle has begun. *T. hudsonicus*, 62. × 100.

ameters of their largest follicles were over 1 mm. The cumulus cells of the one with a vaginal diameter of 10.0 mm were not dispersing, but those of the one with a 9 mm vagina were (Fig. 4.7), hence that animal was probably in behavioral estrus.

Female no. 98 also had unusually numerous corpora atretica and large amounts of fully differentiated interstitial gland tissue in both the cortex and medulla (Fig. 4.6). Primordial and primary follicles were relatively scarce. The thecal gland cells of the ripe follicles were irregularly distributed, in some places three or four cells deep, in others absent (Fig. 4.7). They were large, discoid in shape, appearing rectangular in section, with their broad surfaces parallel to the follicular circumference. Their centrally located nuclei were similarly shaped, almost as thick as the cell but only about one-third its breadth. The cytoplasm was granular. Dilated capillaries were conspicuous among the

Table 4.3 Measurements (in mm) of summer-born red squirrels during their first pregnancy and lactation

Animal number	Date	Greatest diameter		Diameter of largest		Remarks
		Vagina	Uterus	Follicle	C. l.[a]	
			Tubal period			
39	10 Jan.	—	—	0.40	1.00	8–12 cell
99	13 Mar.	10.0	2.0	0.65	1.50	4–8 cell
83	20 Mar.	10.0	2.5	1.00	0.77	Pronuclei
6	20 Mar.	7.0	2.5	0.36	0.75	4-cell
			Embryonal period			
52	26 Mar.	8.0	2.5	0.55	0.90	Primitive streak
51	26 Mar.	7.0	2.5	0.55	0.95	5.5 mm CR[b]
53	26 Mar.	8.0	3.0	0.53	1.30	7.5 mm CR[b]
80	31 Mar.	9.0	3.0	0.57	1.00	Gill slit, 3+ mm
1	9 Apr.	7.0	3.5	0.56	0.84	7.5 mm CR[b]
			Fetal period			
49	26 Mar.	7.0	2.3	0.55	1.00	14 mm CR[b]
			Postpartum period			
79	31 Mar.	10.0	7.5	0.75	1.10	Early lactation
91	20 Apr.	7.0	8.0	0.53	.95	Early postpartum
7	29 May	3.5	1.6	0.63	1.85	Lactation

[a] Corpus luteum.
[b] Crown–rump.

thecal cells, a characteristic of the ripe follicles of other species. Compared with that of most mammals, the thecal gland of the red squirrel and other sciurids is relatively thin, while the interstitial gland tissue is somewhat more plentiful at its fullest development than in most other mammals.

Mammalian ovaries have a well-developed plexus of lymphatic capillaries and vessels in their medullae, and these are often noticeably dilated. Red squirrel ovaries in late proestrus, estrus, and early pregnancy usually have dilated lymphatics. However, this dilatation was erratic, possibly owing to individual variation, but more likely associated with different degrees of hemorrhage at the time the animal was shot, or possibly to different techniques of tissue preservation.

During the tubal and implantation periods of the first pregnancy

There were four animals in this category (Table 4.3). One (no. 83) still had an unruptured follicle 1.0 mm in diameter with dispersing cu-

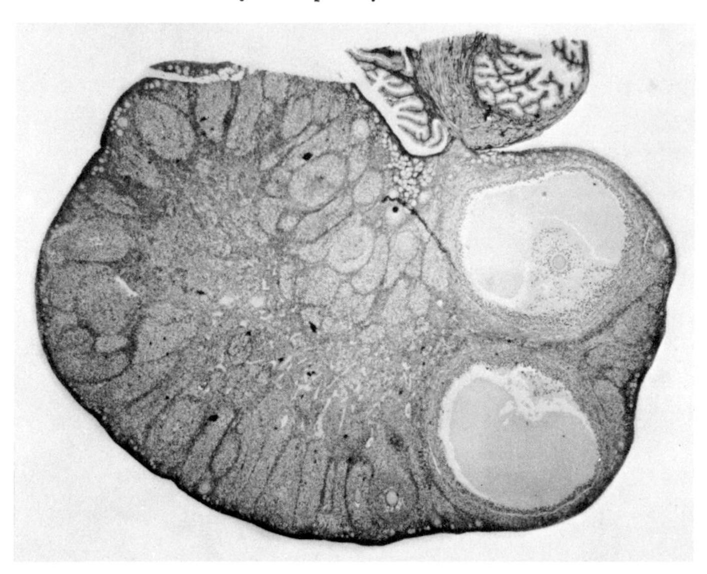

Fig. 4.6 Ovary of a summer-born red squirrel in her first estrus in mid-March, showing two ripe follicles and numerous corpora atretica composed of thecal type interstitial gland tissue. *T. hudsonicus*, 98. × 20.

mulus cells, but pronuclear eggs in the oviduct indicated that it had not completed ovulation and was probably still in estrus. All had smaller vesicular follicles in atresia similar to those present during proestrus and much thecal type interstitial gland tissue.

The diameters of the ruptured follicles during this period, measured between the outer edges of the collapsed follicular epithelial layer, were usually less than those of the ripe follicles in Table 4.2 (but see no. 99, Table 4.3). Through the 4-cell period the thecal gland was conspicuous and there were only occasional cells with pyknotic nuclei or other signs of degeneration, yet by the 8-cell stage it was unrecognizable. A "paraluteal" cell zone, so characteristic of early corpora of many mammals, was not obvious in this group nor in other red squirrels during any phase of the cycle. Number 39, with 8–12-cell embryos, did have an irregular, ill-defined zone of cells which appeared to be somewhat transitional from stromal to luteal cells, but we could not determine whether these were true paraluteal cells (i.e., stromal cells differentiating into luteal cells) or persisting thecal gland cells. Therefore, red squirrel luteal cells are probably derived solely from follicular epithelium, as is true of many

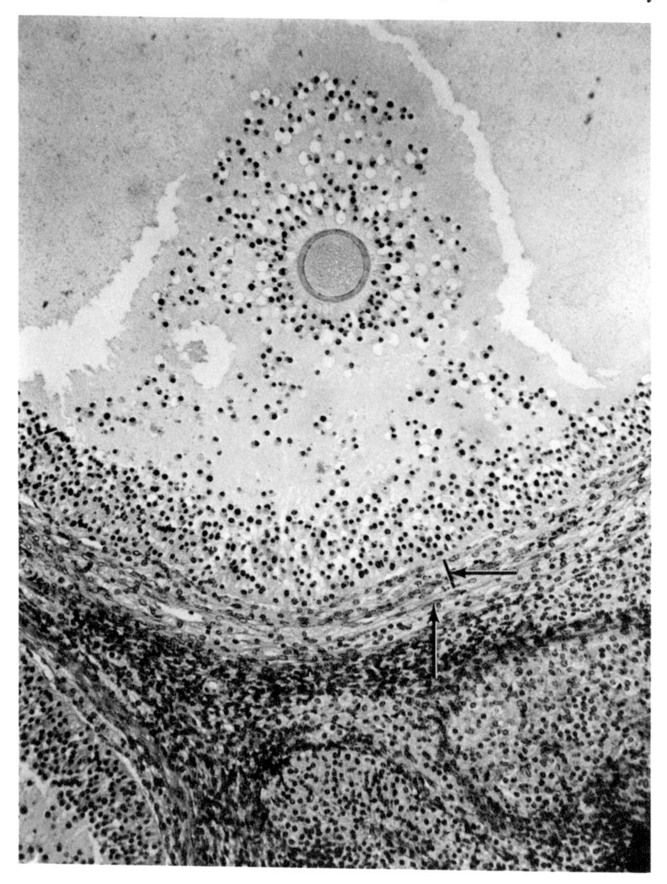

Fig. 4.7 Detail of the red squirrel ovary shown in Figure 4.6. Note the ripe follicle with dispersing cumulus cells, the thin thecal gland (*bar* and *arrow*), and the even thinner and indistinct theca externa (*arrow*). *T. hudsonicus*, 98. × 100.

mammals of this size and smaller. The luteal cells of this period were small, and their cytoplasm was filled with vacuoles of various sizes. Active vascularization of the luteal tissue apparently occurs about the time of implantation, because vessels were just appearing in female no. 39, with 8–12-cell embryos, but they were well developed throughout the corpora lutea of female no. 52 (Table 4.3), which contained embryos in the early primitive streak period.

From the primitive streak through the 7.5 mm CR embryo periods

Five animals comprised this class (Table 4.3). All of their ovaries contained vesicular follicles, most of them in early atresia, with diameters from 0.53 mm to 0.57 mm. Their corpora lutea had lost the small antral remnants present during the late tubal period, and none had any vestige of the thecal gland, nor were there any paraluteal cells. Their luteal cells were larger than in the preimplantation period, and some showed the peripheral vacuolar zone and central homogeneous or granular area so characteristic of mature luteal cells. There was a noticeable decrease in the number of atretic follicles and in the amount of differentiated interstitial gland cells.

Lymphatic dilatation was very erratic in this group. In female no. 80 (gill slit stage embryos) lymphatic capillaries and vessels were extremely dilated, while in female no. 1 (7.5 mm CR embryos) they were not dilated.

During the fetal period of the first pregnancy

Only one female belonged to this class (Table 4.3). The only obvious changes from earlier pregnancy were a decrease in number of atretic follicles and in the amount of differentiated interstitial gland tissue, and the appearance of a thin but definite fibrous capsule around each corpus luteum. The lymphatics were dilated.

During the first postpartum and lactation periods

Three females were in this category (Table 4.3). The stroma tended to be more densely nuclear during this period because of general lowering of the cytoplasmic–nuclear ratio of both the interstitial gland cells and other stromal cells. Atretic follicles decreased in number. The degree of lymphatic dilatation was again erratic.

The great range in diameter of the vagina and uterus in the postpartum and lactating specimens is a reflection of the time available for postpartum involution and also of the difficulty in obtaining a meaningful measurement of the collapsed and flattened uterus.

Ovaries of parous females

In summer during the second proestrus and estrus of the year

The 10 animals in this group (Table 4.4), which were taken between 6 May and 19 July and which had already produced young in the spring, were more alike in their general ovarian morphology than those ob-

Table 4.4 Measurements (in mm) of parous red squirrels in summer during their second proestrus and estrus of the year

Animal number	Date	Greatest diameter		Diameter of largest	
		Vagina	Uterus	Follicle	Corpus luteum
		Proestrus			
94[a]	6 May	3.2	2.8	0.60	1.55
8	4 July	8.0	2.5	0.55	1.40
2	6 July	3.0	1.2	0.61	1.40
66[a]	10 July	3.0	1.7	0.50	1.50
67	10 July	7.0	2.5	0.60	1.10
68[a]	10 July	2.6	1.8	0.49	1.90
122[b]	14 July	7.5	2.5	0.70	1.10
104	15 July	5.0	2.5	0.73	1.32
23	16 July	7.0	3.0	0.50	1.52
88	19 July	6.0	2.8	0.52	1.12
		Estrus			
93	6 May	4.5	2.5	1.00	0.90
120	10 July	10.0	2.5	1.07	1.27

[a] Still lactating.
[b] Slight evidence of the remnants of two corpora albicantia.

tained during their first proestrus (cf. Tables 4.2 and 4.4). They tended to have relatively few atretic follicles and little differentiated interstitial gland tissue, although undifferentiated interstitial gland cells were plentiful around atretic follicles. They also appeared to have markedly fewer primordial, primary, and secondary follicles in their cortices, although, as mentioned above, this could be merely relative to the amount of stroma, for these ovaries were somewhat larger than puberal ones. Their corpora lutea had definite fibrous capsules, usually rather thick. Peripheral nondegenerative type vacuolation of the luteal cells was characteristic with most of the fixatives used. The central cytoplasm in almost every case was distinctly granular, and most animals had from a few to many cells with "foamy" cytoplasm, that is, closely packed small vacuoles of uniform size resembling those of the zona fasciculata of the adrenal. One animal (no. 66) had a few luteal cells with large, clear degenerative type vacuoles which are characteristically numerous in old corpora lutea during the second lactation (see Fig. 4.14).

Some had highly dilated lymphatic vessels; in others they were not dilated. Again this feature was erratic.

The two individuals in estrus had greater amounts of well-differentiated interstitial gland tissue than those in proestrus, but in relation to ovarian size there was noticeably less than in ovaries of nonparous females during their first estrus. The vagina of female no. 93 had the diameter of that of a proestrous animal, but she had several fully ripe follicles and so clearly belonged in the estrus category. She was an example of the occasional wide range that occurs in such features.

In summer and autumn during the second pregnancy and lactation of the year

This category included young females in their second pregnancy as well as older multiparous females in their second pregnancy of the year; it was impossible to distinguish between them on the basis of available data (Table 4.5). Number 11 was a typical example of the early stage of a second pregnancy of the year. A section of one ovary (Fig. 4.8)

Table 4.5 Measurements (in mm) of red squirrels in summer and autumn during their second pregnancy and lactation of the year

Animal number	Date	Greatest diameter		Diameter of largest			Remarks
				Follicle	C. l.[a]		
		Vagina	Uterus		New	Old	
				Tubal period			
70	10 July	9.0	3.2	0.40	1.05	1.90	Uterine morulae
11	11 July	—	—	0.30	0.75	1.20	Pronuclei
71	14 July	8.0	3.0	0.75	1.10	1.45	Uterine morulae?
72	14 July	4.0	3.7	0.64	1.80	2.15	Uterine morulae?
				Embryonal period			
69	10 July	2.3	2.5	0.69	1.12	1.94	2-tube stage
125	23 July	6.5	—	0.80	0.82	1.10	8 mm CR[b]
128	24 July	6.0	—	0.65	0.96	1.17	7 mm CR[b]
				Fetal period			
63	8 Apr.	—	—	0.80	1.00	0.50	32 mm CR[b]
73	14 July	8.0	—	0.76	1.27	1.68	32 mm CR[b]
105	3 Aug.	6.0	—	0.84	1.08	1.50	16 mm CR[b]
				Lactation period			
112	2 Oct.	2.4	1.3	0.30	1.55	1.40	
16	18 Oct.	2.3	1.5	0.26	1.57	1.58	

[a] Corpus luteum.
[b] Crown–rump.

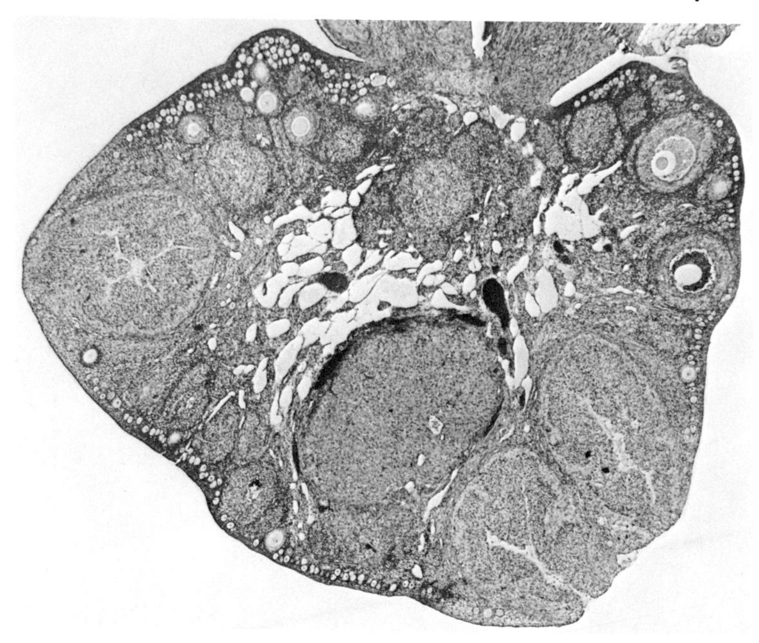

Fig. 4.8 Ovary of a parous red squirrel in mid-July with pronuclear tubal eggs of the second pregnancy of the year. The figure shows one corpus luteum of the previous pregnancy and three new ones, around each of which the zone of thecal gland cells is conspicuous. *T. hudsonicus*, 11. × 16.

shows one corpus luteum of the previous pregnancy and three new ones, each having the outer zone of thecal gland cells (Figs. 4.9 and 4.10). A few degenerating thecal gland cells were still recognizable at the periphery of the new corpora lutea of female no. 70. This animal contained uterine morulae. This is the latest stage in pregnancy in which thecal gland cells are present. The lymph vessels of the medulla were greatly distended. Figure 4.10 shows the corpus margin in the most detail: at this stage the thecal gland cells appear nearly normal, but they degenerate and disappear very soon. Figure 4.11 shows a corpus atreticum with its thecal type interstitial gland cells typical of early pregnancy. Number 63 had an unusually early second pregnancy of the year and is an example of the few exceptionally early or late breeders which almost always appear, even among wild mammals that have clearly defined breeding-season peaks (see Layne, 1954, for similar data).

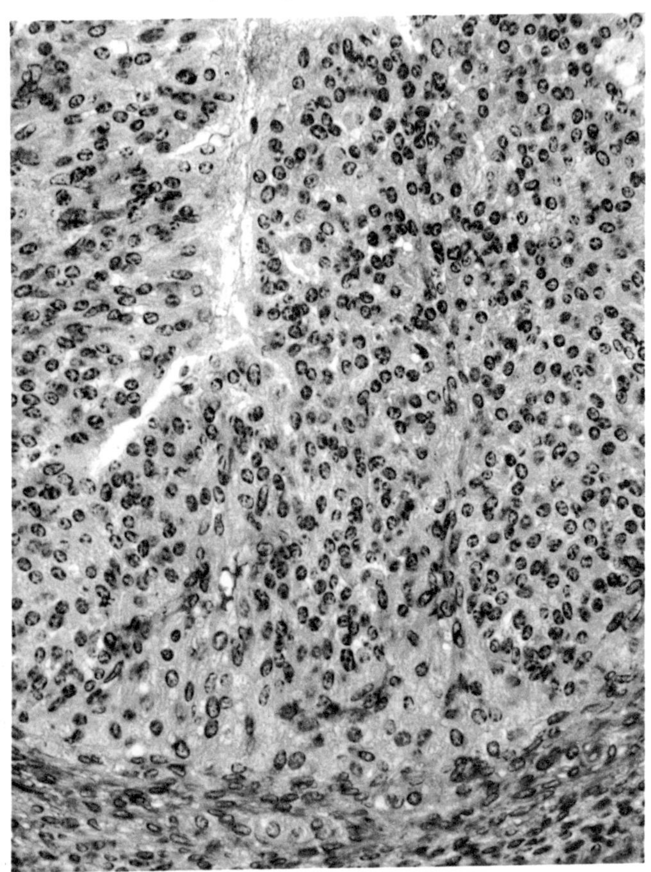

Fig. 4.9 Segment of a new corpus luteum of the red squirrel ovary shown in Figure 4.8. *T. hudsonicus*, 11. × 100.

The first set of corpora lutea persisted through the whole of the second pregnancy and lactation (Figs. 4.12–4.14 and Table 4.5), and the degree to which they maintained their size is striking. With the exception of female no. 63, the corpora were definitely larger than those of the second set until late in the second lactation (no. 16 in Table 4.5) or in the autumn anestrus. The smallness of the old corpora in no. 63 is unusual. Unfortunately, only one ovary of this animal was available. The corpora were clearly of two different ages, so in spite of the early date this animal must have been in the second pregnancy of the year. It is,

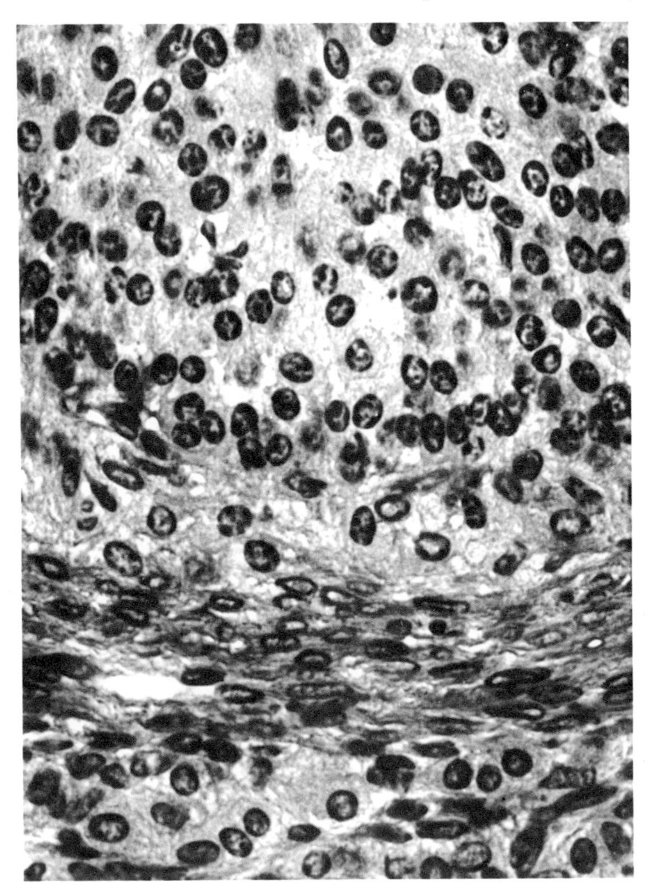

Fig. 4.10 Outer border of a new corpus luteum of the red squirrel ovary shown in Figures 4.8 and 4.9, indicating the still apparently normal thecal gland cells, the theca externa and, below, a few thecal type interstitial gland cells. *T. hudsonicus*, 11. × 600.

of course, possible that the smallness of the older set of corpora could have resulted from premature termination of the first pregnancy and that this in turn may have accounted for the earliness of the second pregnancy. The luteal cells of the first set of corpora were also larger, but late in the second gestation large clear vacuoles began to appear in them. These became very numerous late in lactation and were the chief sign of luteal degeneration (Figs. 4.13 and 4.14). Also during the latter part of this second pregnancy, and especially during lactation, groups

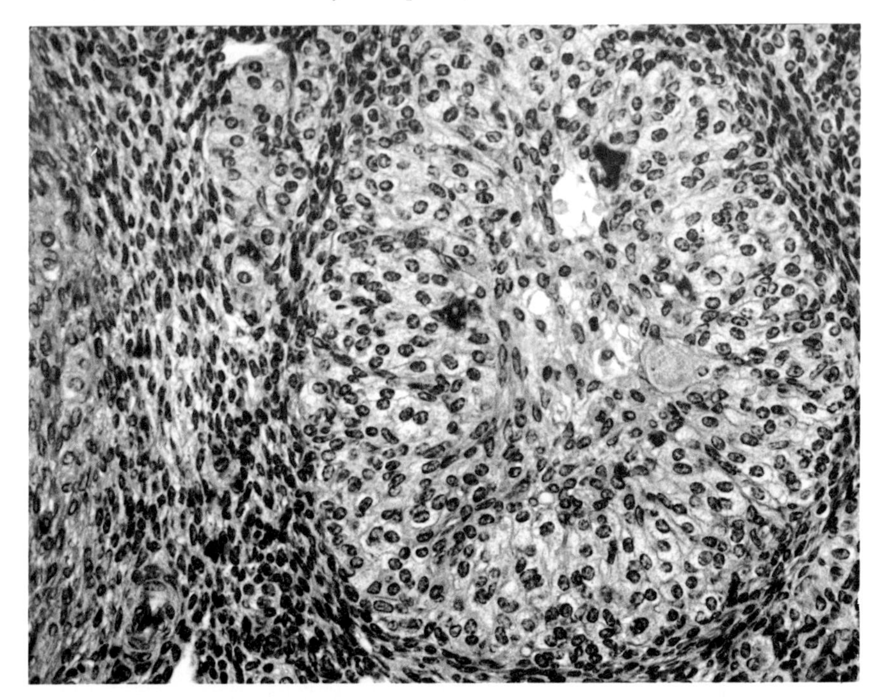

Fig. 4.11 Corpus atreticum with its thecal type interstitial gland cells from the red squirrel ovary shown in Figures 4.8, 4.9, and 4.10. *T. hudsonicus*, 11. × 100.

of outer luteal cells, particularly of the older corpora lutea, herniated into the adjacent lymphatics (Figs. 4.15–4.17). This occurred only in those animals with much-dilated lymphatics; the phenomenon is not clearly correlated with any phase of reproduction, although it was rarely seen in a definitely postlactation individual.

Presumptive polyploid luteal cells were first noticed in our red squirrel ovaries in the corpora of the first set late in the second pregnancy and lactation (Figs. 4.16 and 4.17). More careful observation showed that large, presumably polyploid, cells first appear during the lactation period in the corpora of the contemporary pregnancy, and that they gradually increase, reaching their peak during late lactation of the second pregnancy of the year, the same time at which vacuolar degeneration begins to be conspicuous. (See below, p. 98, for further discussion of gland cell polyploidy.)

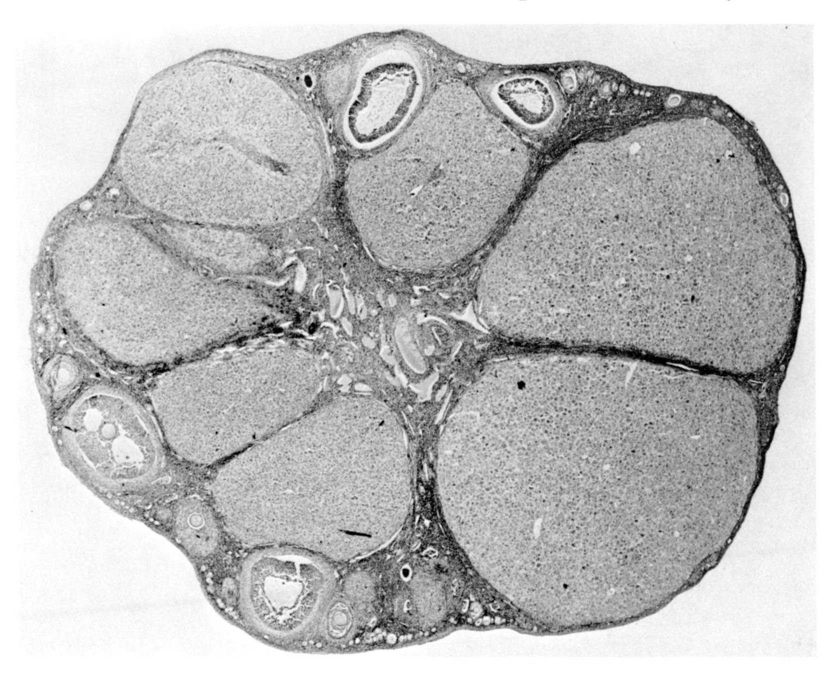

Fig. 4.12 Ovary of a parous red squirrel in July in the uterine morula stage of her second pregnancy of the year, showing two sets of corpora lutea. The corpus at the upper left with the hemorrhagic center is a young one, the rest are from the previous pregnancy. *T. hudsonicus*, 72. × 20.

The functional status of the first set of corpora during the second reproductive cycle and the endocrine basis of persistence during the second period are unanswered questions. Amoroso (1956) reported persistence and some "resurgence" of old corpora lutea during a second pregnancy in the cat. Other than the persistent corpora lutea, there was little detectable difference between the ovaries of this group and those of parous animals during the spring reproductive season.

During the spring reproductive season

Fourteen animals collected between 8 January and 9 July were in this category (Table 4.6). The ones showing the least proestrous changes were known to be parous adults by their laterally compressed uterine horns with an antimesometrial ridge.

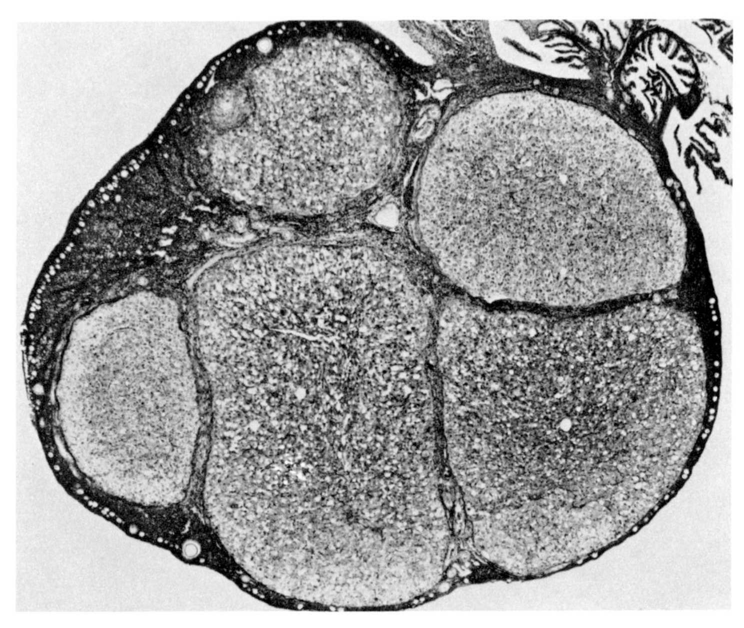

Fig. 4.13 Ovary of a parous red squirrel in mid-October near or past the end of her second lactation period of the year, showing two sets of corpora lutea. Those of the first pregnancy have more vacuolated cells and more and larger polyploid cells. (See Fig. 4.17.) *T. hudsonicus*, 16. × 27.

Ovaries of proestrous and estrous females of this group differed little from those of parous animals entering the summer breeding period, except for the presence of corpora albicantia instead of persistent large corpora lutea from the previous gestation. In an occasional proestrous female (no. 32), small masses of degenerating luteal cells did persist, but in postpartum animals not even corpora albicantia of the previous summer's pregnancy could be recognized. Female no. 50 appears to be an exception, but she had probably resorbed her embryos and so was not comparable to other postpartum females. If the old corpora lutea that persist through a second pregnancy of the year (Table 4.5) are actively functional, one would expect the accompanying new corpora of the second pregnancy to be smaller than corpora of the first pregnancy of the year (Table 4.6). To answer this question one would need more animals and preferably weight data rather than the relatively crude linear measurements we have made. However, certain trends are discussed below on p. 97 (and see Table 4.9).

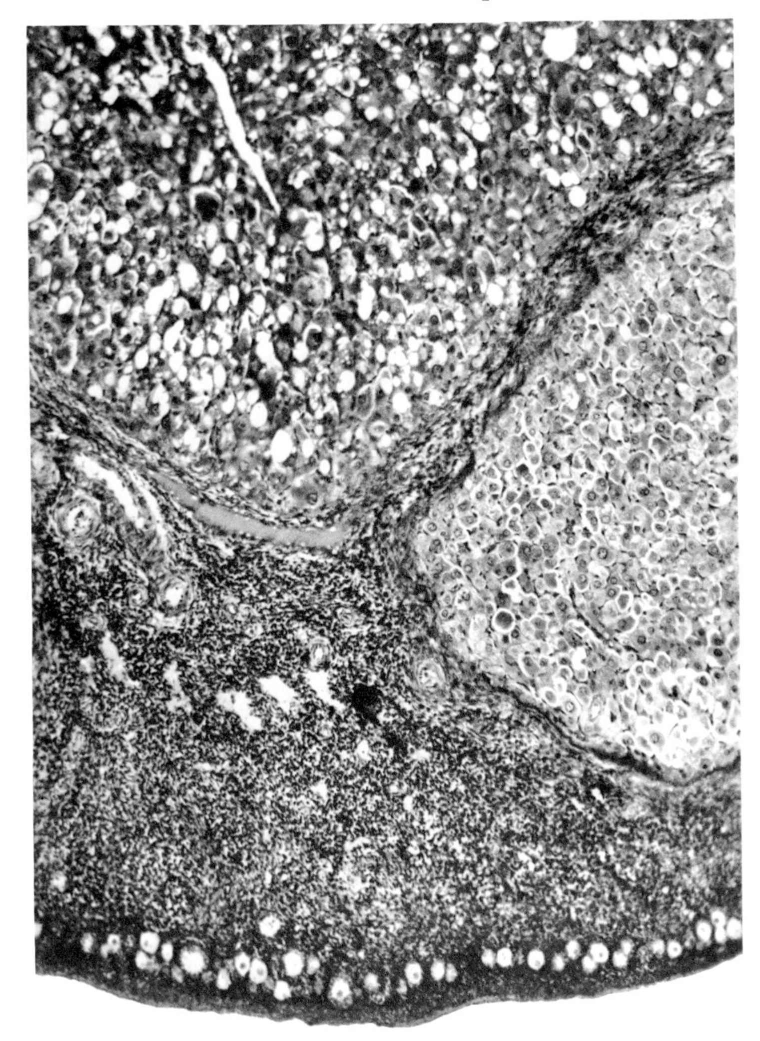

Fig. 4.14 Detail of corpora lutea from the red squirrel ovary shown in Figure 4.13, indicating the marked vacuolation in the older corpus (*upper center*) and the involuted condition of the interstitial gland cells. *T. hudsonicus*, 16. × 100.

There was no difference in the erratic occurrence of lymphatic dilatation. Female no. 56, in late lactation, had luteal cell herniation into the lymphatics.

Ovaries of spring-born females

Prepuberal females before the first proestrus

Thirteen females collected between 9 June and 19 August belong to this group (Table 4.7). In contrast with summer-born prepuberal young collected between 5 September and 3 February (Table 4.1), only five had atretic follicles. Except for some scattered differentiated interstitial cells, believed to be of the fetal type (Fig. 4.3), there was little differentiated interstitial tissue. All had medullary cords containing primary and secondary follicles (Fig. 4.1), and all vesicular follicles were clearly medullary in origin and position (Fig. 4.18).

During proestrus and estrus

There were six proestrous animals in this category (Table 4.8). Although nos. 9 and 126 had vaginal diameters at our arbitrary boundary for the beginning of proestrus, their uterine diameters were large enough

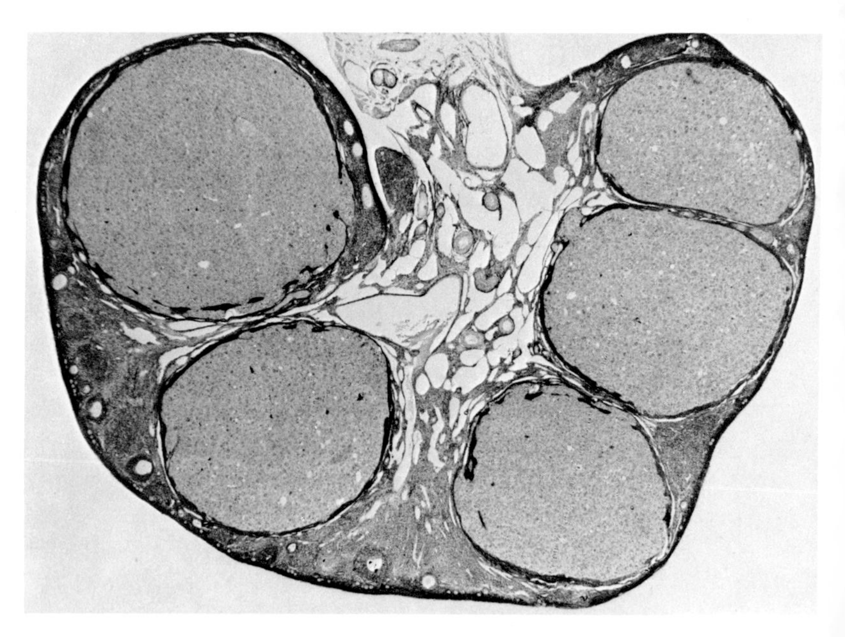

Fig. 4.15 Ovary of a parous red squirrel in October near or past the end of her second lactation period of the year, showing herniation of luteal cells into surrounding lymph vessels and greatly dilated lymphatics of the medulla. (See Figs. 4.16 and 4.17 for more detail.) *T. hudsonicus*, 112. × 21.

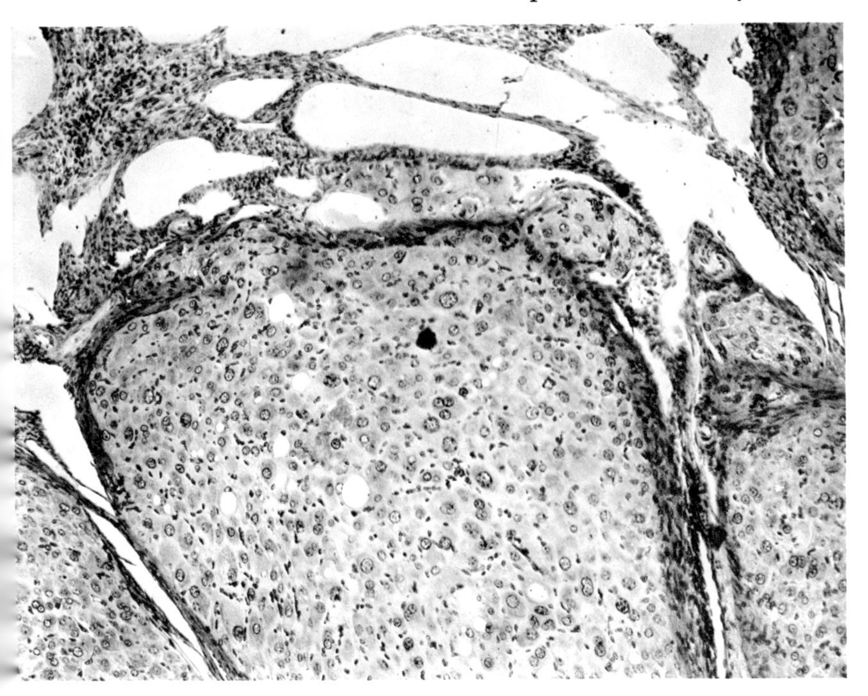

Fig. 4.16 Detail of corpora lutea from the red squirrel ovary shown in Figure 4.15 to point out herniation of luteal tissue into lymph vessels. *T. hudsonicus*, 112. × 100.

to indicate that they were probably definitely in proestrus. All had atretic medullary follicles (Figs. 4.19 and 4.20), much undifferentiated interstitial gland tissue, and a relatively small amount of differentiated, some of which appeared to be of the fetal type. About half had obvious medullary cords containing primary and secondary follicles. All had normal secondary and vesicular follicles which were either connected with medullary cords or clearly in the medulla. Figure 4.21 shows a medium-sized vesicular follicle (female no. 126) with its granulosa herniating into a somewhat dilated rete tubule. The basement membrane of its granulosa, stained with fast green, was clearly continuous with the basement membrane of the rete epithelium. Thus, the connection was in all probability a natural one and not just a herniation caused by the pressure developed during fixation or by rough handling during dissection. The logical explanation is that this follicle originated from a

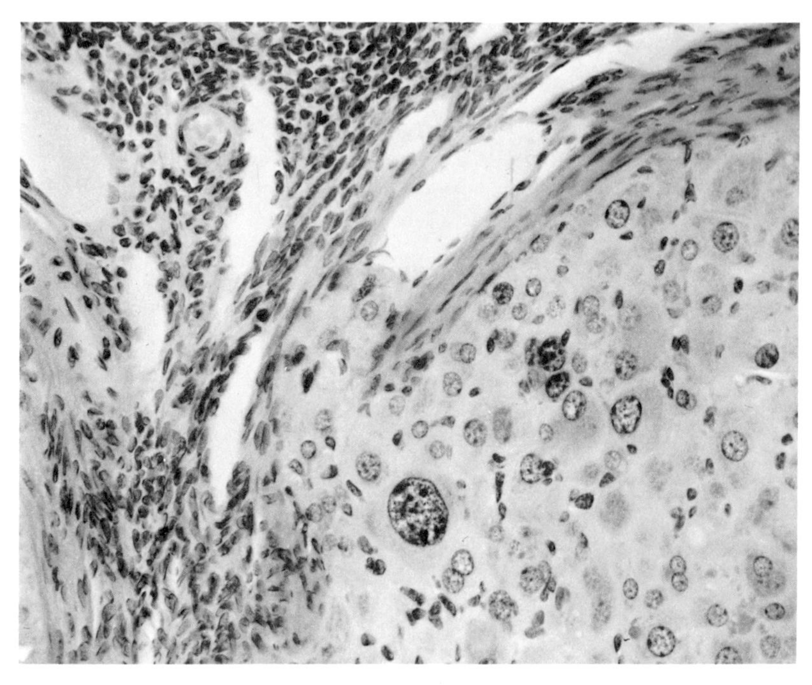

Fig. 4.17 Detail of a corpus luteum from the red squirrel ovary shown in Figure 4.15 to point out the nuclei of different sizes, indicating polyploidy. Also shown is a herniation into a lymph vessel. *T. hudsonicus*, 112. × 200.

medullary tube very close to the tube's connection with the rete and that the opening to the rete allowed the herniation of granulosa cells into its lumen.

The single estrous female had many atretic medullary follicles surrounded by much well-differentiated interstitial gland tissue (Fig. 4.22). This was sometimes so distributed as to suggest its origin not only from the theca interna of atretic follicles but also from the stroma surrounding the medullary cords. The ripe follicles were definitely in the medulla, which was so crowded and distorted by them that the corticomedullary boundary was even more indefinite than usual. It seemed probable, however, that all other medullary follicles were atretic or had completely disappeared and that the numerous remaining normal primary and secondary follicles were all of truly cortical origin. Apparently, only at

Table 4.6 Measurements (in mm) of parous red squirrels during the spring reproductive season

Animal number	Date	Greatest diameter		Diameter of largest		
		Vagina	Uterus	Follicle	C. l.[a]	C. alb.[b]
		Proestrous period				
32	8 Jan.	6.0	1.5	0.63	0.50	+
40	29 Jan.	8.0	1.5	0.50	0	+
41	29 Jan.	3.0	0.8	0.62	0	+
44	3 Feb.	2.5	1.3	0.60	0	+
		Estrous period				
97	11 Mar.	13.0	2.5	0.91	0	+
		Implantation period				
100	11 Feb.	7.0	3.0	0.90	1.30	+
77	17 Mar.	8.0	4.0	0.60	1.00	0
		Embryonal period				
48[c]	26 Mar.	6.0	3.0	0.60	0.96	0
		Postpartum and lactation periods				
50[d]	26 Mar.	7.0	5.0	0.64	0.80	+
85	9 June	2.2	1.4	0.30	1.28	0
84	9 June	2.2	1.0	0.52	1.50	0
87	9 June	2.2	1.5	0.67	0.50	0
116	26 June	3.0	1.8	0.66	1.53	0
56	9 July	2.3	1.6	0.48	1.85	0

[a] Corpus luteum.
[b] Corpora albicantia. All + signs in this column indicate presence.
[c] Embryos were 9 mm CR (crown–rump).
[d] Some evidence of fetuses having been resorbed rather than delivered.

full estrus when about to rupture do the medullary follicles crowd aside the overlying cortical layer of primordial follicles and reach the ovarian surface.

During the first pregnancy

Unfortunately, only one pregnant animal of this category was available (Table 4.8). Her embryos were in the primitive streak period. Except for the young corpora lutea and perhaps an even greater abundance of interstitial gland tissue, there was no appreciable change from the ovary of the estrous phase (Fig. 4.23).

Table 4.7 Measurements (in mm) of spring-born juvenile red squirrels before proestrus

Animal number	Date	Greatest diameter		Diameter of largest follicle
		Vagina	Uterus	
86	9 June	1.0	0.4	0.10
117	2 July	2.0	0.8	0.15
118	2 July	1.3	0.8	0.16
3	6 July	—	—	0.43
57	9 July	1.2	0.4	0.26
58	9 July	1.4	0.9	0.20
102	10 July	1.8	0.8	0.13
119	10 July	1.2	0.8	0.11
10	11 July	1.0	0.5	0.12
124	14 July	1.4	0.4	0.12
103	15 July	1.7	0.7	0.61
89	19 July	1.7	0.6	0.50
60	19 Aug.	1.2	0.6	0.21

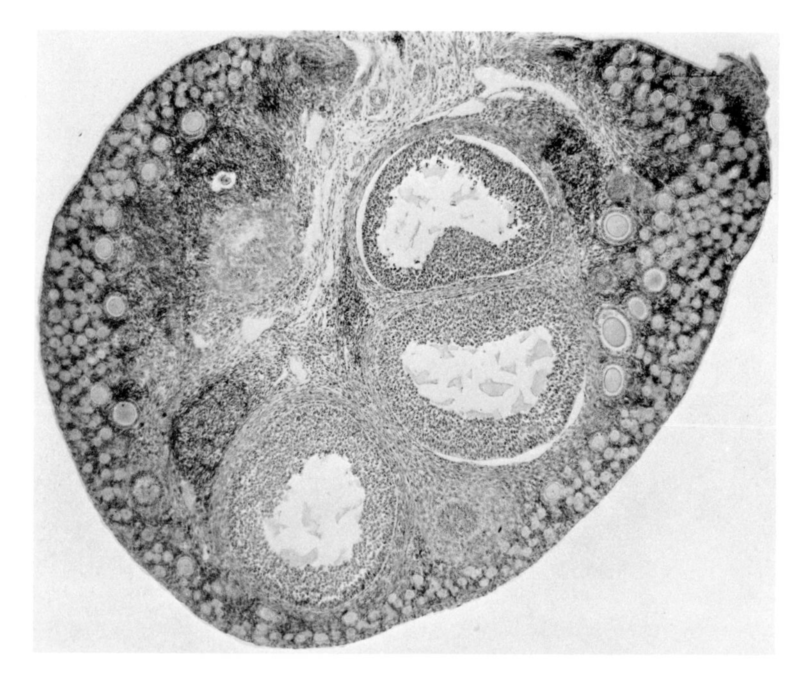

Fig. 4.18 Ovary of a spring-born juvenile red squirrel nearing proestrus. The vesicular follicles are clearly in the medulla. *T. hudsonicus*, 89. × 50.

During the first lactation and anestrus following the summer pregnancy

Six animals were assigned to this category (Table 4.8), because they were taken between 14 September and 4 January and because they had only one set of corpora lutea (or corpora albicantia, no. 5). By all other available criteria they were adults. They had few atretic follicles, almost no differentiated interstitial gland tissue, and relatively few primordial, primary, and secondary follicles. The few early vesicular follicles were very small, 0.40–0.54 mm in diameter. All of these animals could have been adults which had failed to breed during the preceding spring, but there was no indication that this was true. Female no. 5, taken in January, may possibly have been in proestrus, although its vagina and uterus were below the size ordinarily associated with proestrus. However, this one specimen out of the group had very numerous atretic follicles and much well-differentiated interstitial gland tissue in its cortex. Both of these are features which develop during proestrus.

Table 4.8 Measurements (in mm) of spring-born red squirrels during their first reproductive season

Animal number	Date	Greatest diameter		Diameter of largest	
		Vagina	Uterus	Follicle	Corpus luteum
		Proestrous period			
101	10 July	3.0	1.5	0.52	0
9	11 July	2.0	1.0	0.50	0
121	14 July	2.5	1.3	0.85	0
123	14 July	4.0	1.2	0.73	0
126	24 July	2.0	1.0	0.60	0
127	24 July	2.5	1.0	0.56	0
		Estrous period			
107	3 Aug.	12.0	2.5	1.00	0
		Embryonal period			
106[a]	3 Aug.	5.0	2.4	0.70	0.92
		Anestrous period			
13	14 Sept.	3.0	1.5	0.40	1.45
14	14 Sept.	2.0	1.0	0.20	1.35
96	21 Sept.	2.5	1.6	0.27	1.40
61	2 Oct.	—	—	0.50	1.87
115	21 Oct.	2.0	1.4	0.53	1.90
5[b]	4 Jan.	2.0	1.0	0.54	0

[a] In primitive streak stage.

[b] Corpora albicantia present.

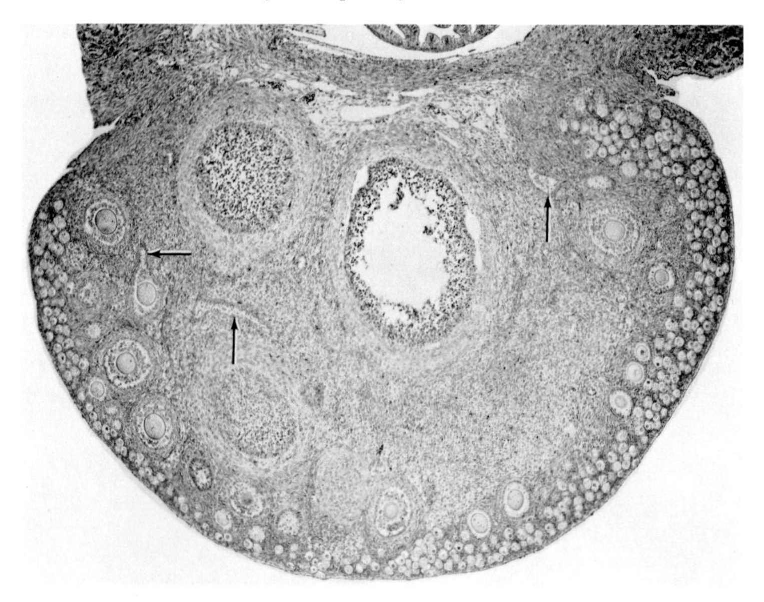

Fig. 4.19 Ovary of a spring-born juvenile red squirrel during proestrus in mid-July, showing tangential sections of two large medullary follicles and portions of medullary cords (*small arrows*). *T. hudsonicus*, 123. × 42.

Discussion and summary

The red squirrel (*Tamiasciurus hudsonicus*) of southern Wisconsin has a midwinter and a midsummer breeding season. Spring-born females breed during the immediately following summer. Summer-born females breed the following winter. The young are born very undeveloped, so their ovaries are comparable with the fetal ovaries of more precocious species. The length of proestrus and estrus is not known, but the dates and frequency of collection of proestrous and tubally pregnant individuals compared with the dates and infrequency of collection of estrous specimens suggest that proestrus probably lasts from two to four weeks and estrus not more than one or two days. There is no evidence to indicate whether red squirrel ovulation is spontaneous or induced. The diameters of ripe follicles averaged approximately 1 mm, and those of corpora lutea of the contemporary pregnancy in mid-gestation ranged from 0.9 to 1.3 mm.

Ovaries of very young juvenile females had distinct thick medullary

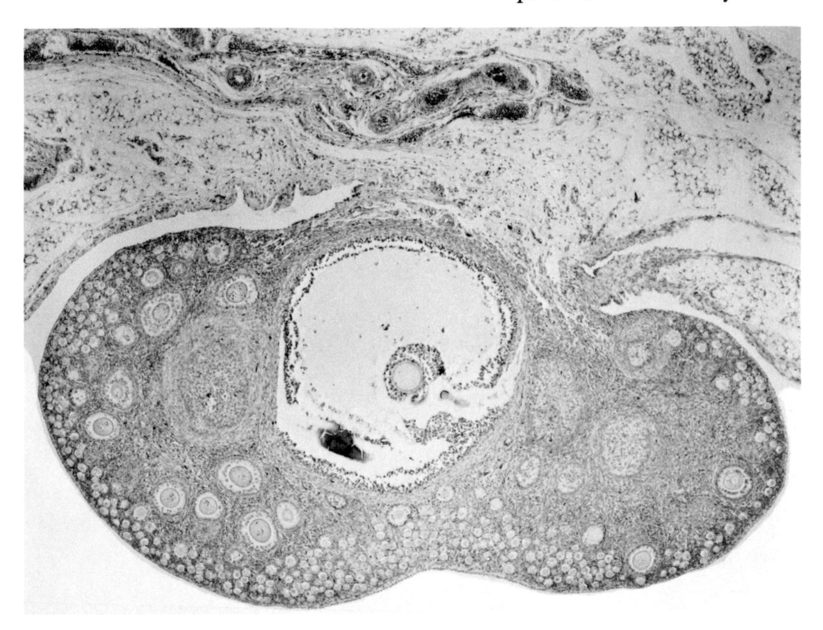

Fig. 4.20 Ovary of another spring-born juvenile red squirrel during proestrus in mid-July, showing a large medullary follicle. The cortex is infantile in character. *T. hudsonicus*, 121. × 42.

cords containing scattered oogonia or oocytes. During the later prepuberal and the proestrous periods, the epithelium around these eggs differentiated to form primary and secondary follicles which sometimes retained connecting portions of medullary cords. In summer-born young, all of these medullary follicles apparently underwent atresia before the first breeding season the following winter. However, in spring-born young, although most medullary follicles became atretic, some matured and obviously provided the only ripe follicles of their first breeding season in the immediately following summer. All follicles of parous females appeared to be of cortical origin, and only rarely were even small remnants of medullary cords recognizable.

Loosening and separation of cells, i.e., edema of the cumulus, were typical of all fully ripe follicles. The thecal gland of mature follicles was relatively thin, only three to four cells in thickness. It was often absent in spots. Its individual cells were typical of those of most mammals: large, discoid, and with granular cytoplasm. After follicular rupture, a prom-

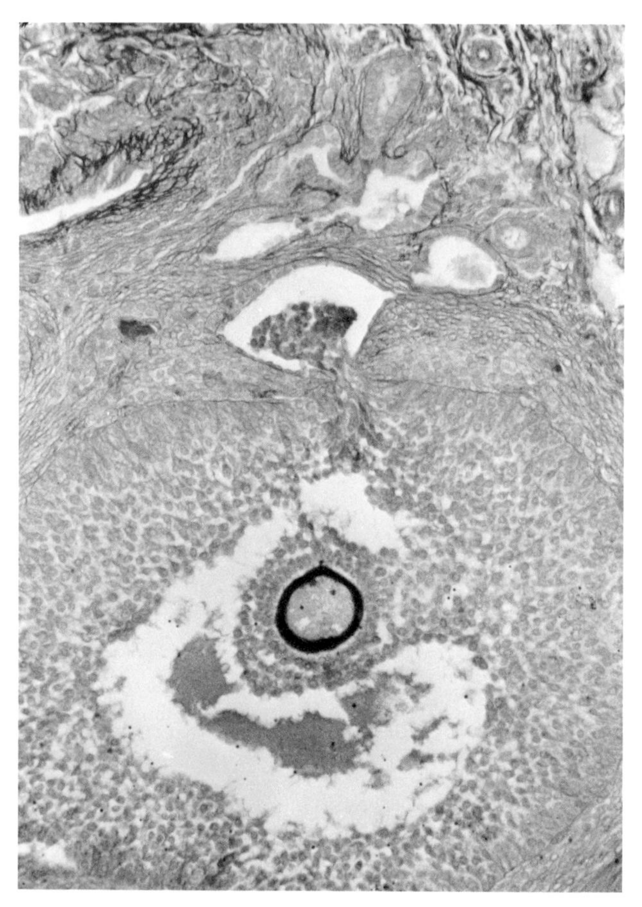

Fig. 4.21 Medullary follicle with its granulosa herniating into the lumen of the rete. A spring-born juvenile red squirrel during her first proestrus in late July. *T. hudsonicus*, 126. × 155.

inent thickened band of thecal gland persisted during the early tubal period and then disappeared, except for one instance (female no. 70, Table 4.5) in which a small amount was detected in an animal with uterine morulae.

Since very slight signs of paraluteal cells were found, it was assumed that follicular epithelium was probably the sole source of luteal cells. This seems to be generally characteristic of mammals of this size and smaller. Corpora lutea of females which bred in the summer had become corpora albicantia by the beginning of the spring gestation.

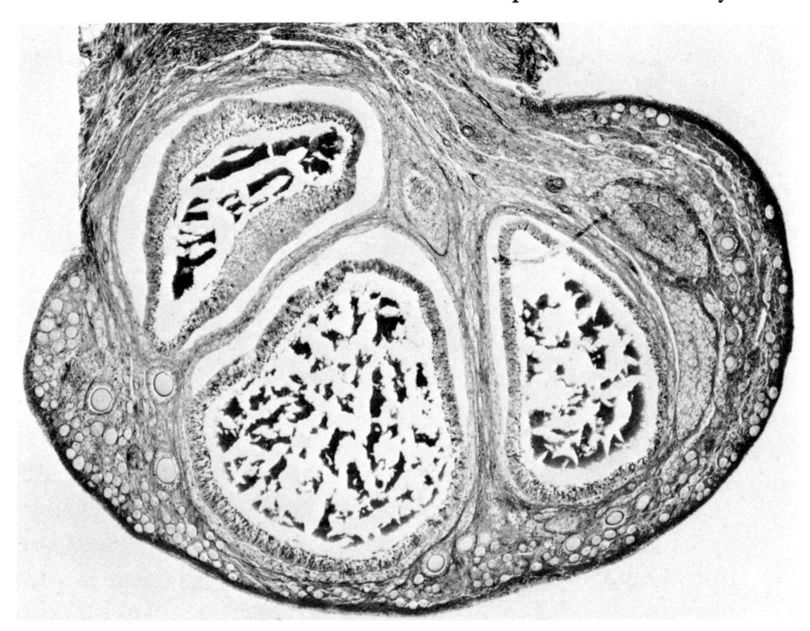

Fig. 4.22 Ovary of a spring-born red squirrel during her first estrus in early August, showing three ripe medullary follicles and, at the right, two corpora atretica, presumably derived from medullary follicles. (Severe shrinkage artifact in the large follicles.) *T. hudsonicus*, 107. × 40.

The persistence of corpora lutea of the late winter gestation through the succeeding summer gestation and lactation raises a number of questions about the physiology of reproduction in this species and about the corpora lutea themselves. Our data on the corpora (condensed in Table 4.9) are admittedly crude, but they do indicate two trends which should be checked more carefully in species showing a similar pattern. They suggest (1) that the corpora continue to enlarge throughout most of the lactation period and (2) that those of a first pregnancy of the year remain as large or larger than those of a succeeding pregnancy, at least until well into the autumn anestrous period. Animals entered the summer proestrus while still lactating, with their spring set of corpora showing no evidence of cell degeneration. Signs of luteal cell degeneration in the form of large cytoplasmic vacuoles did appear in some females late in pregnancy, but vacuolation was not advanced until late in lactation when the second set of corpora also began to show the same phenomenon.

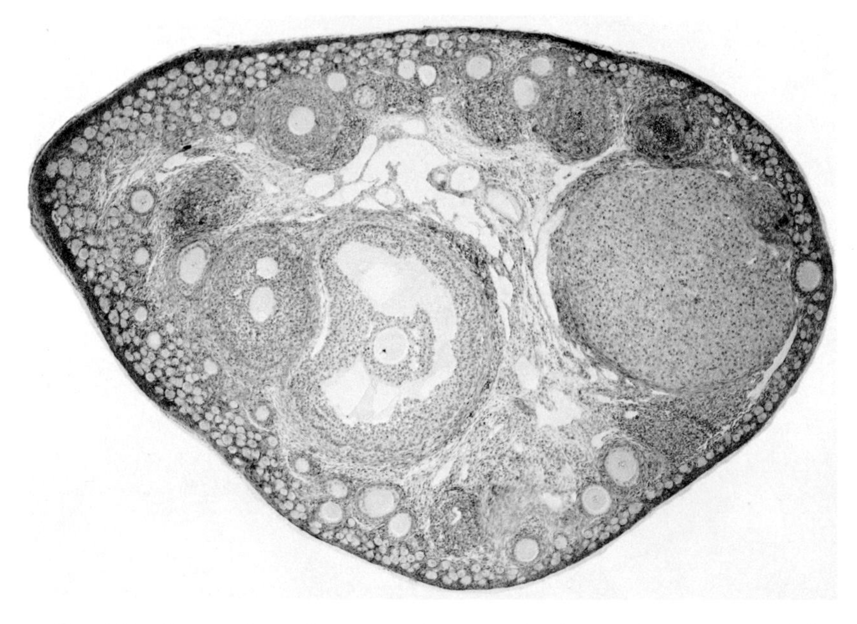

Fig. 4.23 Ovary of a spring-born red squirrel in the primitive streak period of her first pregnancy in early August. The corpus luteum was presumably derived from a medullary follicle. A large medullary follicle in very early atresia is present. *T. hudsonicus*, 106. × 38.

Luteal cell nuclei of three or more size categories in old corpora indicated that polyploidy probably occurs in these glands. Polyploidy is known to occur in mammalian liver and brain and, of course, notably in the salivary gland cells of dipterous insects. Since all of these, including luteal cells, are secretory, it may be that polyploidy is in some way correlated with the rapid synthesis for cell export, such as is known to occur in these organs. Corpora lutea might be favorable subjects for study of this phenomenon, as activity of these cells can be controlled to some extent by luteotrophic hormones both in vivo and in vitro.

Interstitial gland tissue of the fetal type was present in the medullary regions of ovaries of juveniles before the onset of follicular degeneration. After atresia had set in, it was impossible, with the techniques used, to distinguish between cells of this type and the numerous scattered interstitial gland cells probably derived from the thecae internae of atretic follicles. Interstitial gland tissue subjectively seemed to increase in

Table 4.9 Average diameters (in mm) of corpora lutea in red squirrels at different periods of the reproductive cycle

Reference and season	Proestrus and estrus	Uterine embryos primitive streak to 9 mm CR[a]	Fetuses 14 mm CR[a] to term	Postpartum, lactation, and anestrus
	Summer-born in 1st pregnancy			
Table 4.3; spring	—	1.00 (5)	1.00 (1)	1.30 (3)
	Parous females in 2d breeding of the year			
Tables 4.4–4.5; summer	1.34 (12)	old 1.40 (3) new 0.97 (3)	old 1.23 (3) new 1.12 (3)	old 1.49 (2) new 1.56 (2)
	Parous females in 1st pregnancy of the year			
Table 4.6; spring	—	0.96 (1)	—	1.24 (6)
	Spring-born in 1st pregnancy			
Table 4.8; summer	—	0.92 (1)	—	1.59 (5)

Note: Number in sample in ( ).
[a] Crown–rump.

amount as follicular atresia increased during proestrus, estrus, and tubal pregnancy, and to decrease during later pregnancy and lactation to a minimum during late lactation and anestrus. However, it was not measured, and even subjectively seemed to vary rather erratically. Even more erratic was the degree of its cellular differentiation. Ovaries near the time of estrus and during tubal and early uterine pregnancy certainly had the greatest number of fully differentiated interstitial gland cells and those during late lactation and anestrus the least. These conditions exist in most other mammals.

Five

The human ovary

The best detailed description of the adult human ovary is that of Watzka (1957). We have drawn heavily on his data and for confirmation of our own observations, and we wish to acknowledge our indebtedness to him here rather than by numerous citations which would clutter the text. In this chapter, we attempt to describe briefly the ovary during the different periods of life from birth to old age and at the various phases of the menstrual and pregnancy cycles. A summary of the development of the ovary is contained in Chapter 3.

Ovary of the newborn

The ovary of the newborn baby is basically like the infantile ovaries of mammals whose young are born in a relatively precocious condition, but is more advanced than those of such altricial newborn as the rat and rabbit. It normally contains one or more vesicular follicles with diameters of 3–5 mm. Because of their location in the medulla, they usually do not cause prominent bulges on the surface. Probably these follicles are derived from medullary cords, but there is no direct proof of this, and presumably they are stimulated to enlarge by gonadotropic hormones. It is well known that the female neonate's cervical region is markedly enlarged and that the mammae of both sexes often secrete "witch's milk"; however, since this cervical hypertrophy and mammary activity disappear within a few days, they are in some way dependent upon intra-uterine placental relationships. Medullary vesicular follicles, usually atretic ones, continue to occur sporadically throughout childhood. This, of course, is evidence of some gonadotropin production by the child. Both gonadotropins and steroid hormones are known to reach the fetal blood stream in the placenta. These could be maternal hormones transmitted unchanged through the interhemal membrane, or maternal hormones modified in the trophoblast of the membrane, or they could be strictly synthetic products of the trophoblast. Probably all three situations occur. For recent contributions to these problems, see Diczfalusy (1962), Diczfalusy, Pion, and Schwers (1965), and Jost (1969).

Most of the large follicles of neonatal and infant ovaries which we have examined, or which are described in the literature, are in some stage of atresia, and are usually surrounded by a zone of well-differentiated interstitial gland cells derived from their thecae internae (Mossman, Koering, and Ferry, 1964; Wagenen and Simpson, 1965) (Figs. 5.1 and 5.2). Some follicular atresia may start days or possibly even weeks before birth, but the process becomes very marked at or soon after birth when the placental circulation is stopped and the infant's supply of maternal and placental hormones is exhausted.

Except for the large vesicular follicles, ovaries of the human newborn and infant resemble those of most other infant and juvenile mammals.

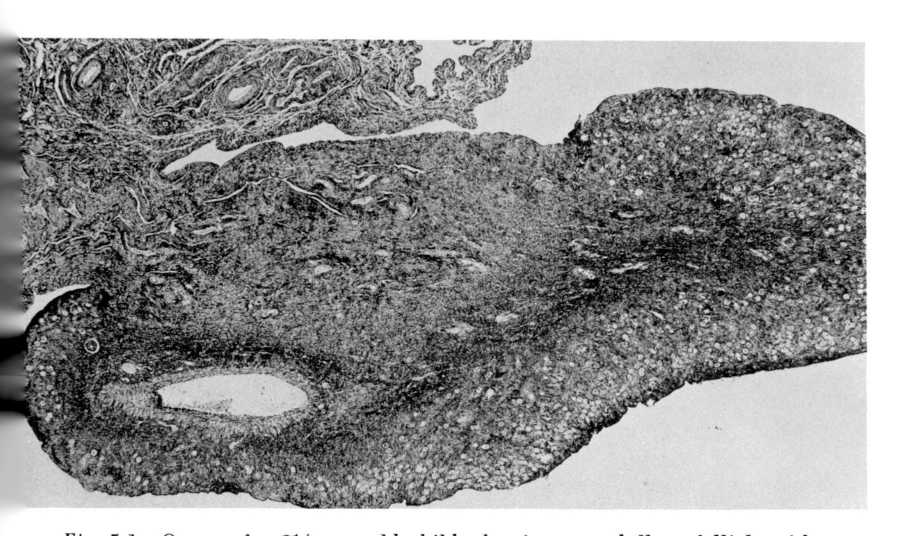

Fig. 5.1 Ovary of a 2½-year-old child, showing a medullary follicle with thecal type interstitial gland tissue. See Figure 3.1 for more detail of the cortex of this ovary. The ovary at this time is very similar to that of a late fetus or newborn. Human, 143L. × 18.

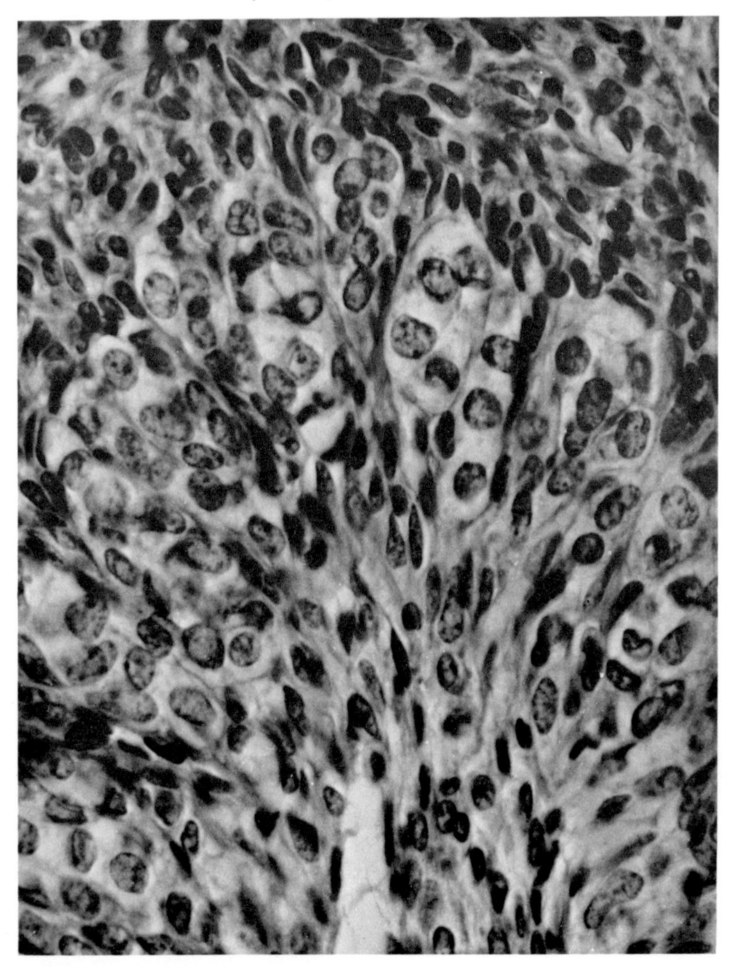

Fig. 5.2 Thecal type interstitial gland cells of the atretic follicle shown in Figure 5.1. Human, 143L. × 600.

Their cortex is relatively thick and can be considered to consist of three intergrading zones packed with "ova" (see Figs. 3.1 and 5.1). In the innermost of these three zones, each oocyte is enclosed by a simple low columnar epithelium. These are primary follicles. The intermediate zone has an even denser population of oocytes, but they are enclosed by simple squamous epithelium to form primordial follicles. The narrower surface zone has a few oocytes and oogonia, some actually in the surface epithelium. These are the "naked" ova, often occurring in groups or associated with irregular masses, or cords, of epithelium (cortical cords and "egg nests"). Most of these naked ova are early primary oocytes, but some are still oogonia (Oehler, 1951). The surface epithelium is simple cuboidal to simple squamous with occasional indentations opposite or attached to the cortical cords and egg nests (Oehler, 1951; Watzka, 1957). There is no distinct tunica albuginea, and the stroma of the cortex is very cellular with relatively few reticular and collagenic fibers.

The medulla contains the larger blood and lymph vessels together with the vesicular and atretic follicles and the interstitial gland tissue of the atretic follicles. There are also occasional primary follicles and scattered, narrow, inconspicuous segments of what have been interpreted as medullary cords. However, Forbes (1942) rarely found medullary cords in ovaries of fetuses longer than 280 mm CR and never in infants or children. The relation of the medullary follicles to the cords has not been established in human ovaries, but it is likely that a thorough study would show that these follicles originally develop from the cords, as they do in the juvenile chipmunk (*Tamias striatus*) and red squirrel (*Tamiasciurus hudsonicus*) (see Chapter 4, p. 88). This probably takes place late in fetal life rather than in the newborn.

A distinct rete lies in the cephalic portion of the hilus and adjacent mesovarium. Its anastomosing tubules are lined by a simple columnar to squamous nonciliated epithelium. The tubule diameter, including the epithelium, usually ranges from 30 to 60 μ. The lumina are about one-third as wide. With the usual histological techniques, stained slides show nothing in the lumina except an occasional bit of cellular debris.

Ovary of childhood

The extent to which new vesicular follicles continue to develop during infancy and early childhood is uncertain, but it is obvious that there is much individual variation. Simkins (1932), Potter (1963), and Valdes-Dapena (1967) all report them to be common throughout childhood. As in other mammals, the closer to puberty, the larger the cortical follicles, with due allowance for probable cyclic periods of follicular growth. Wagenen and Simpson (1965) show a well-developed corpus atreticum with differentiated interstitial gland tissue in what appears to be the medulla of an ovary of a 10-year-old, and we have seen similar medullary atretic follicles in the ovaries of children of 13 months and 2½ years (Mossman, Koering, and Ferry, 1964). It seems improbable that these could have persisted from the neonatal period. Therefore we surmise that vesicular medullary follicles develop frequently, probably periodically, during infancy and at least early childhood.

Kraus and Neubecker (1962) found corpora atretica with "luteinized thecae," which are thecal type interstitial gland tissue, in "almost all" of 121 children from birth to 14 years. These must have been medullary in the infants and cortical in the older children.

It is unfortunate that no thorough investigation of the human ovary during this period has ever been made. For such a study to be reliable, certain preliminary conditions are essential: 1) reliable histories of the individuals to eliminate specimens which are in any way abnormal because of certain types of illness, drug or hormone treatment, or genetic defects manifesting themselves in abnormalities of structure or function of the genital system; 2) well-prepared material, which means relatively quick fixation and good microscopic and ultramicroscopic techniques; 3) an adequate number of comparable specimens from each age to make possible an estimation of normal individual variability and the detection of cyclic changes which probably start long before puberty becomes obvious. Until such a study is made, our knowledge of this period of ovarian development must be considered superficial and our hypotheses tentative.

Ovary during puberty

Our own material for this period is scanty, and the literature is again inadequate. What we do provide represents gleanings from various sources and must therefore be considered just as tentative as that of the preceding section (Simkins, 1932; Potter, 1963; Wagenen and Simpson, 1965; Valdes-Dapena, 1967).

It is generally accepted that girls usually begin to menstruate before they ovulate, and that anovulatory cycles are actually the rule during the first few months after the menarche. However, pregnancies which occur before menstruation begins prove that ovulation sometimes does precede the menarche. It follows, then, that the puberal ovary may show very large atretic follicles and much other follicular atresia during the middle of the menstrual cycle, but no ripe or recently ruptured follicle such as is normal in the fully mature woman, and that after midcycle there may be no corpora lutea. That such ovaries are far more common during early puberty than during active reproductive life is a reasonable assumption, but we know of no studies that prove this.

It is probable that the potential for producing medullary follicles is exhausted before puberty, and that the follicles which ripen or approach that condition during puberty are produced in the cortex. Also, during childhood and early puberty the ovary increases in size to several times the linear dimensions it had during infancy. Because hyperplasia of the cortical stroma accounts for most of this increase, cortical primary and secondary follicles tend to become widely separated and appear to have become greatly reduced in number. All estimates of the actual number of oocytes indicate that there is a reduction at this time to about 190,000, half the number in the newborn, but the relative stromal increase makes this reduction appear even greater.

Ovary during the mature reproductive years

Man is a relatively large mammal, and the human ovary is in many ways characteristic of that of other large mammals, especially of those that ovulate only one or two eggs at a time. The ovary of a mature woman averages about 35 mm in length, 20 mm in depth, and 15 mm in thickness, and weighs about 4 g. It easily ranges from half to double this weight. Except for the distortion by very large follicles or corpora lutea, it is essentially almond-shaped, not lobed, and relatively smooth-surfaced. Its boundary with the broad thick mesovarium and with the proper ligament is not sharp, either grossly or microscopically. Cortical tissue may extend several millimeters into what grossly appears to be the proper ligament, and smooth muscle bundles (Hansen, 1957) often penetrate into the hilus and medulla from the mesovarium. McNeill (1931) described aberrant ovarian tissue and a corpus luteum in the proper ligament of the ovary close to the uterus.

The cortex resembles very cellular connective tissue and has delicate reticular and collagenic fibrils. Widely scattered primary and secondary follicles lie near the indistinct tunica albuginea, which usually ranges in thickness from 30 to 200 μ. Vesicular and atretic follicles and corpora albicantia, usually also widely spaced, occur in the deeper cortex and along the indistinct corticomedullary border. Only the larger follicles and well-developed corpora lutea extend through the entire thickness of the cortex and produce bulges on the ovarian surface. Hence, the general impression one gets from most sections of mature human ovaries is of a great amount of cellular connective tissue (the cortex), of blood and lymph vessels embedded in fibrous tissue (the medulla),

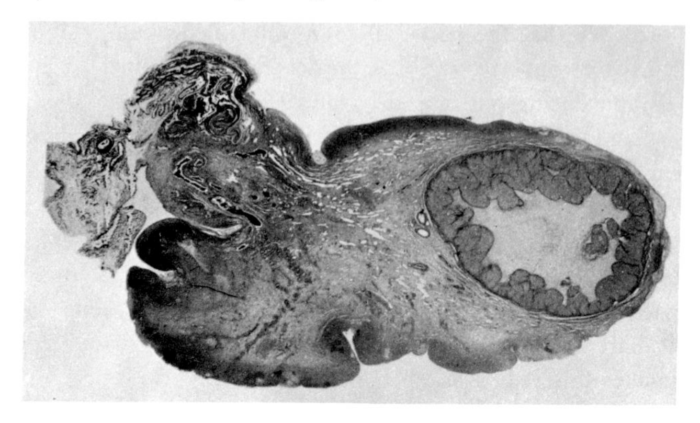

Fig. 5.3 Typical section of an ovary of a woman in about the 15th week of pregnancy. Note the cellular (dark) cortex with relatively few cortical structures. The corpus luteum is still hollow and growing by accretion from paraluteal cells. It has bulged deeply into the medulla. Human, 51. × 2.8.

and of a surprising scarcity of follicles and other typical ovarian elements (Fig. 5.3). Anatomically, the ovary is one of woman's most uninteresting features!

There is nothing unusual about the follicles of a human ovary in their entire history. Ripe follicles are commonly 10–15 mm in diameter and have a moderately developed thecal gland. Upon rupture, the follicular epithelium begins to lutealize, the thecal gland cells disappear quickly, presumably by degeneration as in other mammals, and within three to five days the immediately adjacent stromal cells begin to hypertrophy to form paraluteal cells (Fig. 5.4). These differentiate into luteal cells and add to the size of the corpus by accretion. (See G. W. Corner, Jr., 1956, for a detailed description of corpora lutea of this period.) Late in gestation paraluteal cells are usually not present, although we have seen some cells resembling them in one corpus luteum at eight months of pregnancy.

The corpora lutea of menstruation and of pregnancy have been described many times, and the literature is reviewed in detail by Watzka (1957). Gillman and Stein (1941) considered the total amount of luteal tissue to be greatest from about the middle of the second month to the end of the fifth month of pregnancy. However, they believed that cytological and physiological involution begins during the second month.

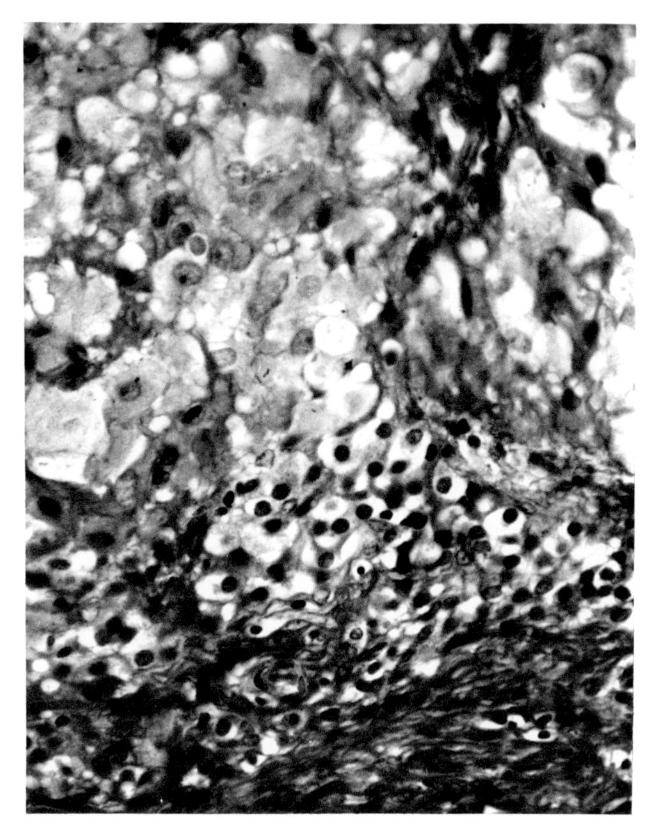

Fig. 5.4 Border of the corpus luteum of Figure 5.3, showing paraluteal cells intergrading with true luteal cells centrally (*above*) and with stromal cells peripherally (*below*). Human, 51. × 375.

Several glandular cell types have been described in the human corpus luteum, partly on a histochemical basis (White et al., 1951; Nelson and Greene, 1953, 1958). They are probably all simply various states of one basic luteal cell. There is no good evidence of two definitive types of luteal cells such as occur in artiodactyls and cetaceans.

Crisp, Dessouky, and Denys (1970) studied the ultrastructure of human paraluteal ("theca lutein") and granulosa luteal cells from two corpora lutea of a ruptured tubal pregnancy estimated to be of the 9th week, and from corpora of three nonpregnant women on days 15, 21, and 23 of the menstrual cycle. The granulosa luteal cells differed from paraluteal cells in having a more homogeneous and electron-lucent nuclear matrix, larger and more pleomorphic mitochondria, more numerous isolated Golgi complexes, and more abundant whorls of both granular and agranular endoplasmic reticulum. They also had numerous bundles of 50 A filaments and patches of microvilli bordering both inter- and intracellular canaliculi. They mentioned no transitional types.

Degeneration of corpora lutea results in distinctive corpora albicantia, each composed of a dense "cottony" mass of very fine collagenic fibers with very few cells (Fig. 5.5). Similar ones are found in apes and many other large animals, including the artiodactyls. In artiodactyls, they contain much yellow pigment, but in man they are white, hence the name "albicans." They probably persist for many months, or even years, for ovaries of older women show many more than those of young women. The fine collagenic fibers appear to be remnants of the delicate capsule and trabecular system of the corpus luteum. Many very small corpora albicantia are also present in the human ovary. These are derived from

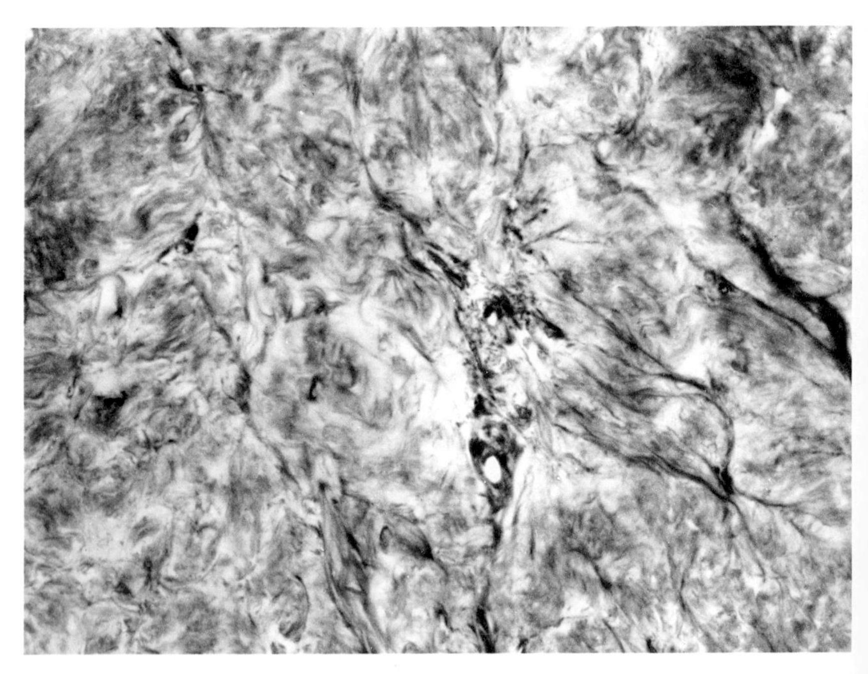

Fig. 5.5 Detail of the cottony fine collagenic fibers of a large corpus albicans derived from a corpus luteum. Human, 51. × 310.

the basement membranes of the follicular epithelium of atretic follicles (Fig. 5.6). Such accessory corpora albicantia are often called "glassy membranes" while their membranous nature is still visible. Eventually, they have the same microscopic structure as the primary corpora albicantia. It seems probable that, regardless of origin, these fibrous bodies are slowly resorbed; otherwise they would accumulate in far greater numbers than can be accounted for by the already large number that do occur in postmenopausal ovaries.

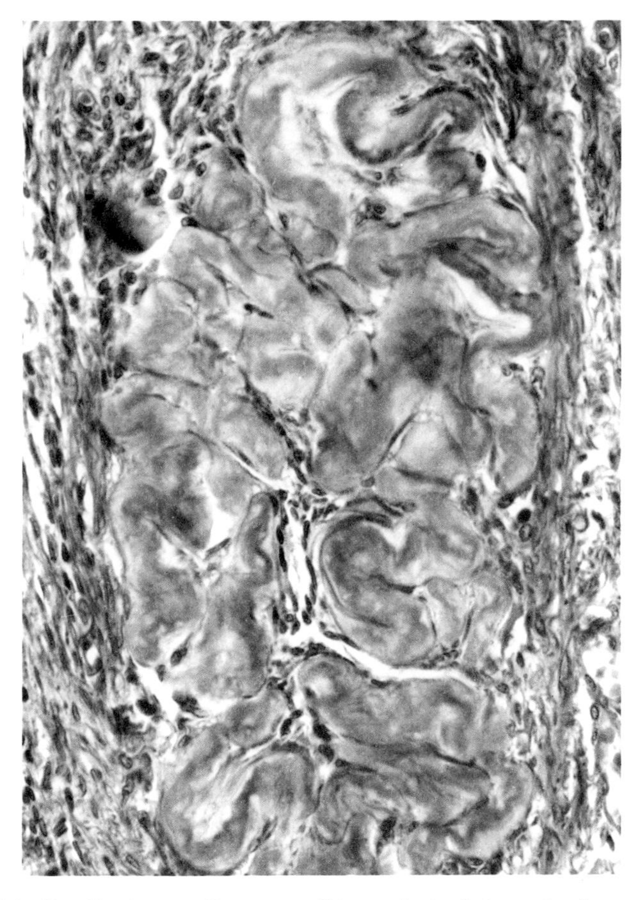

Fig. 5.6 Detail of a small corpus albicans derived from the basement membrane of the follicular epithelium of a collapsed atretic follicle. Fibroblasts that had entered the follicle are still present in the center. Human, 51. × 310.

Atretic follicles of all sizes, presumably derived from follicles of all stages, are common. The atretic vesicular follicles and probably the larger atretic secondary ones give rise to corpora atretica, that is, to zones of thecal type interstitial gland tissue surrounding each degenerating follicle (see Chapter 6, pp. 168–84). Waves of atresia and hence of interstitial gland tissue abundance occur near each ovulation period. Probably an even greater wave of atresia and of interstitial gland abundance occurs late in gestation (Mossman, Koering, and Ferry, 1964; Govan, 1970). It is at these periods that the gland cells appear to be largest and most differentiated, and therefore probably most active functionally. As in other mammals, this thecal type interstitial gland tissue forms whenever atresia of larger secondary and vesicular follicles takes place, hence it is present in significant amounts at least periodically from birth to many years after menopause. Savard (1968) summarized the results of studies on tissue slices of "stroma" from human ovaries incubated with acetate-1-^{14}C. Several steroids were synthesized, principally 4-androstenedione, dehydroepiandrosterone, and testosterone, but also significant amounts of estradiol-17β and estrone. No doubt this tissue contained thecal type interstitial gland cells, since atretic follicles were seen in histological samples of it. Although these experiments do not demonstrate the specific functions of human thecal type interstitial gland tissue in vivo, they do indicate that it has an important and probably complex part to play in the basic physiology of woman, particularly so since it is present throughout her life.

Only one other type of interstitial gland tissue is known to occur in the human ovary, the adneural type ("sympathicotroph," or "Berger," cells) (see Fig. 6.85), which is discussed in Chapter 6, pp. 201–3. Small numbers of these glandular cells, which closely resemble testicular interstitial gland cells, occur in or adjacent to the nerves of the mesovarium and ovarian hilus from the fetal period to old age (Kohn, 1928), and are best developed during fetal life and after puberty, especially during pregnancy (Wallart, 1927). They seem to be a constant feature of human and other primate ovaries. Hilus cell tumors are often composed of similar cells and are masculinizing, thus indicating that normal adneural gland cells are probably androgen secretors.

Probable gonadal adrenal tissue was seen in the hilus of only one of our ovary series, perhaps because we did not section enough of the mesovarium. However, "adrenal rests" and adrenal cortex type tumors are known to occur. Some hilus cell tumors may well arise from gonadal adrenal tissue instead of adneural cells.

Accessory corpora lutea derived from atretic follicles have been observed several times in human ovaries. Usually, they consist of a small area of lutealized granulosa of a large atretic follicle. We have recorded two instances of larger luteal bodies involving at least half of the wall of a large atretic follicle (Figs. 5.7 and 5.8). These were about one-third to one-half the diameter of the primary corpus and could easily be seen and mistaken for normal corpora, although microscopically there was no evidence that the follicles had ever ovulated. In all cases thecal type interstitial gland tissue is also differentiated around the nonlutealized portion of the follicle (Fig. 5.9). Accessory corpora lutea may occur near the primary corpus, but we have observed an instance of one being in the opposite ovary, so there is no reason to believe that their

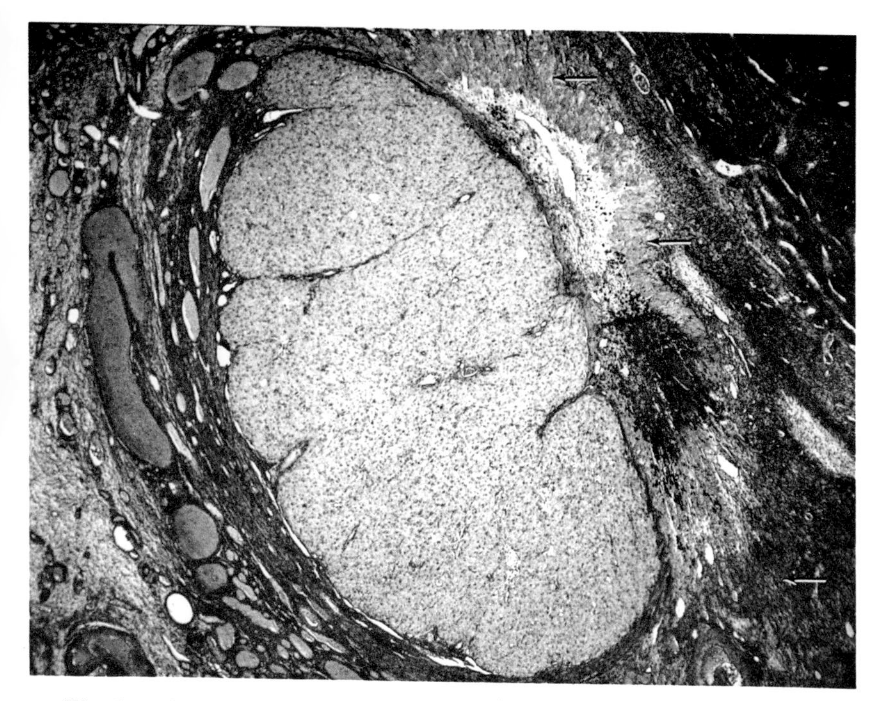

Fig. 5.7 Section through an atretic follicle, part of the wall of which has formed an accessory corpus luteum and part of which has formed the typical radiating columns of thecal type interstitial gland tissue (*arrows*). Human, 119R. × 25.

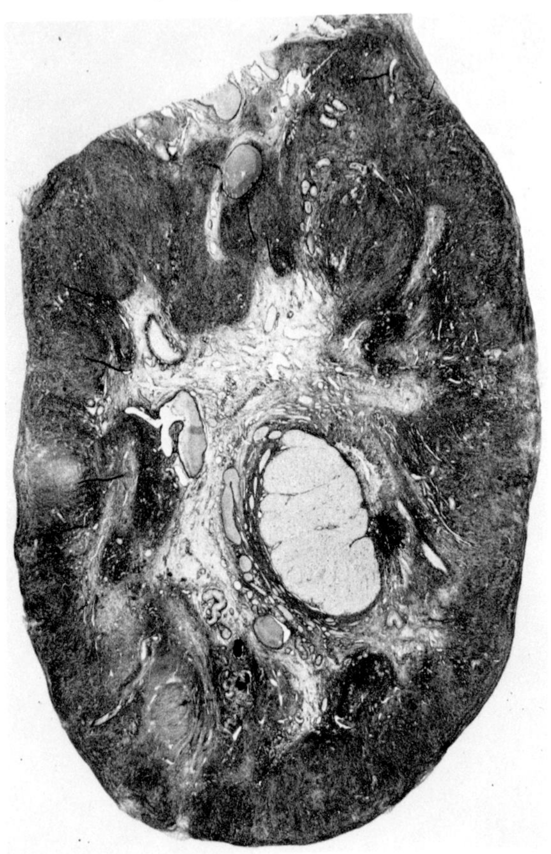

Fig. 5.8 The whole section shown in Figure 5.7 to indicate the relative size and position of the accessory corpus. Note the dense cellular cortex typical of human ovaries. Human, 119R. × 7.7.

presence is in any way dependent upon the direct influence of the primary corpora.

A common peculiarity of human ovaries, probably a form of mild pathology, is the presence at random spots on the surface and in the outer cortex of so-called ovarian decidual cells (Fig. 5.10) (Greene and Nelson, 1952; Israel, Rubenstone, and Meranze, 1954). Some consider these related to endometriosis. We have some evidence that they are correlated with adhesions. We doubt that they are cytologically or func-

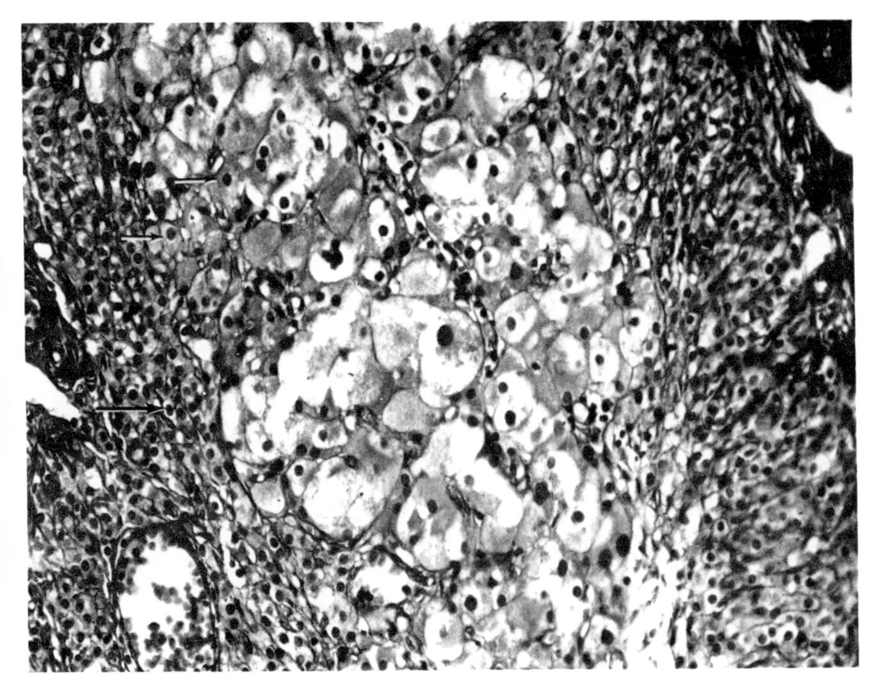

Fig. 5.9 Section of a very small accessory corpus luteum, showing centrally located luteal cells intergrading with paraluteal cells (*small arrows*) and these with thecal type interstitial gland cells (*large arrow*). Human, 51. × 250.

tionally homologous to uterine decidua, although they do somewhat resemble it.

The rete and epoophoron ductules of adult human ovaries are relatively small and inconspicuous, and we have never seen medullary cords in human ovaries.

The postmenopausal ovary

The ovaries of women many years past the climacteric still contain follicles ranging in stage from primary to early vesicular. However, it is hard to find in them a vesicular follicle which is not already in atresia, although obviously a few growing follicles must occur to account for the degenerating ones. Thecal type interstitial gland tissue is usually fairly plentiful and often appears to be well differentiated, even in ovaries of women of 70 years or more. Presumably, it is to some extent functional.

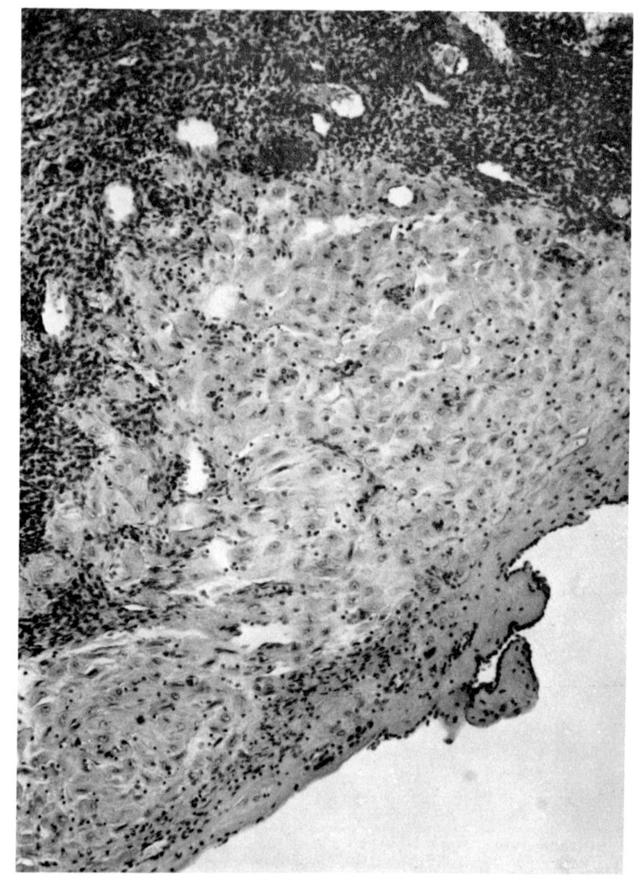

Fig. 5.10 Ovarian "decidua" at the surface of an ovary and near a point of adhesion. Human, 142R. × 95.

It is the one glandular tissue found in human ovaries from the late fetus and neonate into old age, but unfortunately its glandular function is unknown.

Corpora albicantia are characteristically numerous in postmenopausal ovaries. It is possible that many are remnants of those formed a few years previously from degenerating corpora lutea, but more are probably derived from relatively recent atretic follicles.

Hence the human postmenopausal ovary is not the completely inert nonfunctional fibrous mass that many have considered it to be. Like the ovary of childhood, the postmenopausal ovary should be investigated as to its possible endocrine functions and especially as to its potentiality for such function, if properly stimulated.

Summary

Watzka (1957) has provided an excellent and very detailed description of the adult human ovary.

The ovary of the newborn normally contains vesicular medullary follicles, most of which are in atresia. These are often surrounded by thecal type interstitial gland tissue. The cortex is packed with oogonia, "naked" oocytes, and primordial and primary follicles.

During childhood the medullary follicles appear to continue to develop, undergo atresia, and produce thecal type interstitial gland tissue. Apparently follicular activity also gradually increases in the cortex. At puberty, vesicular and ripening follicles are present in the cortex, but have disappeared from the medulla. However, a complete account of the human ovary through childhood and puberty has never been written.

The adult human ovary varies greatly in size. It is amygdaloidal in shape with a long thick mesovarium. Histologically, it is not sharply delimited, as cortical tissue may extend several millimeters into its ligaments, especially into the proper ligament. The cortex consists of a very cellular connective tissue stroma in which follicular elements are usually widely separated and apparently scarce. Ripe follicles have a diameter of 10–15 mm and are surrounded by a moderately developed thecal gland. The evidence indicates that upon ovulation the differentiated thecal gland cells degenerate and that within a few days paraluteal cells (really lutealizing stromal cells) appear around the lutealizing mass of former follicular epithelium. Thus, the corpus luteum is derived primarily from follicular epithelium, supplemented by the accretion of lutealizing stroma. Corpora lutea eventually regress to corpora albicantia composed of "cottony" masses of fine collagenic fibers with very few cells. These "white bodies" are better developed in the apes and man than in any other group of mammals, and no doubt derive their name from observations on the human ovary. In man and apes larger atretic follicles also give rise to corpora albicantia identical in structure to those derived from corpora lutea but smaller. These probably account for many of the numerous corpora albicantia characteristic of menopausal ovaries.

Follicular atresia with the formation of corpora atretica of thecal type interstitial gland tissue is common in human ovaries at midcycle and particularly during the last weeks of gestation. An occasional large atretic follicle undergoes partial lutealization, thus forming accessory corpora lutea which may be one-third to one-half the diameter of the primary corpora and so be easily confused with them.

Adneural gland tissue is probably a constant feature of the mesovarium and hilus of human ovaries. Probable gonadal adrenal tissue has also been seen in this area. Both occur in such small amounts that they may have little functional significance; however, the masculinizing hilus cell tumors that frequently occur in this area could be derived from either of these cell types.

The rete and epoophoron of the mature human ovary are relatively small, and medullary cords are probably absent.

The postmenopausal ovary usually contains many small atretic follicles and considerable quantities of thecal type interstitial gland tissue. Thus it may retain some endocrine function. It also contains numerous corpora albicantia, presumably mainly derived from atretic follicles.

Six

Comparative morphology of specific ovarian tissues and structures

Surface epithelium and tunica albuginea

We are certain of only one exception to the rule that the mammalian ovary is suspended by the mesovarium. In pregnant mouse deer (*Tragulus javanicus*), the ovary is attached directly to the uterine wall: there is no mesovarium. However, this is a secondary condition due to pregnancy, for nonpregnant specimens do have a very short mesovarium. We have no data to show whether or not this species is an exception to the previous statement that pregnancy does not greatly alter the position of the ovaries. Other than the usually narrow line of attachment to the mesovarium, the ovary is covered on all surfaces by a serous membrane — the visceral peritoneum, or serosa (see Fig. 1.1). In the armadillos (Dasypodidae) and horses (Equidae), the line of attachment to the ovary is very broad, while in most carnivores (Canidae, Ursidae, Procyonidae, Mustelidae, Pinnipedia) the mesovarium is a short, thick stalk and the bare area of the ovary is therefore roughly circular.

The serosa of any mammalian organ typically consists of a surface layer of simple squamous epithelium, the mesothelium, plus a thin layer of irregular fibrous connective tissue. Either of these layers may vary in character in different organs and in different species. The fibrous layer may be very thin and so intimately fused with the underlying fibrous capsule of the organ as to be indistinguishable from it. At an early embryonic stage the mesothelium consists of a single layer of mesenchymal cells flattened on their coelomic surfaces, but still having stellate processes on their deep sides (Gruenwald, 1942). At a later embryonic period the cells usually become low columnar, all their stellate cell processes disappear, and a basement membrane develops which separates them from the underlying connective tissue. They may remain permanently more or less columnar, especially in areas where there is little movement of one peritoneal surface against another surface, as near the hilus of the gonad.

In mammalian embryos the gonads first appear as a longitudinal ridge on the ventromedial surface of each mesonephros during the period when the limb buds become obvious. At first these ridges are merely a thickening of the mesothelial cells, which assume a more columnar shape. Later, primordial germ cells move into the ridges. Hyperplasia of these and of the mesothelium and the underlying mesenchyme builds up the definite gonadal bodies. The tendency toward a columnar surface epithelium persists into adult life in the case of the ovary, especially near the hilus. By and large, however, the bulk of the surface epithelium of mammalian ovaries ranges from a simple low columnar to a simple squamous condition, the latter being perhaps the most common. Occasionally, more than one layer of cells occurs, in most cases as local areas of irregular pseudostratification. This is more common in young females. Such a condition is usually associated with oogonia or developing primary oocytes embedded in the surface epithelium. However, in the adult eastern mole (*Scalopus aquaticus*) large areas of pseudostratification are normal. In the juvenile eastern chipmunk (*Tamias striatus*) much of the ovary is normally covered with an irregularly pseudostratified low columnar epithelium (Fig. 6.1). In large areas it is quite uniformly two cells in thickness — the deeper ones larger and rounded with clear cytoplasm, the superficial ones fitted over the others to form a smooth surface. The superficial cells have uniformly basophilic cytoplasm, and send processes between the others to the basement membrane. The significance of these two layers is unknown. The deeper layer closely resembles oogonia and very early oocytes.

Ovarian surface epithelial cells are always cuboidal or columnar in areas of formation of cortical cords or tubules, even in adult ovaries, whether or not there is evidence of new formation of oogonia in the area. The cells in these areas are almost always embryonic, that is, pluripotential or undifferentiated, with a basophilic cytoplasm and a low ratio of cytoplasmic volume to nuclear volume. The surface epithelium of the adult ovary of the porcupine (*Erethizon dorsatum*) is columnar except at the stigmata of young corpora lutea. In juveniles of this

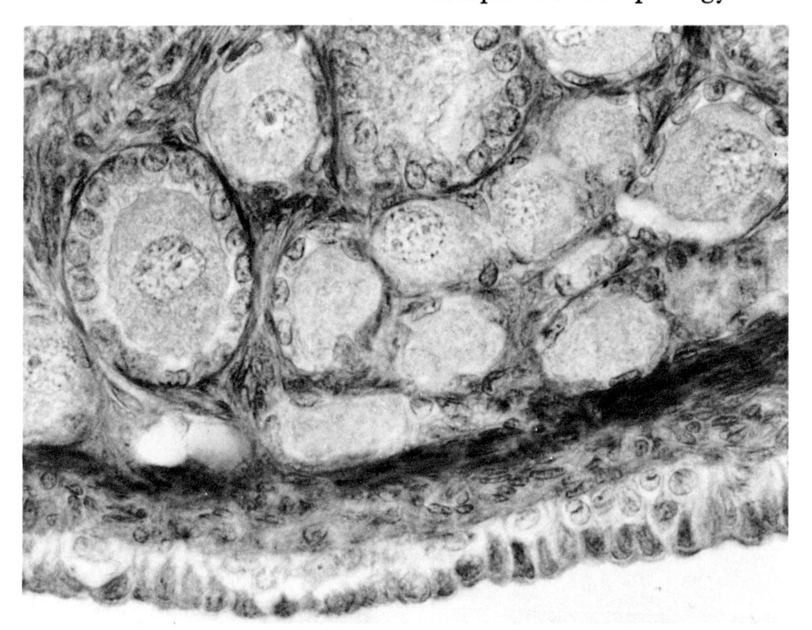

Fig. 6.1 Two-layered (pseudostratified) ovarian surface germinal epithelium of a juvenile eastern chipmunk (*Tamias striatus*); the basal layer may be oogonia. Note also the tunica albuginea and primordial and primary follicles. *T. striatus*, 62. × 620.

species, oocytes, probably abortive ones, commonly occur in the surface epithelium at any point on the ovary (Fig. 6.2). Likewise in this species, ingrowths of cortical tubes may occur anywhere on the ovarian surface, most frequently in juveniles (Fig. 6.3) (Mossman, 1938).

Sometimes there is no clear distinction between the surface epithelial cells and underlying cells, especially in small mammals, such as shrews (Soricidae) and mice (Muroidea), which lack a definite tunica albuginea (Fig. 6.4). In such cases, no basement membrane is visible, at least by light microscopy. The embryonic nature of both the surface and deeper cells seems obvious in these cases where the conditions are much the same as in the fetal ovaries of the several species described by Gruenwald (1942). As pointed out above, he demonstrated that at an early stage the only obvious difference between the surface mesothelial cells and the subjacent stromal cells was the flattened coelomic surface of the former. It seems probable that new surface epithelium can regenerate directly from the stroma, but this has never been directly proven. Even in ovaries with a usually distinct fibrous tunica albuginea,

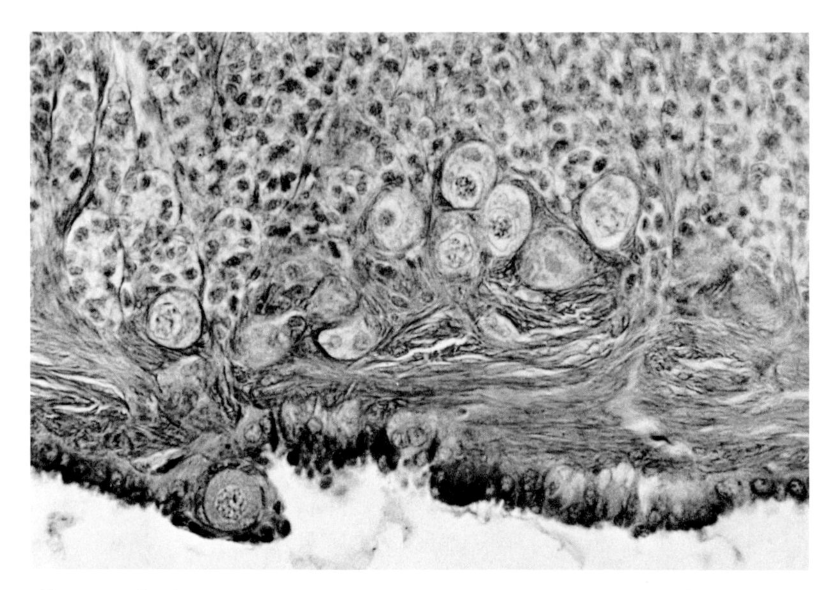

Fig. 6.2 Surface area of the ovary of a juvenile porcupine (*Erethizon dorsatum*), showing a primordial follicle in the surface epithelium and numerous basal lightly stained cells that resemble oogonia. The tunica albuginea is fibrous. In the background of the cortex is stromal type interstitial gland tissue. *E. dorsatum*, 2. × 375.

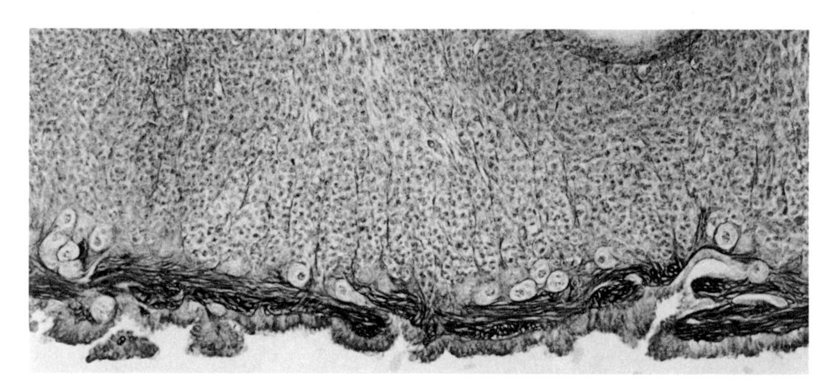

Fig. 6.3 Outer cortex of the ovary of a juvenile porcupine (*Erethizon dorsatum*) with epithelium-lined cortical tubules perforating the thin fibrous tunica albguinea. *E. dorsatum*, 2. × 150.

such as those of the porcupine, areas can be found where the surface epithelium seems to be in no way set off from the underlying cells except by its position.

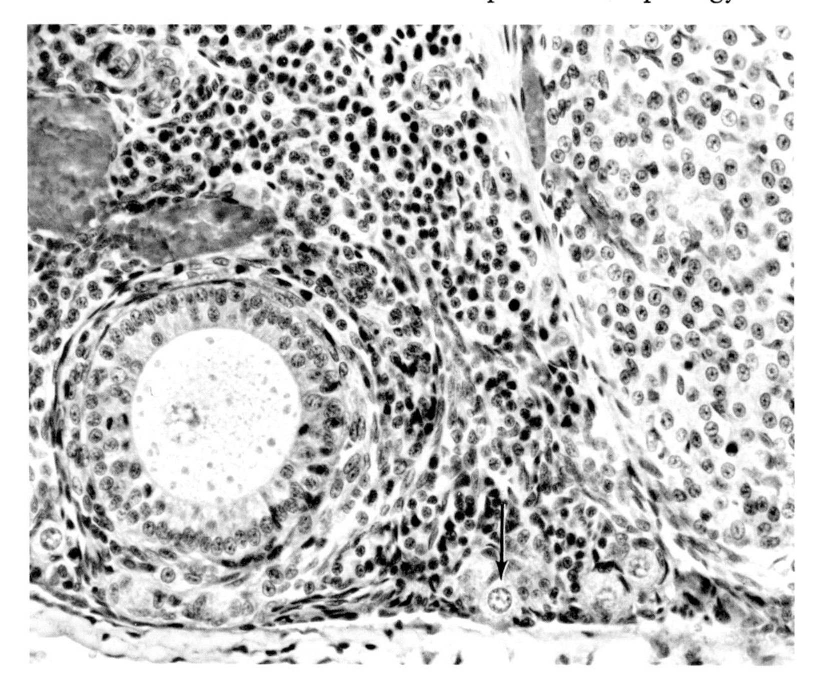

Fig. 6.4 Outer cortex and surface of the ovary of a woodland jumping mouse (*Napaeozapus insignis*), showing the lack of a definite surface epithelium and tunica albuginea. This condition is common in many areas of very small mammalian ovaries. Because the wall of the ovarian bursa is in almost direct contact with the ovarian surface, there is little chance that the surface epithelium was artificially abraded. Stromal type interstitial gland tissue fills most of the space between the corpus luteum and the early secondary follicle. The arrow indicates an oocyte that appears to lack a complete follicular epithelium and might therefore be considered a "naked ovum." *N. insignis*, 12R. × 300.

The fibrous layer of the normal serosa is generally inconspicuous, especially on organs having a heavy fibrous capsule. In ovaries of very small mammals, such as shrews and mice, there is often no sign of it or of a tunica albuginea. In mammals the size of rats, an indistinct subepithelial fibrous layer is usually identifiable, at least in some areas of the ovary. This is commonly considered to be the tunica albuginea. In large animals, the porcupine, for example, there may be a distinct subepithelial fibrous zone or capsule (Fig. 6.3), certainly an analog, if not the homolog, of the testicular tunica albuginea. In the human ovary, the follicles and other cortical structures in cortical zone stroma may be so scattered that there appears to be no qualitative difference throughout the depth of the cortical zone, and hence no distinct fibrous layer or tunica albuginea. However, more typically, especially in young women, the surface epithelium is underlain by a zone of rather fibrous tissue ranging from about 200 to 1,500 μ in thickness (Fig. 6.5). This grades rather abruptly into the outer cortex, which is noticeably more cellular and less fibrous. Most of the primary follicles occur in the outer portion of the cortex, only sparingly in the fibrous zone. Certainly this fibrous zone is analogous to the tunica albuginea. However, it is seldom clearly distinct from the underlying cortex, so if one considers it homologous to the tunica albuginea, one must admit that it is much less differentiated as a fibrous capsule than is that of the testis or of many other mammalian ovaries.

We must conclude, then, that the serosa of the adult mammalian ovary

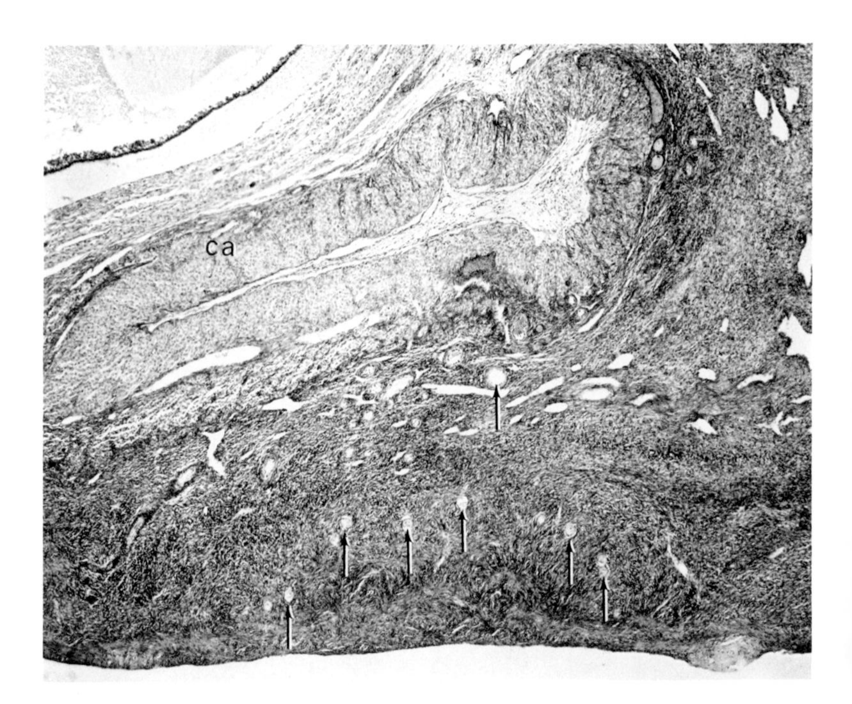

Fig. 6.5 Ovarian cortex of a 39-year-old woman in about the 14th week of pregnancy to show its cellular nature, the absence of a distinctly demarcated tunica albuginea, and the scarcity of follicles (*arrows*). A large corpus atreticum (*ca*) is present. Human, 142L. × 40.

is relatively embryonic. The surface epithelial cells of very small mammals may be mesenchymal in nature, except for their flattened coelomic surfaces, and they are commonly proliferative, even in adult mammals. A subepithelial connective tissue supporting layer is seldom differentiated as a distinct layer. Although the serosa of the ovary, in whatever form it occurs, is continuous with that of the rest of the peritoneal cavity, it actually bears a much more intimate morphological and physiological relation to the organ as a whole than does that of other organs, with the possible exception of the adrenal cortex whose capsule cells continue to differentiate into cortical cells. Even the adult ovary retains much of its embryonic character, its epithelium having a more or less continuous and formative relationship with it. For this reason, the epithelial covering of the ovary is regarded as different from the mesothelium of the rest of the coelom, and in view of its function in gonadal development it should be called *surface germinal epithelium* (*epithelium germinale superficiale*). It is confusing and inexact to call it simply the "germinal epithelium," for that would apply to any epithelium in some way related to the formation of germ cells. Since germ cells are found in the surface epithelium, in the cortical and medullary epithelial cords, and rarely even in the rete and epoophoron tubules (see pp. 127–36), germinal epithelium is widespread throughout the ovary and even to some extent in the mesovarium. One could argue with good reason for also including follicular epithelium in this category. Since follicular epithelium is often derived directly from the stroma, which is itself of mesenchymal origin, as is the surface epithelium and medullary cord epithelium, a good case exists for considering that there is no basic difference between epithelial and stromal elements of the ovary. We hold this view.

Invaginations of the surface epithelium known either as egg cords or tubules, or as cortical cords or tubules, were described by Pflüger (1863) in ovaries of the fetal and newborn dog. These structures have since been found to be common in adult ovaries of many carnivore families (Canidae, Ursidae, Procyonidae, Otariidae, Phocidae), some of the primates (Lemuroidea), and rodents (Rodentia). Their extreme development in the gray seal (*Halichoerus grypus*) was first described by Harrison-Matthews and Harrison (1949). (See also Harrison, 1950.) In this species, epithelium-lined tubes and clefts perforate the definite tunica albuginea at frequent intervals and dilate just beneath it and parallel to the surface into broad, thin crypts. These epithelium-lined spaces are so numerous as to appear to almost separate the tunica albuginea and surface epithelium from the underlying cortex. A similar but less extreme condition occurs in the adult raccoon (*Procyon lotor*) (Fig. 6.6). The presence of ova of various sizes in or associated with the deeper epithelium of these cortical sinuses strongly suggests that this is a region of neo-oogenesis. Similar relationships of ova to much narrower and more infrequent cortical cords have been found in the slow loris (*Nycticebus*) by Duke (1967) and Butler and Juma (1970). They also occur in the adult ovaries of many other mammals, including the grizzly bear (*Ursus horribilis*) (Fig. 6.7), the black bear (*Ursus americanus*), and the porcupine (see Fig. 6.3) (Mossman, 1938). Oocytes have been seen in the epithelium of the surface of the ovaries of many mammals, including *Nycticebus*, the porcupine, the dog, the weasels (*Mustela*), and one of the palm civets (*Paradoxurus*).

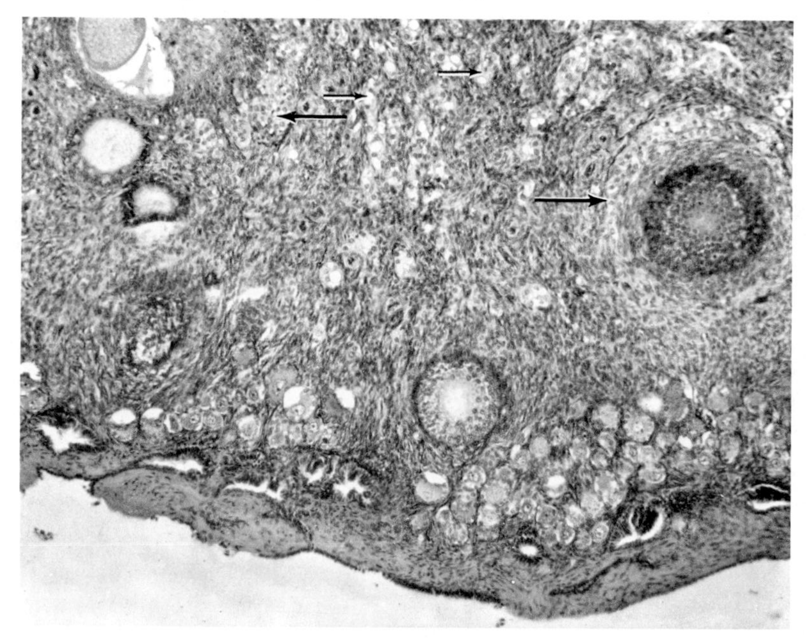

Fig. 6.6 Cortex and surface of the ovary of a raccoon (*Procyon lotor*) pregnant with late gill slit stage embryos. The epithelium-lined cortical tubules are piercing the fibrous tunica albuginea and expanding to form subsurface crypts. A few strands of medullary cord type interstitial gland cells lie in the stroma (*small arrows*). Thecal type interstitial gland cells are also present (*large arrows*). *P. lotor*, 2. × 100.

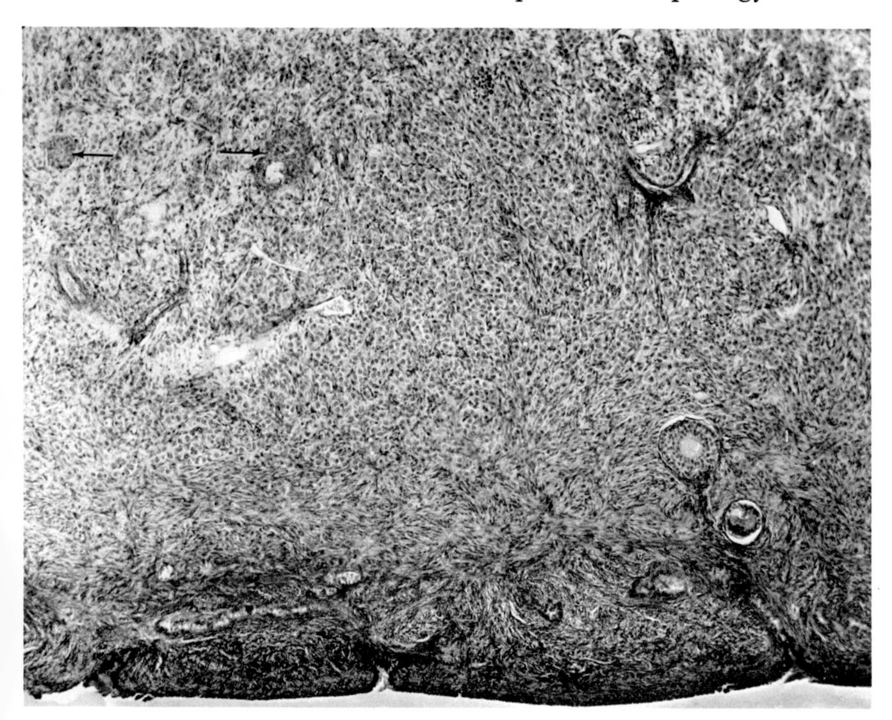

Fig. 6.7 Surface and outer cortex of the ovary of a nonpregnant parous grizzly bear (*Ursus horribilis*). Cortical tubules, or cords, pierce the poorly delimited tunica albuginea and terminate in primary and secondary follicles. The cortex and medulla are filled with interstitial gland of both the thecal and stromal type, here chiefly the latter. Two remnants of medullary cords can be seen (*arrows*). *U. horribilis*, 1. × 75.

Regions of the ovary

The surface epithelium, the tunica albuginea when present, and a thick zone of relatively cellular stroma containing oocytes, follicles, corpora lutea, and interstitial gland tissue make up the ovarian cortex. The inner core, or ovarian medulla, always contains the larger blood and lymph vessels, all or part of the ovarian rete, medullary cords if they are present, usually some interstitial gland tissue, and in juveniles of many species medullary follicles as well. Hyraxes (Hyracoidea) have the only known adult mammalian ovary in which the border between medulla and cortex is really sharp (Fig. 6.8). In all others, not only is there much

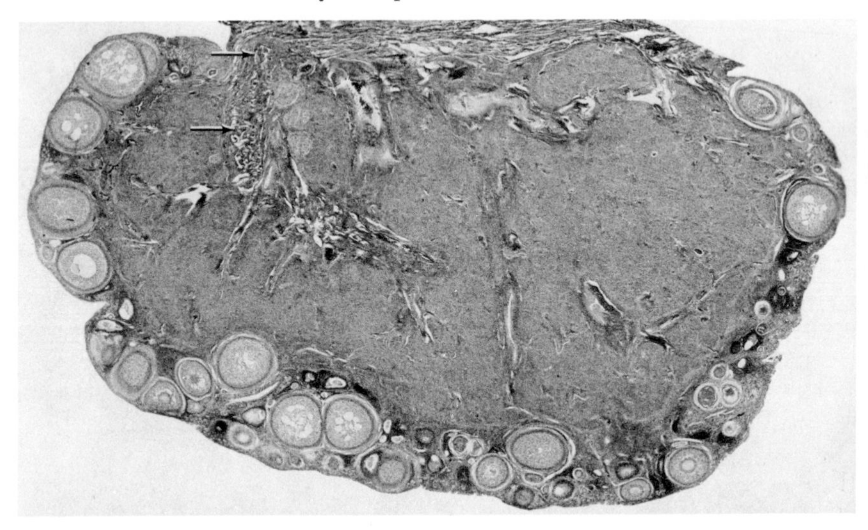

Fig. 6.8 Ovary of one of the rock dassies (*Procavia*), probably a nonparous adult. It shows the distinct border between cortex and medulla, as well as the broad attachment of smooth muscle at the hilus, the extensive rete (*arrows*), and the massive interstitial gland tissue of the medulla. *Procavia*, BP940. × 11.

interdigitation of the zones grossly, but microscopically the boundary is also very arbitrary. Large follicles and corpora lutea commonly extend deep into the medulla (Figs. 5.3 and 6.9–6.15), while medullary cords in some species extend into the cortex (Figs. 6.16 and 6.17). Degenerating corpora lutea and late stages of atretic follicles seem to sink gradually from the cortex into the medulla before they disappear. This is actually the result of the continuing growth of the superficial zone of the cortex (see Chap. 7, pp. 272–73). We therefore feel that the anatomical distinction usually emphasized between medulla and cortex is of relatively little significance. However, the terms are useful to indicate loosely the topographical position of the structures.

The classical concept of the ovarian cortex and medulla is that the medulla is derived from the primary sex cords and other portions of the indifferent stage of the embryonic gonad, while the cortex is derived from the surface epithelium of the indifferent stage by means of a second period of proliferation which forms the outer cortical elements. In some mammals, there does appear to be a distinct break between the original proliferation of the primary gonad and the second proliferation of cortical elements. However, in others the formation of the cortex seems to be simply a continuation of the primary proliferation, so that long after sex differentiation many cortical cords may be found which are continuous with medullary cords, although many are not. Developmentally, then, the difference between cortex and medulla is not distinct. Follicles form in both portions, but only in a few species do medullary follicles become functional. Also, since in most species various forms of interstitial gland tissue occur in both medulla and cortex, the medulla has to be recognized as a functional part of the ovary. Probably the two portions are in the process of evolving as functionally separate entities, but the evidence for or against this evolution has never been compiled.

Usually, the ovarian hilus is relatively long and narrow. It is only at or near the cephalic end of the hilus that the entrance of the large ovarian blood and lymph vessels is located. Here also are the ovarian rete and the efferent ductules of the ovary, which connect the rete to the duct of the epoophoron. However, the position of the rete varies in different species. Typically, it is entirely within the ovary, but may be almost entirely outside in the mesovarium. This area is comparable to the mediastinum testis and can logically be called the mediastinum ovarii. This is a fairly useful term to distinguish the cephalic part of the medulla from the rest, for it is in the cephalic part that the all-too-common rete cysts and hilus cell tumors originate (see p. 31).

Ovarian germ cells and follicles

Because the details of growth, maturation, and fertilization of the mammalian ovum have been of critical importance chiefly to geneticists and cytologists, it has become an established custom among anatomists, embryologists, and physicians to refer to all stages of the female germ cell from oogonium to zygote as ova, or eggs. In fact, human embryologists long ago fell into the questionable practice of speaking of even the early conceptus, up to at least the third month, as an ovum. This is a regrettable inaccuracy, yet because of its frequent use and real convenience it will no doubt remain in accepted usage for many years. For details of oogenesis and the cytology of female germ cells, the reader should consult Austin (1961), T. G. Baker (1963), Peters and Levy (1966), and Zamboni (1970).

Naked ova

Ova are numerous in the ovaries of adult mammals. Some of these appear to lie naked in the stroma (see Figs. 6.1, 6.2, and 6.4), or interspersed among epithelial cells of cortical cords (see Fig. 6.3) or tubules,

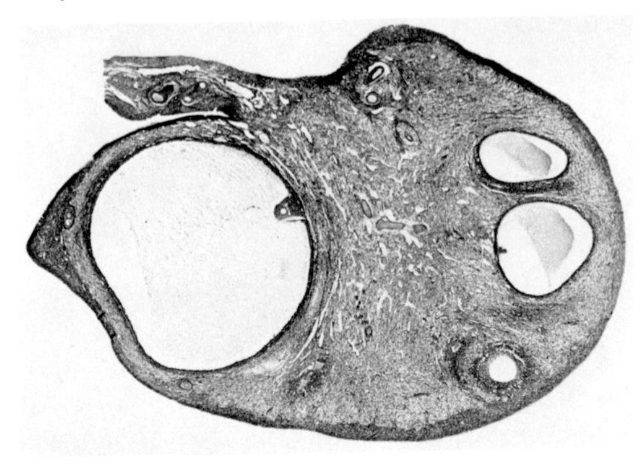

Fig. 6.9

Fig. 6.10

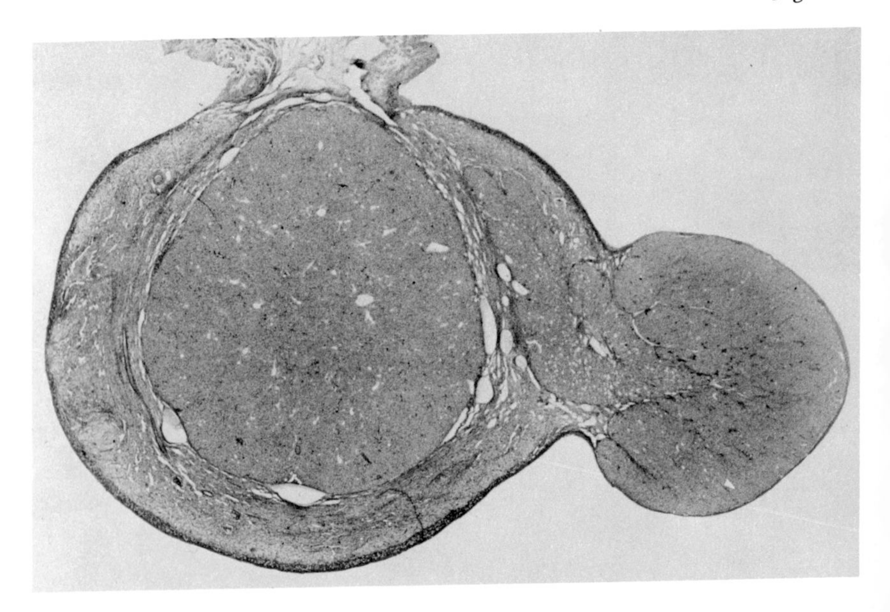

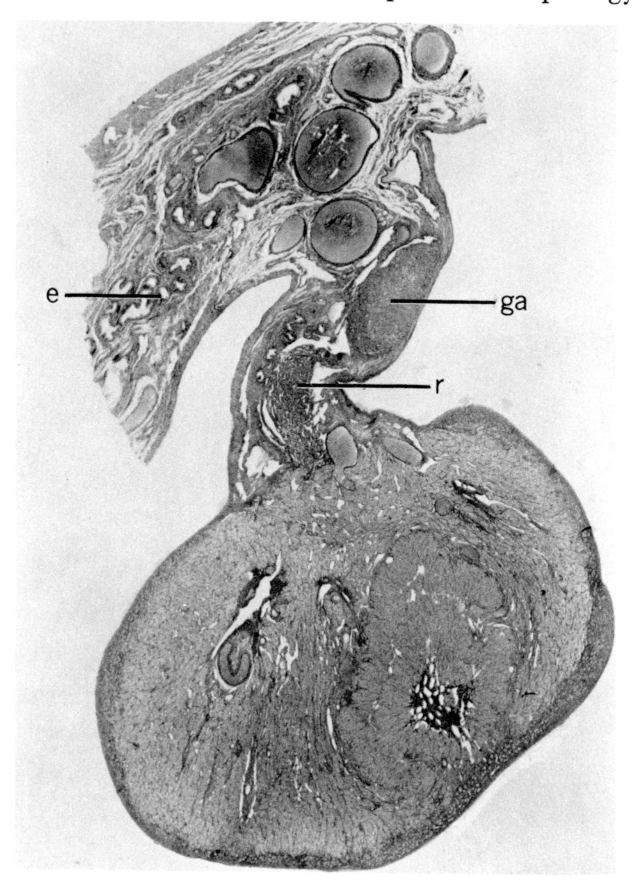

Fig. 6.11

Figs. 6.9–6.11 Sections of the left (Fig. 6.9) and right (Figs. 6.10 and 6.11) ovaries of one of the pangolins (*Manis*) in mid-gestation. Note the very pronounced differences between each ovary of a pair and between different areas of the same ovary, as well as the difference in appearance and degree of invasion of the medulla between an everted corpus luteum and the more usual type. The cumulus is attached at the deep side of the two larger follicles (Fig. 6.9). Figure 6.11 also shows the large rete (*r*), the epoophoron tubules (*e*), and the gonadal adrenal body (*ga*). *Manis*, 1. Figs. 6.9 and 6.10, × 8; Fig. 6.11, × 12.

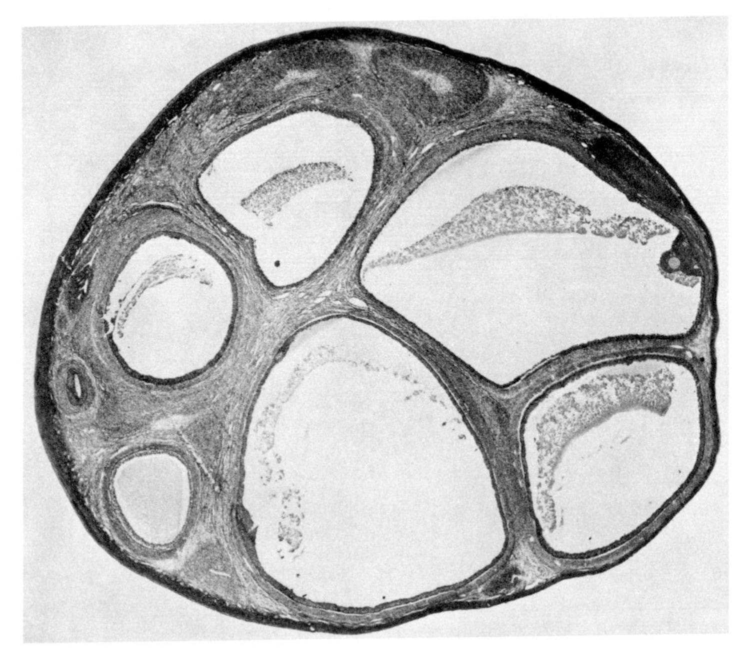

Fig. 6.12 Ovary of a springhare (*Pedetes capensis*) in late gestation, showing large follicles encroaching on the medulla and two large corpora atretica (*top center*) with their zones of thecal type interstitial gland tissue. *P. capensis*, 11. × 13.

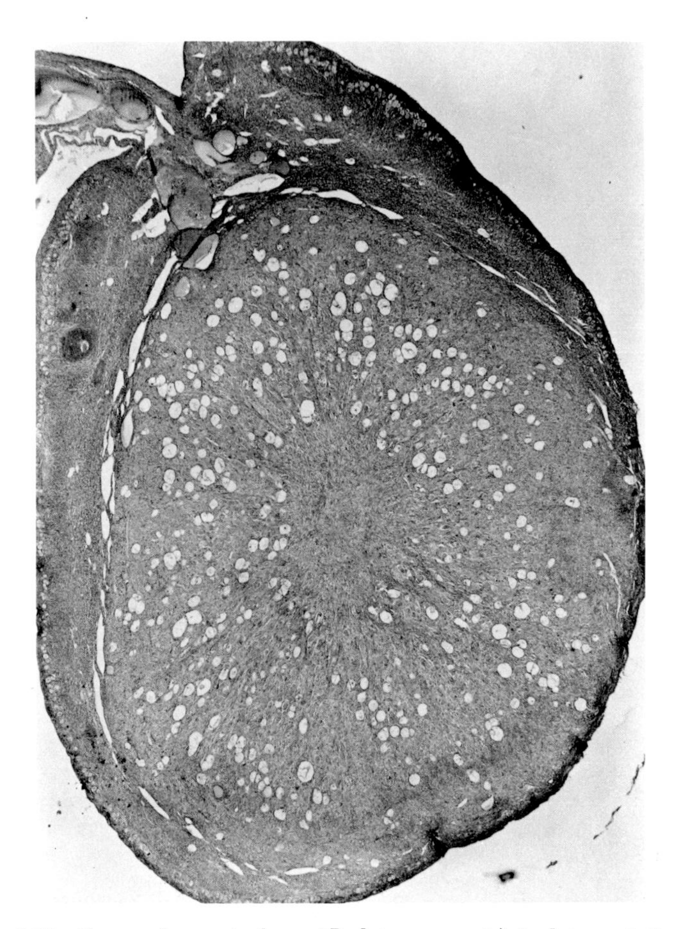

Fig. 6.13 Ovary of a springhare (*Pedetes capensis*) in late gestation, showing the relatively great size of the corpus luteum and its large vacuoles of degeneration. *P. capensis*, 10. × 11.

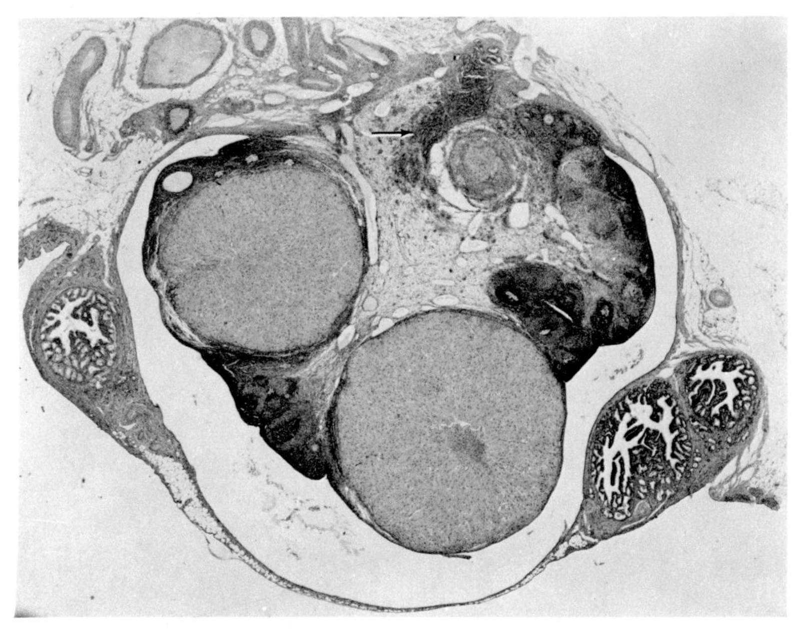

Fig. 6.14 Ovary of a European ferret (*Mustela putorius*), showing the relatively thin cortex (dark) and the extent to which the corpora lutea distort it. The rete extends well into the medulla (*arrow*), and medullary cord remnants appear as scattered black specks in the otherwise weakly stained connective tissue of the medulla. *M. putorius*, A84. × 16.

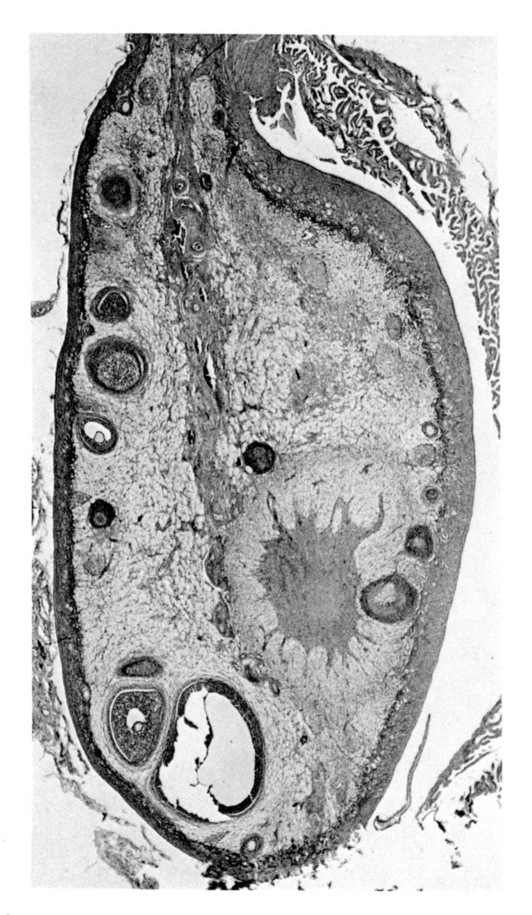

Fig. 6.15 Ovary of a springhare (*Pedetes capensis*), showing the very distinct fibrous outer zone of the cortex contrasted with the inner zone and medulla, both of which are occupied by massive amounts of thecal and stromal type interstitial gland tissue. *P. capensis*, 134R. × 16.

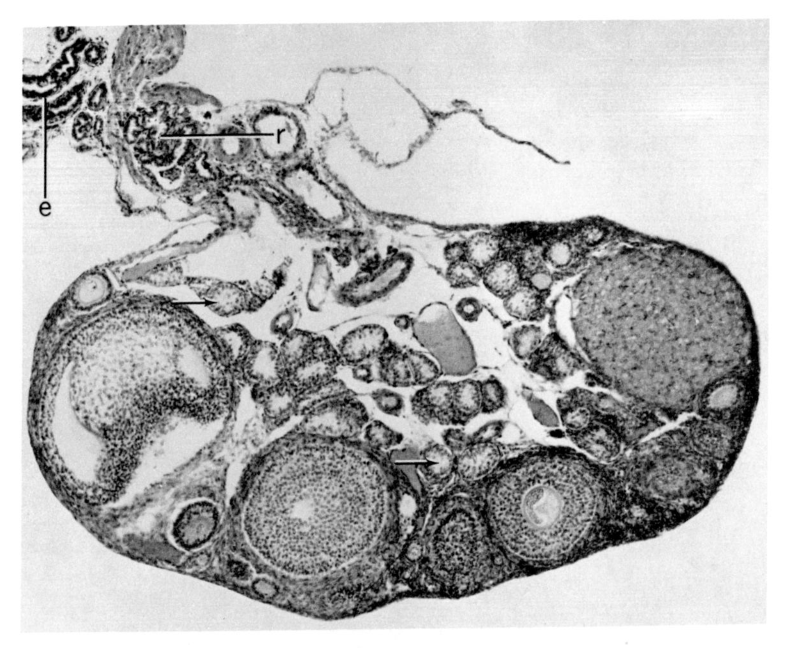

Fig. 6.16 Ovary, mesovarium, and mesosalpinx of a vagrant shrew (*Sorex vagrans*) in mid-gestation, showing medullary cords ("testis cords") (*arrows*), rete (*r*) and epoophoron (*e*). *S. vagrans*, 4. × 77.

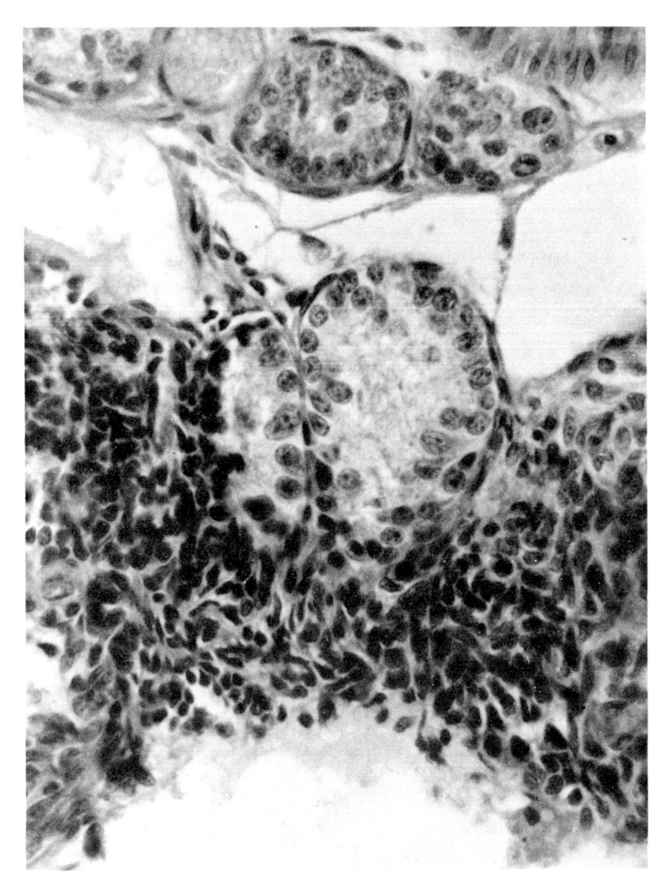

Fig. 6.17 Cortical cords continuous with surface tissue of the ovary of a vagrant shrew (*Sorex vagrans*) pregnant with limb bud embryos. *S. vagrans*, 13. × 490.

or even in the surface epithelium (see Fig. 6.2) or that of the rete (Figs. 6.18–6.21) or epoophoron (Fig. 6.22). When these germ cells are about the same size as the surrounding epithelial cells and barely distinguishable from them, they are likely to be still in the oogonial stage. However, most have enlarged noticeably beyond the size of the epithelial cells, and their nuclei are in the prophase of the first meiotic division. These are therefore early but still-growing primary oocytes.

Nonvesicular follicles

As we have seen in Chapter 2, when the oocytes have reached a diameter of 20–40 μ, they are enclosed either in a definite sheath of simple squamous epithelium to become primordial follicles or in simple columnar epithelium to form primary follicles. By the time the primary oocytes have reached a diameter of around 100 μ, they are usually surrounded by a noncellular membrane, the zona pellucida, and by stratified columnar or partly pseudostratified (Lipner and Cross, 1968) epithelium of

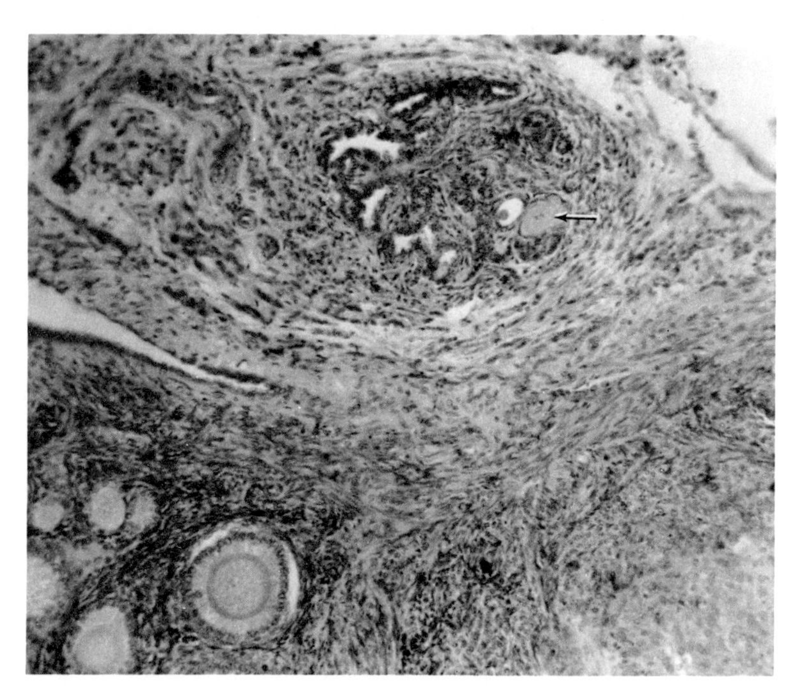

Fig. 6.18 An ovum in the rete (*arrow*) of a nonparous gray squirrel (*Sciurus carolinensis*) in proestrus. *S. carolinensis*, 36. × 120.

from two to many layers. These are then considered to be secondary follicles.

Vesicular and ripe follicles

Typically, but not in all mammals (see Chapter 2, p. 41), large fluid-filled spaces form among the follicular epithelial cells of the secondary follicles and coalesce into the single follicular cavity, or antrum, leaving the oocyte and its several surrounding layers of epithelial cells to form the cumulus oophorus, a hillock eccentrically attached to the epithelial wall of the now-hollow vesicular follicles. The vesicular follicles are usually subdivided arbitrarily into classes according to their diameters. However, ripe follicles are especially distinct because the cumulus becomes edematous and its cells tend to swell and disperse just before ovulation (Figs. 4.6, 4.7, 6.23–25).

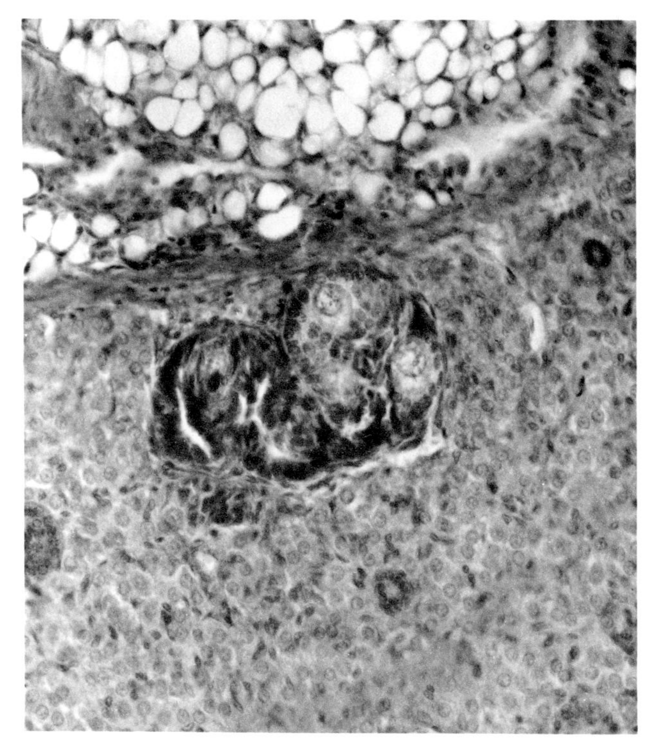

Fig. 6.19 Two ova in the rete of a star-nosed mole (*Condylura cristata*) near full term. *C. cristata*, 13. × 250.

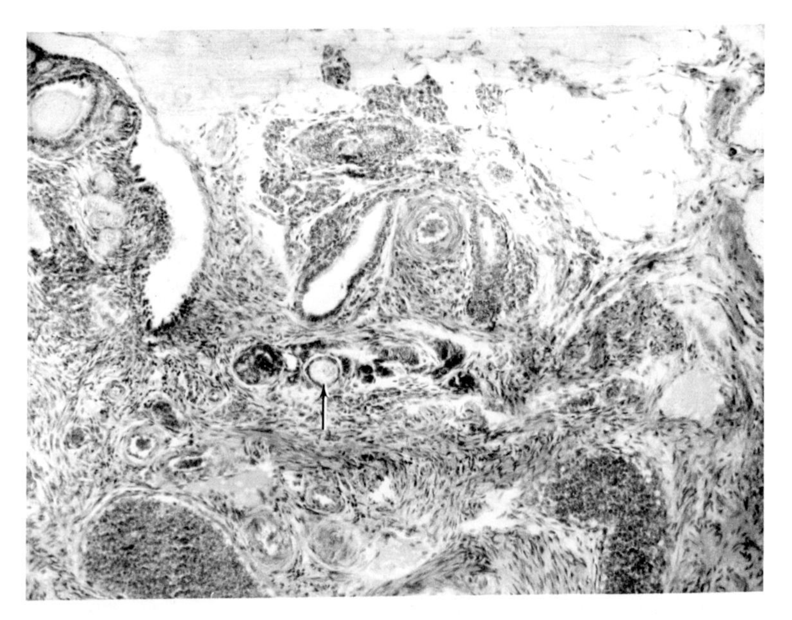

Fig. 6.20 An ovum (*arrow*) in the rete of a 13-lined ground squirrel (*Spermophilus tridecemlineatus*) in lactation. *S. tridecemlineatus*, 90. × 100.

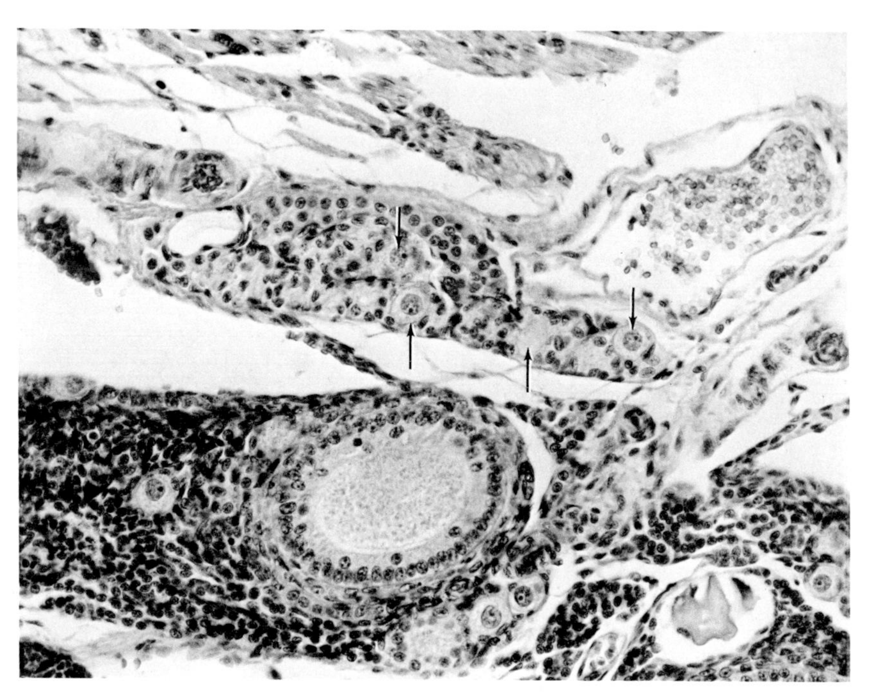

Fig. 6.21 Ova (*arrows*) in the rete of a parous jumping mouse. *Zapus*, 1. × 250.

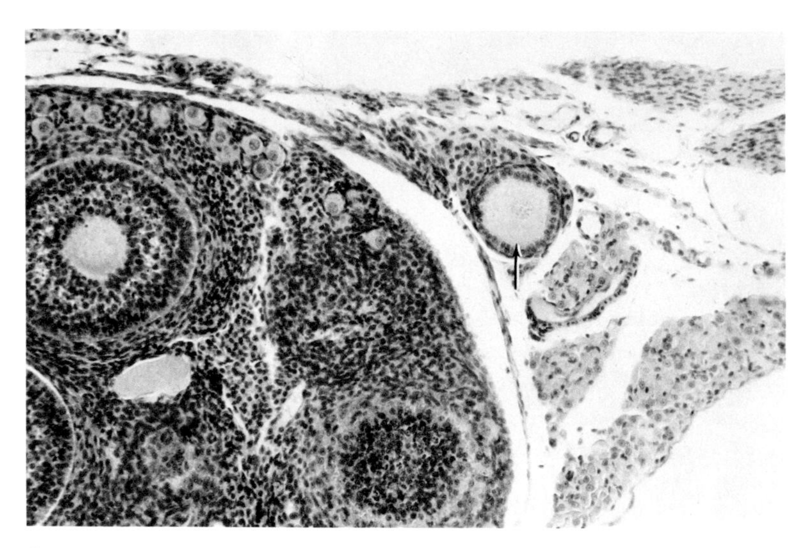

Fig. 6.22 Ovum (*arrow*) in a primary follicle in the epoophoron of a non-parous jumping mouse. *Zapus*, 3. × 130.

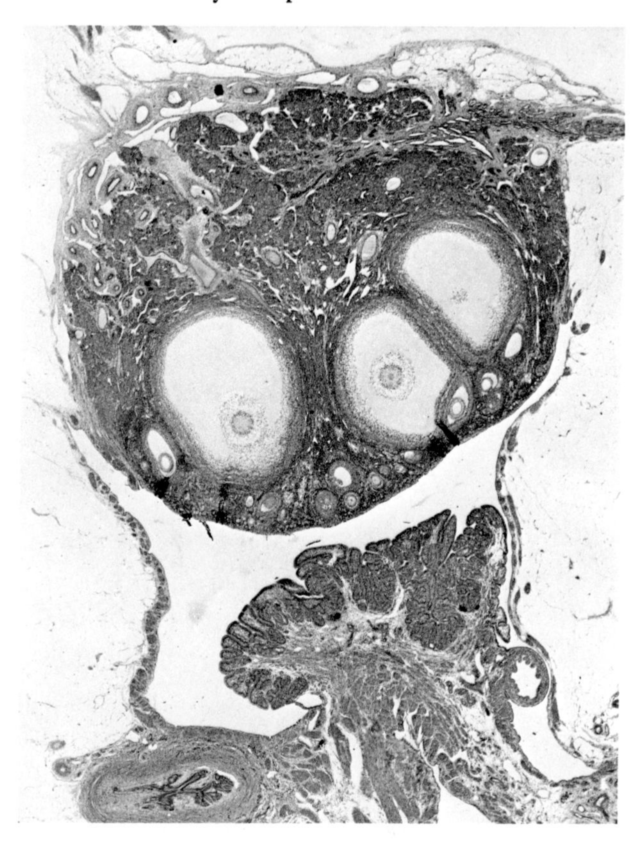

Fig. 6.23 Ovary of a striped skunk (*Mephitis mephitis*) in estrus, showing typical ripe follicles. The medulla is largely occupied by interstitial gland tissue. Cumulus cells are beginning to disperse (see Fig. 6.24). *M. mephitis*, 4. × 16.

Finer subdivsions of the four major categories of follicles have been made, but there is no general agreement on the criteria for these and, as a rule, no real need for such further subdivisions. In fact, many workers would combine the primordial and primary types under the single term "primary follicles."

Atretic follicles

Degeneration may set in, as pointed out in Chapter 2, at any stage up to that of a practically ripe vesicular follicle. Such degenerating or

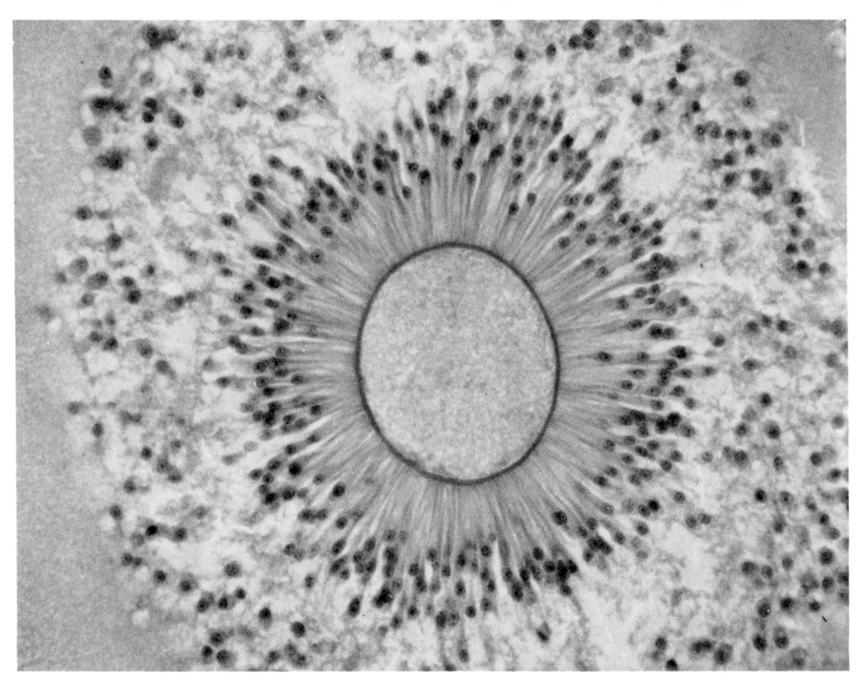

Fig. 6.24 Cumulus with corona radiata, zona pellucida, and oocyte from the ovary of the striped skunk (*Mephitis mephitis*) shown in Figure 6.23. The elongated cells of the corona radiata are prominent, and the cells of the rest of the cumulus are dispersing. *M. mephitis*, 4. × 250.

atretic follicles present a wide variety of appearances, and in adult ovaries are almost always far more numerous than all normal secondary and vesicular follicles combined. For further discussion of these, see below, p. 166 and Chapter 7, pp. 237–40.

The oocyte

As noted above, the diameter of the primary oocyte is between 20 and 40 μ when the primary follicle stage is attained. Chromatinic material within the primary oocyte's centrally located nucleus (15–25 μ in diameter) is rather uniformly dispersed, but there are clumped chromatinic strands in the pangolins (*Manis*), the domestic rabbit, and the guinea pig. After ordinary histological fixation, the oocyte cytoplasm is usually uniformly granular in appearance. One or more vacuoles, some

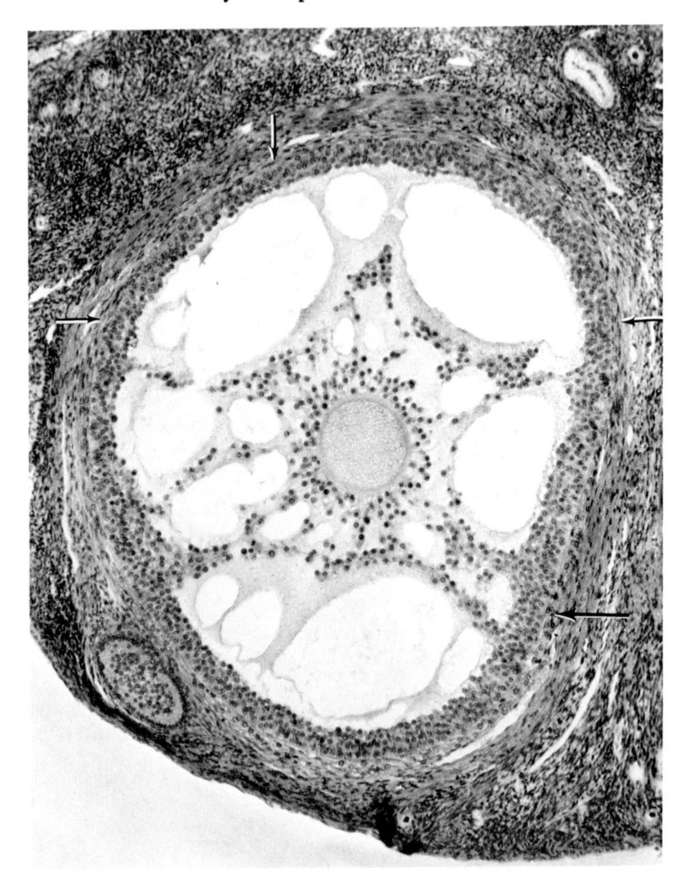

Fig. 6.25 The ripe follicle of a pika (*Ochotona princeps*) in estrus. This is typical, as well, of the appearance of the ripe follicles of rabbits and hares (Leporidae), which also have several epithelial cell trabeculae connecting the rather centrally located cumulus with the follicular wall. The thecal gland (*large arrow*) is thin and interrupted in some places (*small arrows*). The zona pellucida is invested by corona radiata cells, but the cells of the rest of the cumulus are dispersing in preparation for ovulation. *O. princeps*, 12. × 100.

approaching the size of the nucleus, may be seen in the cytoplasm of some oocytes in the ovaries of the pangolins, the cat, and the moose (*Alces alces*). The significance of this vacuolation is uncertain. The so-called yolk nucleus (yolk nucleus complex; Balbiani's vitelline body) is remarkably clear in the oocytes of primordial and primary follicles in gliding lemurs (*Cynocephalus*), tree shrews (*Tupaia*), and *Nycticebus.* No special techniques are necessary to demonstrate the complex in these species. In many other species, the yolk nucleus is only faintly visible or not visible at all with routine techniques. The yolk nucleus undergoes an apparent fragmentation, with peripheral dispersion of the fragments, as the oocyte and follicle increase in size. A single centriole is often visible in the midst or vicinity of the yolk nucleus, especially in oocytes of the primordial and primary follicles. Adams and Hertig (1964), Hertig and Adams (1967), Hertig (1968), and Zamboni (1970) have published excellent papers on the fine structure of the oocytes and follicles in the guinea pig and man. Hertig (1968) paid special attention to Balbiani's body. (See also p. 221.)

Ovulation and fertilization

As pointed out in Chapter 2, it is not until the last few hours before ovulation that a primary oocyte starts the final nuclear processes which culminate in its division into a secondary oocyte and into the first polar cell (polocytus primus), along with reduction of chromosomes to the haploid number. This first maturation division is usually completed just before ovulation. The second maturation division ordinarily occurs after the egg is shed and has been penetrated by the sperm cell. Hence, strictly speaking, most mammalian eggs, so far as now known, do not become true ova until after the sperm cell has penetrated them.

Exceptions to the above do occur. O. P. Pearson and Enders (1943) said that the eggs of the domesticated silver phase of the red fox (*Vulpes fulva*) are ovulated as primary oocytes and give off the first polocyte near the middle section of the oviduct. Apparently, sperm penetration and second polocyte formation then proceed as in other mammals. Much more striking exceptions were described by Strauss (1938, 1939, 1964) in two genera of Tenrecidae (tenrecs and Madagascar "hedgehogs") — *Setifer* and *Hemicentetes.* The ripe follicles of these animals have no antrum (Fig. 6.26). Spermatozoa in large numbers penetrate the very thin and somewhat loosened ovarian surface epithelium and the theca overlying the follicle, enter the mildly edematous epithelial tissue of the ripe novesicular or secondary follicle (Figs. 6.27 and 6.28) and reach the zona pellucida, through which one of them presumably perforates and then fertilizes the ovum. Because at least one polocyte is present before the sperm enters, the ovum at this time is a secondary oocyte, as it is in other mammals at the time of ovulation. Since there is no follicular

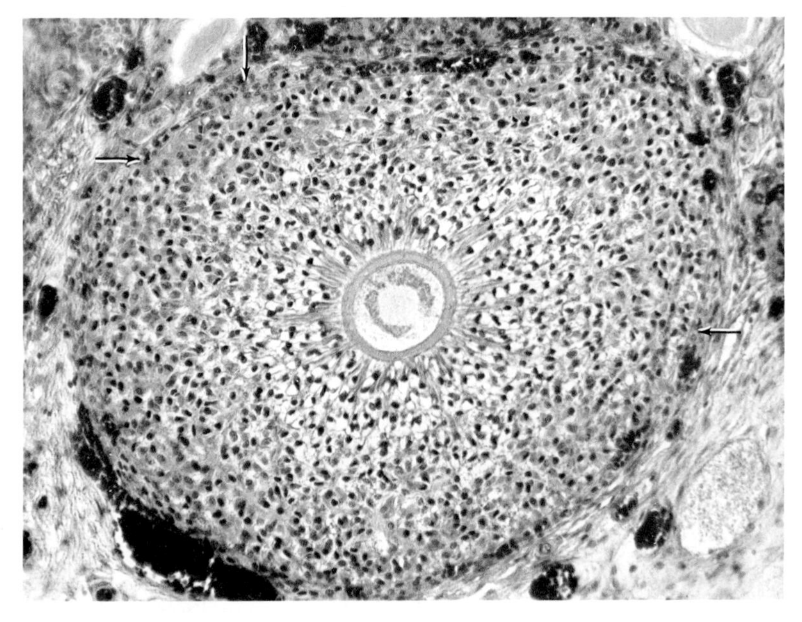

Fig. 6.26 Ripe follicle of a large Madagascar "hedgehog" (*Setifer setosus*). There is no antrum. The thecal gland (*arrows*) is thin and occasionally interrupted. *S. setosus*, ser. 60. × 150. (Courtesy of Fritz Strauss.)

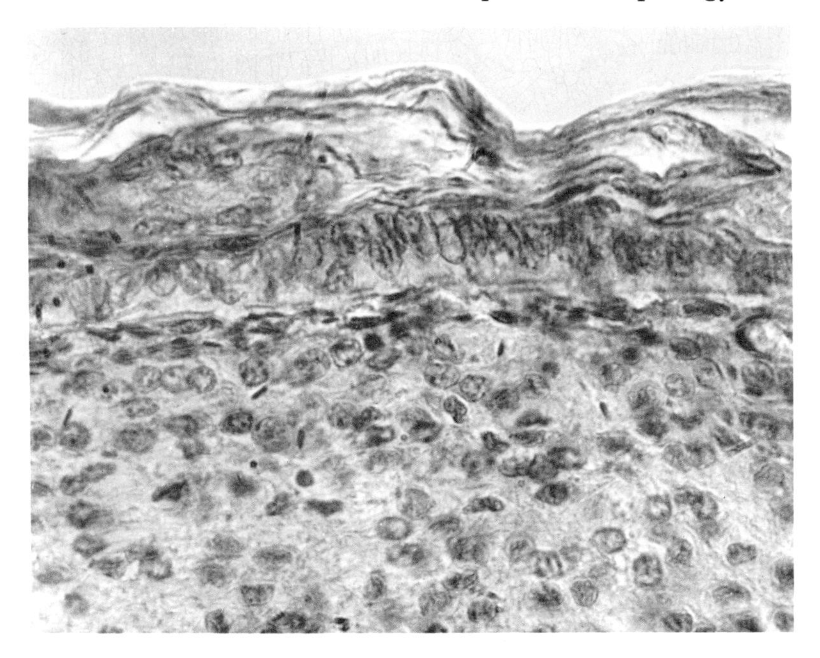

Fig. 6.27 Surface of the ovary and a ripe follicle of a large Madagascar "hedgehog" (*Setifer setosus*), showing spermatozoa that have penetrated into the follicular epithelium. *S. setosus*, ser. 36a. × 600. (Courtesy of Fritz Strauss.)

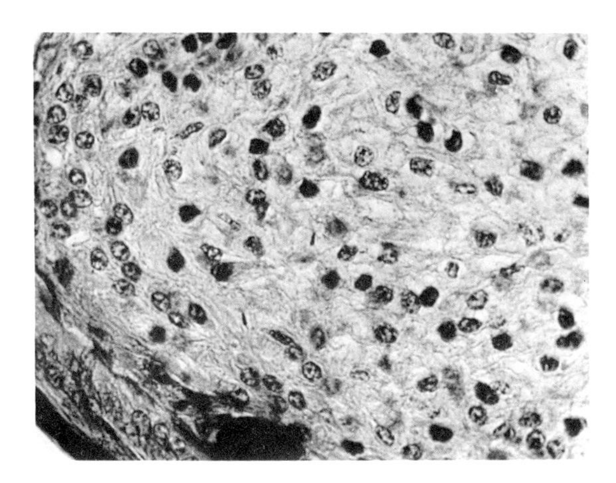

Fig. 6.28 Spermatozoa that have penetrated among the epithelial cells of a ripe follicle of a large Madagascar "hedgehog" (*Setifer setosus*). *S. setosus*, ser. 36. × 395. (Courtesy of Fritz Strauss.)

liquor, ovulation does not involve a "bursting" of the follicle; rather there is a lateral shifting of the epithelial cells from between the ovum and the ovarian surface, possibly augmented by contraction of the theca externa, until the ovum with its zona lies bare of epithelium on the surface of the everted follicle, or potential corpus luteum (Figs. 6.29 and 6.30). The ovum then detaches and passes into the ovarian bursa and oviduct.

The moon rat (*Echinosorex gymnurus*) also has ripe follicles without an antrum. It is unknown whether ripe follicles with no antrum occur

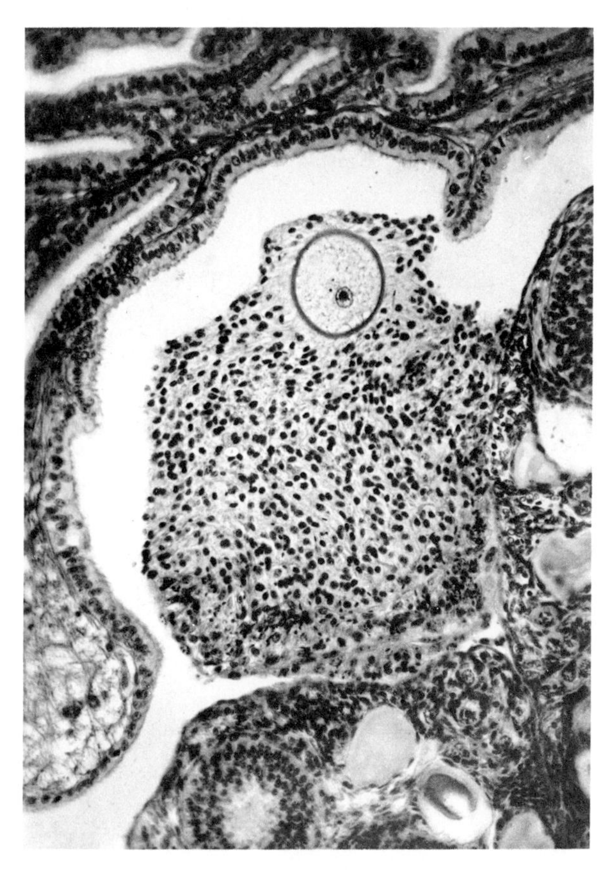

Fig. 6.29 Fertilized egg in the process of ovulation from the antrumless follicle of a large Madagascar "hedgehog" (*Setifer setosus*). *S. setosus*, ser. 63. × 150. (Courtesy of Fritz Strauss.)

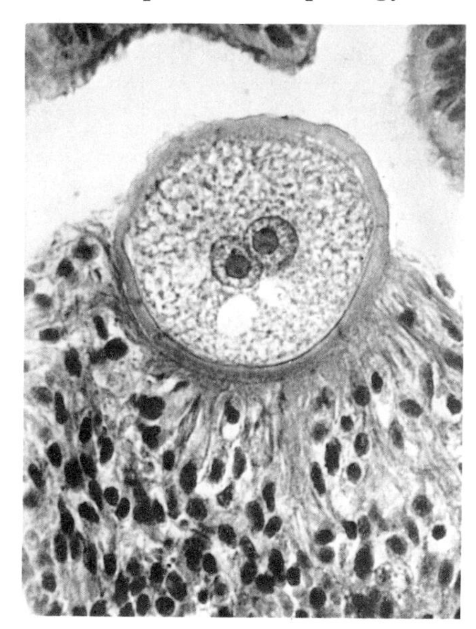

Fig. 6.30 Pronuclear egg of a large Madagascar "hedgehog" (*Setifer setosus*) about to be shed from the follicle into the ovarian bursa. *S. setosus*, ser. 62. × 500. (Courtesy of Fritz Strauss.)

in any mammalian families other than the Tenrecidae and Erinaceidae (moon rats, gymnures, and hedgehogs). In many Soricidae, Talpidae (moles), and Microchiroptera (bats), the antrum of ripe follicles is very small and often crossed by trabeculae or divided by septa (Figs. 6.31–6.35). Accidental retention of the cumulus and oocyte in ruptured vesicular follicles does happen, and fertilization of such an egg while still in the ovary accounts for some cases of "ovarian pregnancy," but so far *Setifer* and *Hemicentetes* are the only mammals known that normally have intrafollicular fertilization.

The follicular capsules

We have already observed in Chapter 2 that in mammalian ovaries an encapsulating zone of capillaries and modified stromal cells, the theca interna, always develops adjacent to the follicular epithelium of secondary and vesicular follicles, and that immediately outside of this another more fibrous and relatively avascular theca externa may develop, especially around the larger secondary and vesicular follicles. The ovaries of mammals the size of the common rat and smaller frequently show no sign of this outer zone, and even in larger mammals it may be present only around the larger vesicular follicles. Possibly this theca ex-

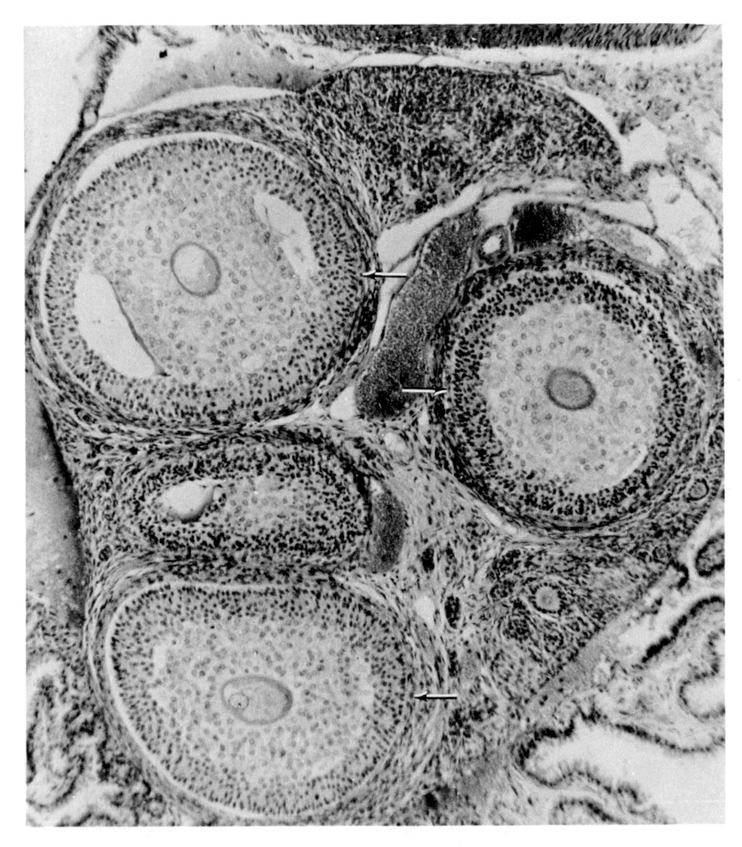

Fig. 6.31 Ovary of a vagrant shrew (*Sorex vagrans*). The apparently normal ripe follicles are characteristic of follicles with very small antra. This animal, however, had gill slit stage embryos. Thin zones of thecal gland (*arrows*) are present. *S. vagrans*, 6. × 100.

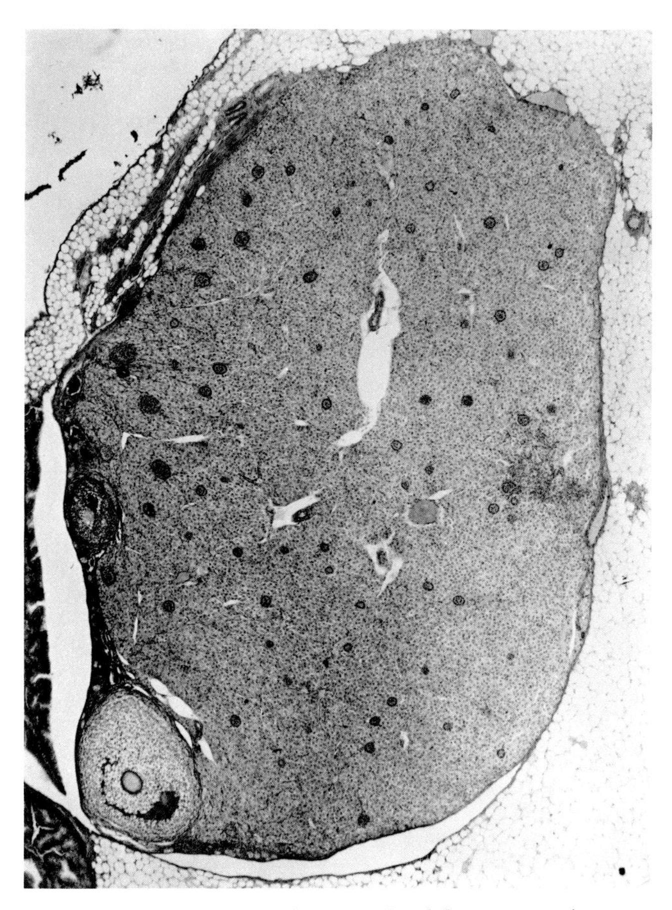

Fig. 6.32 Ovary of a star-nosed mole (*Condylura cristata*) in estrus, showing the very small antrum. Most of this ovary is gonadal adrenal type interstitial gland tissue with scattered epithelial spheroids, which are remnants of the medullary cords. *C. cristata*, 1. × 40.

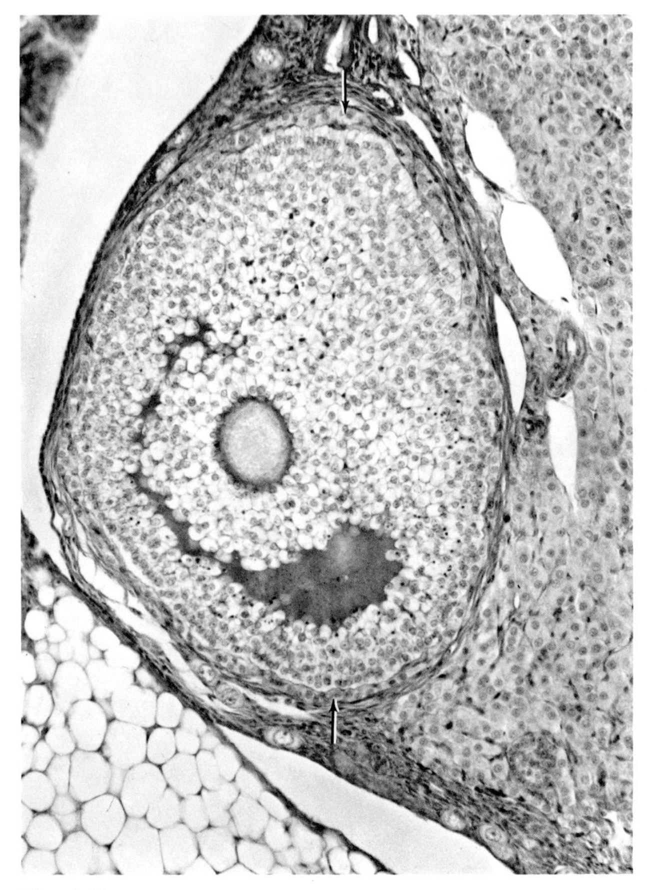

Fig. 6.33

Fig. 6.33 Detail of follicle and interstitial gland tissue in the ovary of the star-nosed mole (*Condylura cristata*) shown in Fig 6.32. The thecal gland (*arrows*) is thin and interrupted. *C. cristata*, 1. × 158.

Fig. 6.34 Ovary of a big brown bat (*Eptesicus fuscus*) in February, showing an apparently ripe follicle. Note the central position of the oocyte, the hypertrophied cumulus cells, and the small isolated antral spaces between the cumulus and the remainder of the follicular epithelium. *E. fuscus*, 6. × 100.

Fig. 6.35 Apparently ripe follicle of a big brown bat (*Eptesicus fuscus*) in December. Note the small isolated antral spaces. *E. fuscus*, 7. × 250.

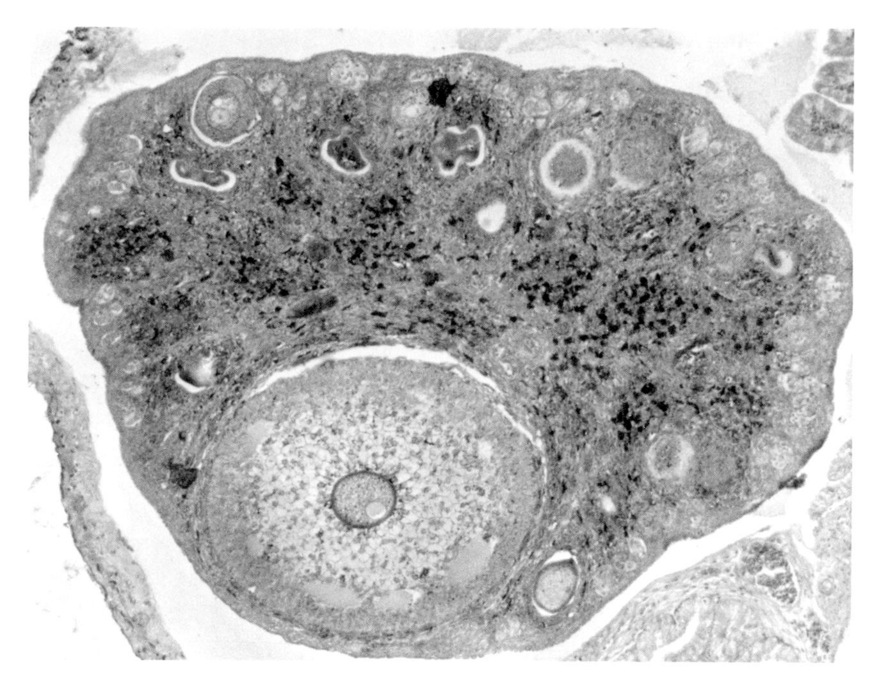

Fig. 6.34

Fig. 6.35

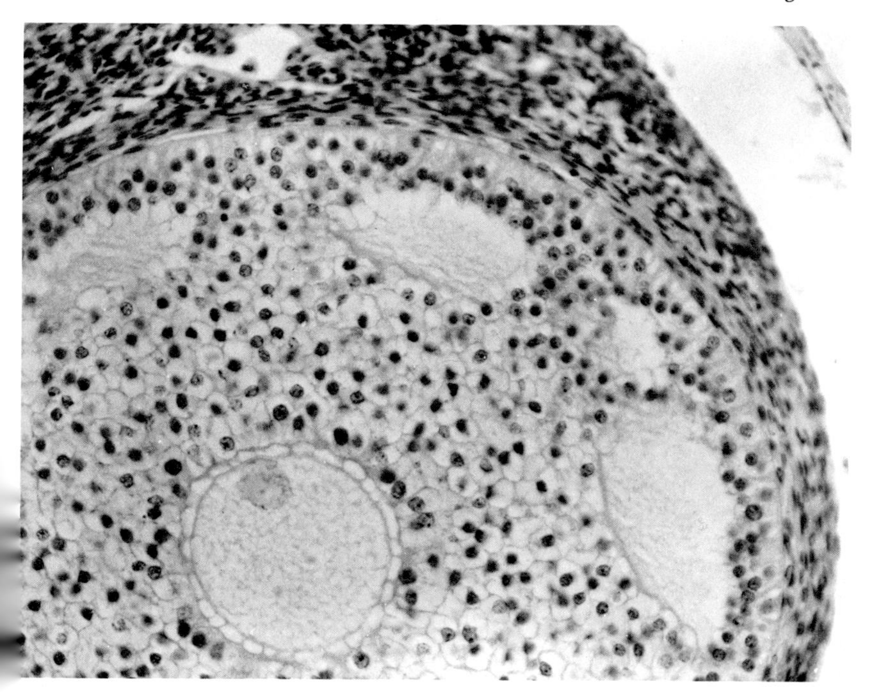

terna is merely the result of rapid follicular expansion, causing a re-arrangement of the surrounding cellular connective tissue of the stroma. However, there is noteworthy evidence in a few species (man, rabbit, rat, cat) of contractility of some cells of the theca externa and general stroma. This may possibly aid follicular rupture at ovulation (Claesson, 1947; Hansen, 1957; Lipner and Maxwell, 1960; Rocereto, Jacobowitz, and Wallach, 1969).

The theca interna is of major significance. Its capillaries nourish the follicle and egg. After ovulation, they sprout among the differentiating luteal cells to furnish the blood supply of the corpus luteum. Some of the stromal cells of the theca interna differentiate into the thecal gland cells of ripening follicles, or interstitial gland cells, if the follicle undergoes atresia.

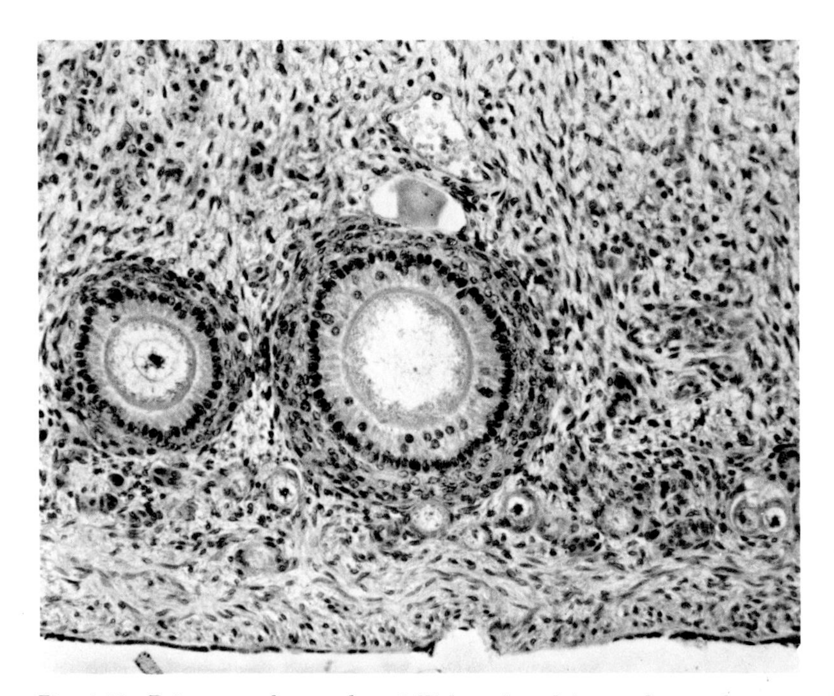

Fig. 6.36 Primary and secondary follicles of a plains pocket gopher (*Geomys bursarius*), showing early theca interna. Between the larger follicle and the venule is a remnant of zona pellucida in a completely atretic follicle. *G. bursarius*, 13. × 195.

Thecal gland

In at least one group of rodents (Geomyoidea), the theca interna can be recognized even around larger primary follicles (Fig. 6.36), but usually it is not obvious until the secondary follicle period. At first its cells simply appear to be rounded stromal cells among which mitosis is active. Hyperplasia continues without much change in the size or appearance of these cells until well into the vesicular follicle period. Then, as

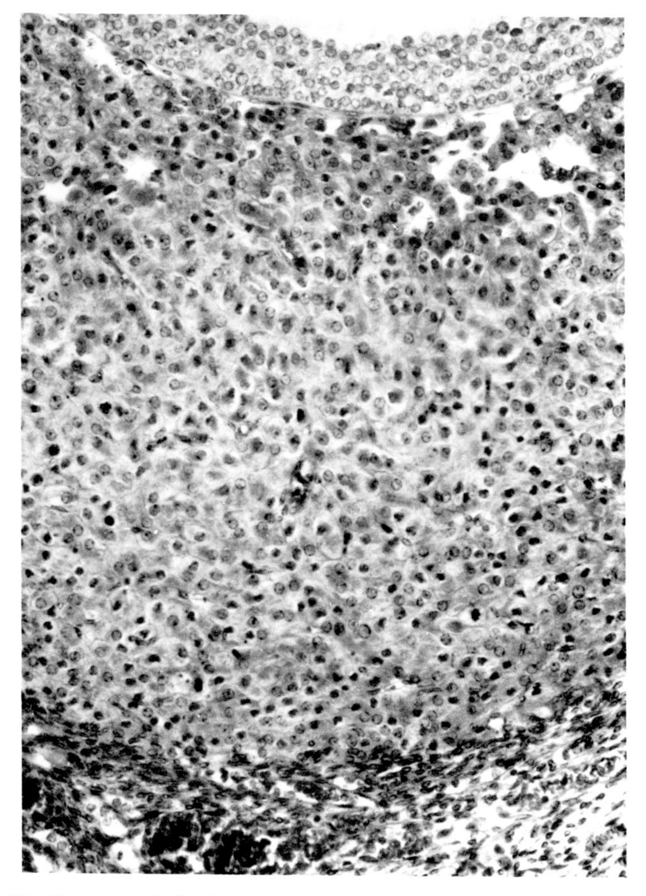

Fig. 6.37 Segment of the developing thecal gland of a plains pocket gopher (*Geomys bursarius*). Note the more highly differentiated cells centrally and the incorporation of relatively undifferentiated stromal cells peripherally. *G. bursarius*, 4. × 198.

the final enlargement and ripening of the follicle begins, the thecal cells next to the follicular epithelium develop more and more granule-bearing cytoplasm and cease dividing. This glandular differentiation proceeds outward, while more dividing stromal cells are added peripherally (Fig. 6.37). Accretionary growth in thickness and centrifugal differentiation stops near or at the time of estrus. Thus, depending upon the species, a glandular parenchyma is built up, ranging from only a few cells in thickness (Fig. 6.38) to a zone as thick as the follicular diameter, as in the plains pocket gopher (*Geomys bursarius*) (Fig. 6.39). Even the early theca interna contains a capillary net, and as the epithelioid glandular tis-

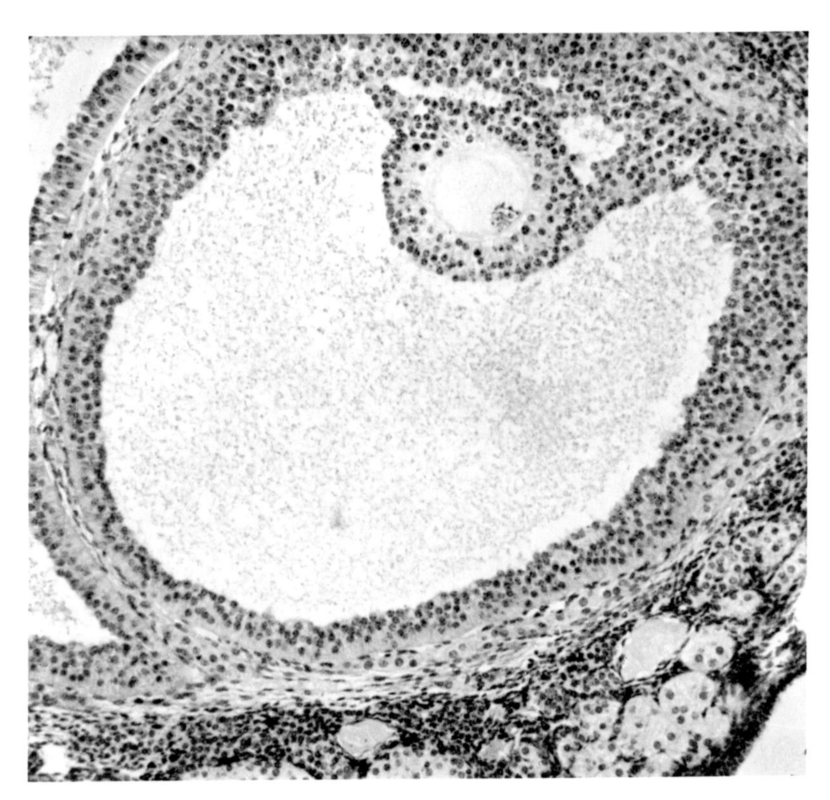

Fig. 6.38 Nearly ripe follicle of an estrous long-tailed weasel (*Mustela frenata*), showing the thin interrupted zone of thecal gland cells. Fluid-filled spaces in the attached portion of the cumulus presumably result from the process of detachment of the cumulus. *M. frenata*, 43. × 145.

sue differentiates these capillaries become more dilated. They serve not only as nutritive vessels for the follicular epithelium, but also as a vascular plexus with relations to the epithelioid cells typical of that of an endocrine gland (Fig. 6.40). This, then, is the thecal gland. Although very slightly developed in some species, it is a constant feature, as far as known, of the ovaries of all mammals in late proestrus and estrus (Mossman, 1937; Stafford, Collins, and Mossman, 1942) (Figs. 4.7, 6.25, 6.26, 6.31, 6.33, and 6.37–6.44).

In all species examined, the maximum glandular differentiation of the theca interna is reached at estrus. Soon after estrus the glandular cells begin to degenerate, both in the case of ovulated follicles and of large unruptured follicles which become atretic during this period (Fig.

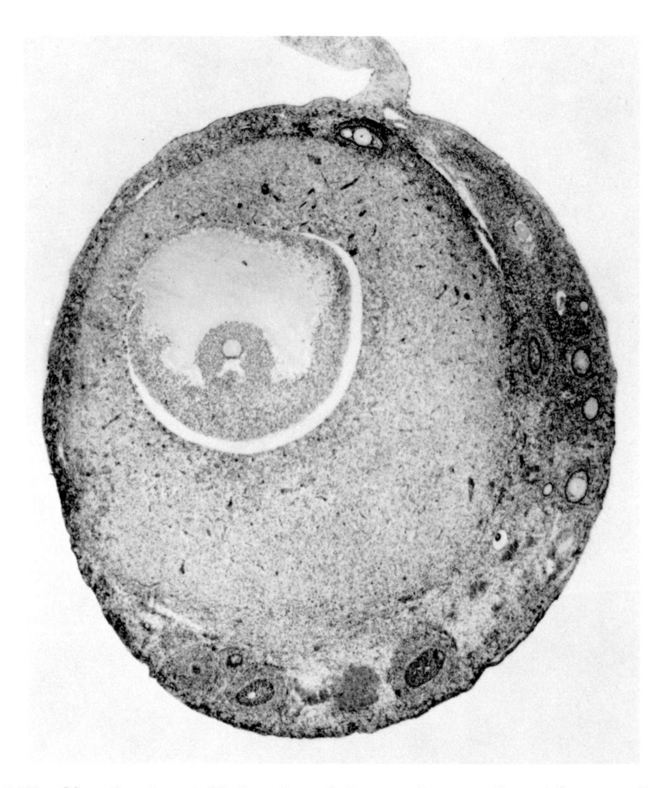

Fig. 6.39 Nearly ripe follicle of a plains pocket gopher (*Geomys bursarius*), showing the massive thecal gland. *G. bursarius*, 2. × 30.

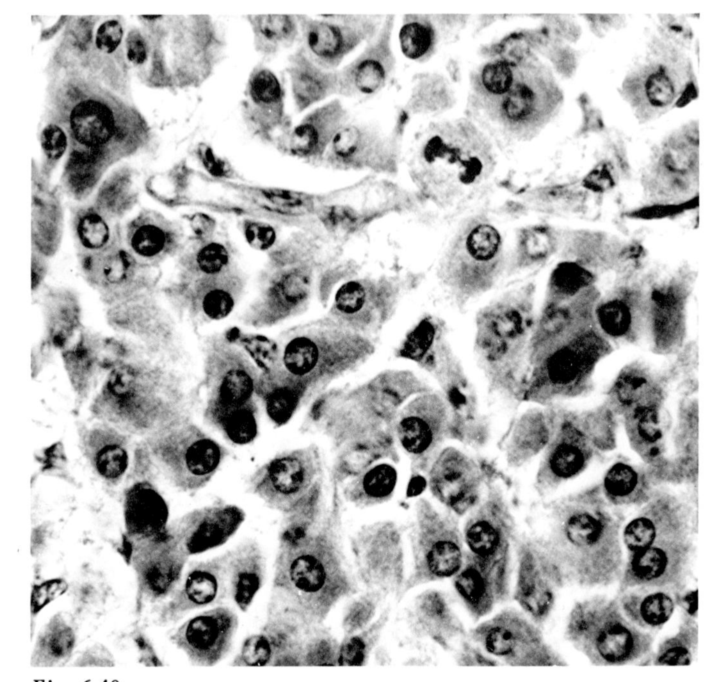

Fig. 6.40

Fig. 6.41

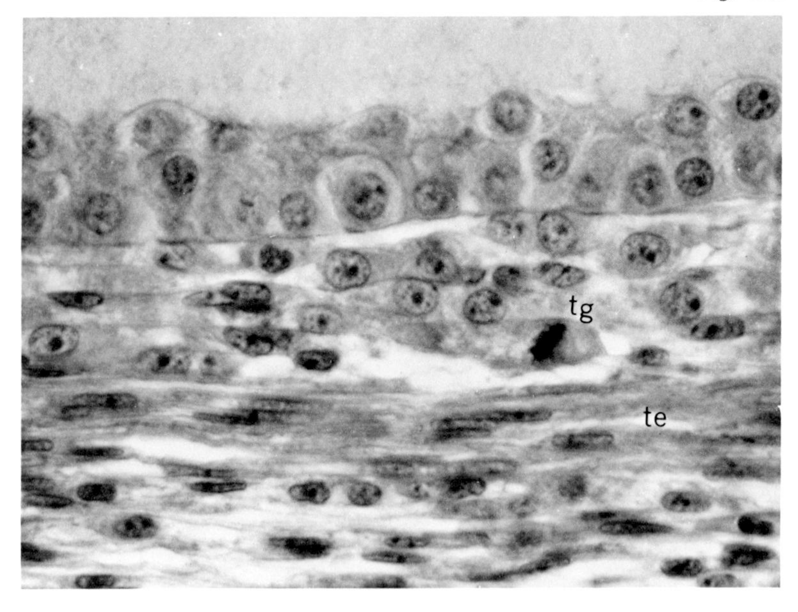

Fig. 6.42

Fig. 6.40 Fully differentiated thecal gland cells of a plains pocket gopher (*Geomys bursarius*). *G. bursarius*, 3. × 290.

Fig. 6.41 Portion of the wall of the nearly ripe follicle of a porcupine (*Erethizon dorsatum*) in estrus to show the thecal gland (*tg*) and theca externa (*te*). *E. dorsatum*, 26. × approx. 600.

Fig. 6.42 Nearly ripe follicle of one of the pangolins (*Manis*), showing the moderately thick thecal gland, the most common condition in mammals. *Manis*, 1. × 155.

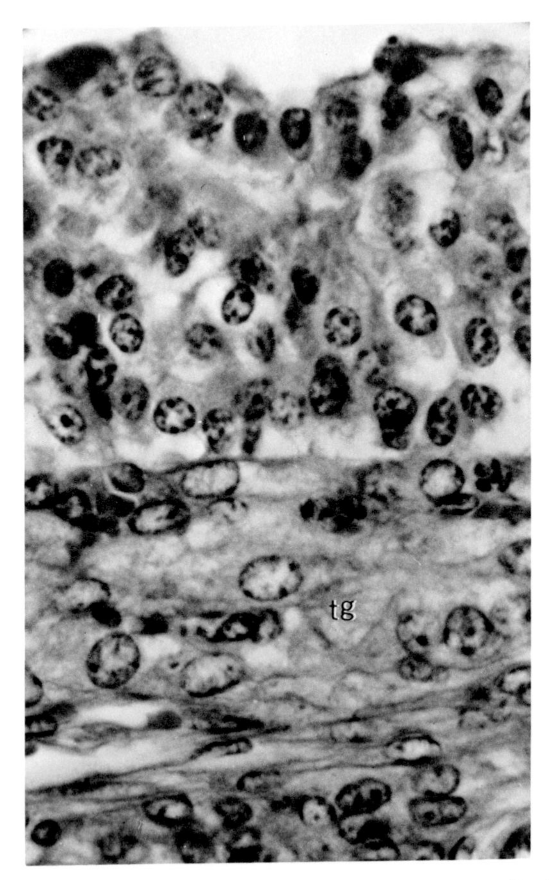

Fig. 6.43 Thecal gland (*tg*) of a guinea pig (*Cavia porcellus*) in late proestrus. *C. porcellus*, 42L. × approx. 700.

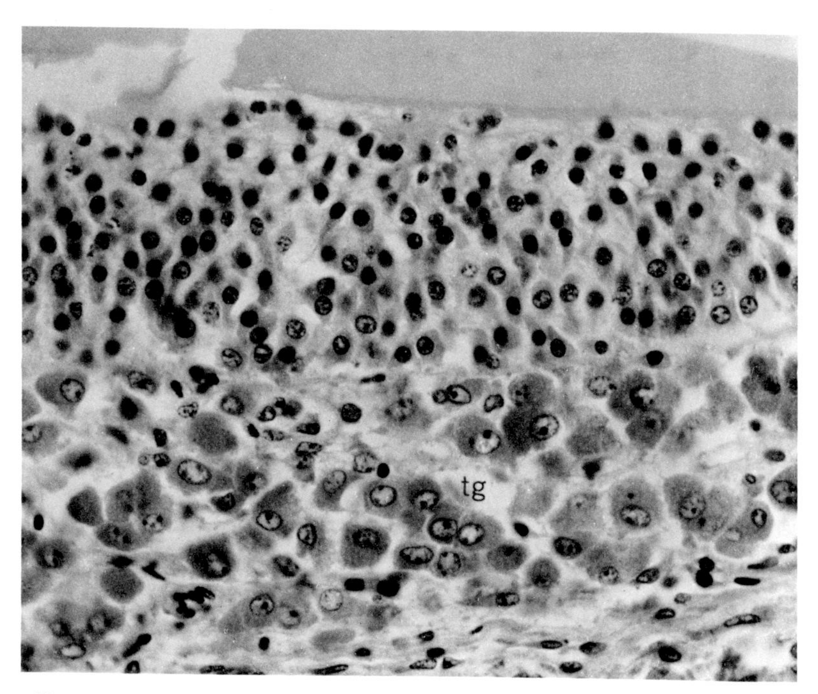

Fig. 6.44 Thecal gland (*tg*) of a presumably ripe follicle of a mare (*Equus caballus*). *E. caballus.* × 400.

6.45). By the time the embryos have reached the uterus, the differentiated thecal gland tissue has disappeared. The outer, less-differentiated cells of the gland may remain and differentiate into paraluteal and eventually into luteal cells, thus contributing to the accretionary growth of the corpus luteum.

There has been much controversy as to whether the thecal gland does contribute to the corpus luteum. We have seen no evidence that *differentiated* thecal gland cells ever do this. However, there is no doubt that

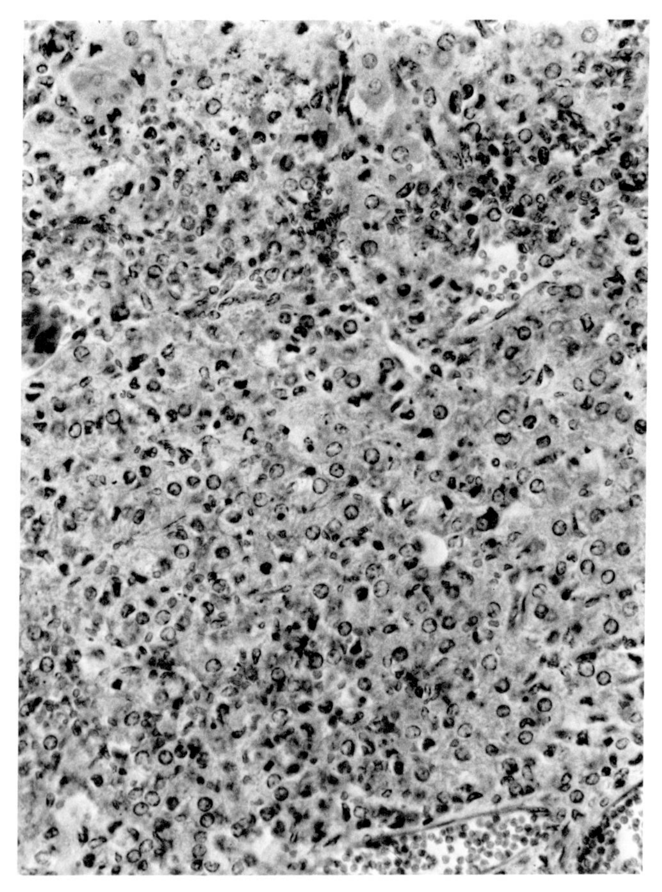

Fig. 6.45 Degenerating thecal gland surrounding an early corpus luteum of a plains pocket gopher (*Geomys bursarius*). Luteal cells are at the extreme top. *G. bursarius*, 13. × 200.

the corpora lutea of some mammals are for a time surrounded by paraluteal cells. As a rule, these show every gradation from undifferentiated stromal cells to typical luteal cells (Figs. 6.46 and 6.47). The relatively undifferentiated outer cells of the thecal gland may well be the first to become paraluteal cells, the still-more-peripheral stromal cells undergoing paraluteal differentiation after the thecal gland itself has degenerated.

The glandular nature of the theca interna of maturing vesicular follicles in the "mole, shrew, and weasel" was indicated in the illustrations of MacLeod (1880) and in the human ovary by Meyer (1911) and Ruge (1913), but apparently none of them thought of it as a gland. This is not surprising, since the idea of ductless glands in the ovary was first seriously considered after the publications on the corpus luteum by Fraenkel and Cohn (1901), Fraenkel (1903), and Loeb (1906). Meyer (1924) thought of the theca interna as being nutritive to the follicular epithelium. Strassmann (1961) theorized that the theca interna, by producing a "cone" at the side of the follicle nearest the ovarian surface, built a "pathway" of fragile tissue through which the ripening follicle reached the surface and was able to rupture. Found in only a few species, these cones are almost always associated with atretic follicles; they are therefore interstitial gland tissue, not thecal gland. Thus, it seems to us that Strassmann's theory is completely illogical and unsupported by the facts.

Mossman (1937) saw the highly developed theca interna of the plains pocket gopher. Because of its large size, obvious glandular morphology, and definite correlation with the time of proestrus and estrus (see Figs. 6.39 and 6.40), he could not fail to recognize its glandular nature and its obvious significance. Corner (1938) favored the thecal gland as the source of estrogens. Belt and Pease (1956) showed that thecal gland cell mitochondria had the tubular cristae characteristic of steroid secretors. Dubreuil (1957) considered it to be an estrogenic gland ("glande estrogene") induced from the stroma by the follicular epithelium. Moricard (1958) confirmed the observations of Belt and Pease on the tubular cristae of the mitochondria and demonstrated the numerous lipid droplets related to the Golgi apparatus that occur in other steroid secretors. Falck (1959) demonstrated estrogenic action of the walls of ripe follicles containing thecal gland cells by making contiguous intraocular transplants with vaginal mucosa. While these and several histochemical

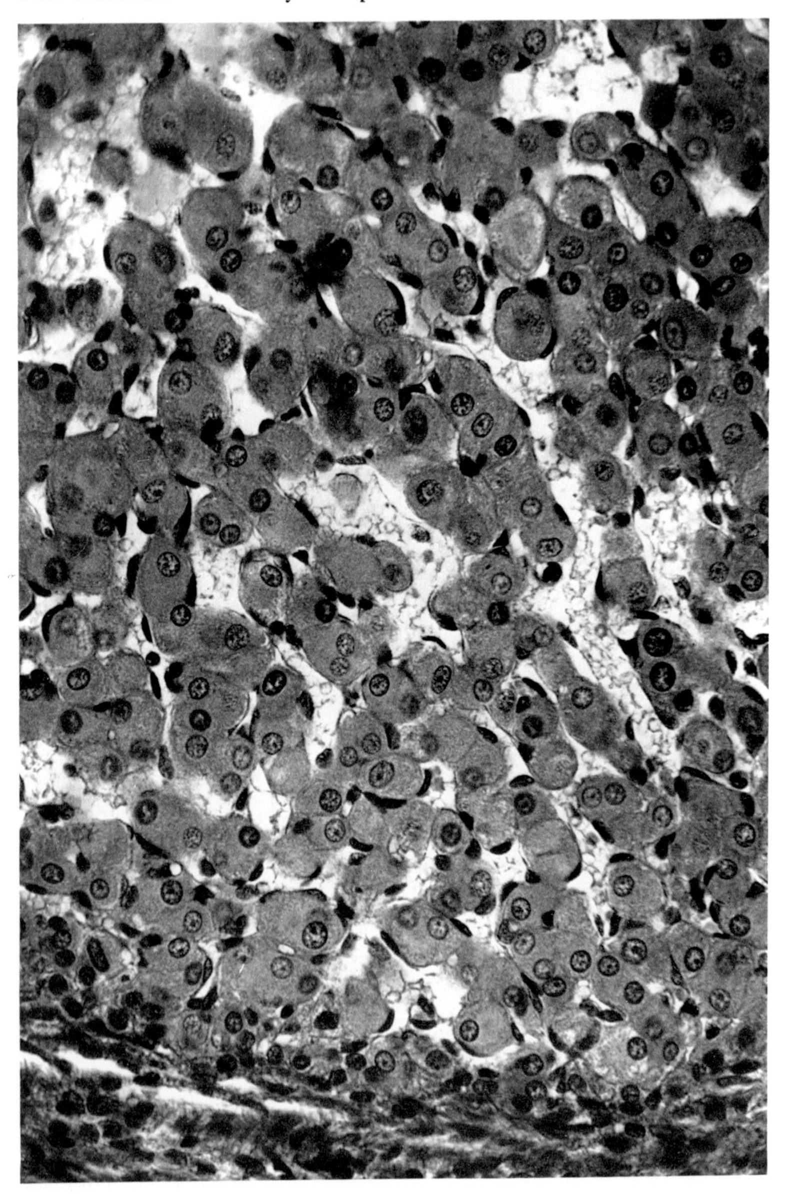

Fig. 6.46 Segment of a primary corpus luteum of a porcupine (*Erethizon dorsatum*) with an embryonic disc embryo. Note the narrow paraluteal zone below. *E. dorsatum*, 29. × 400.

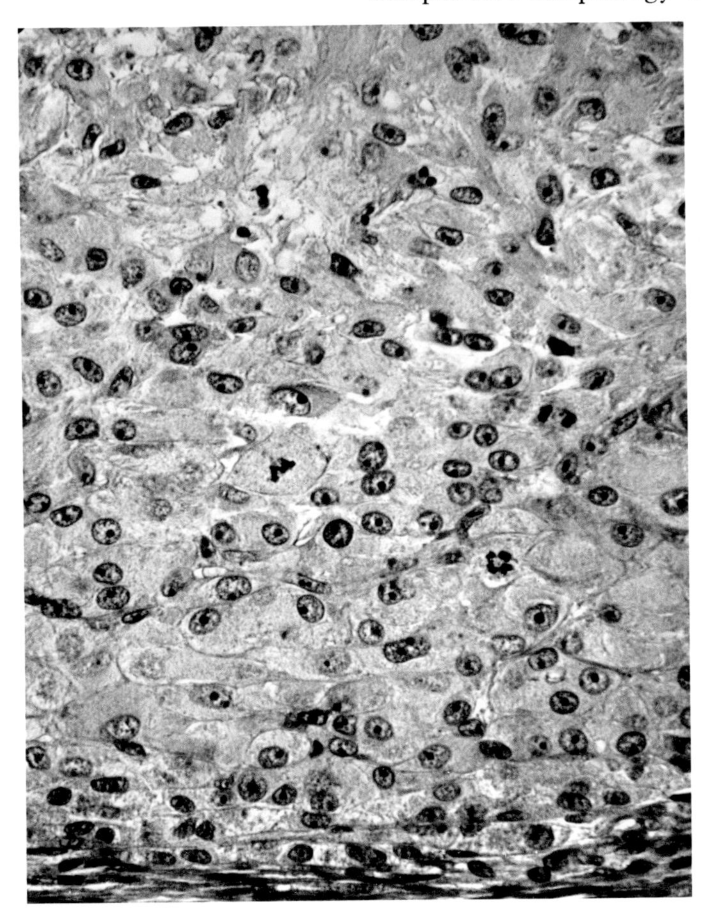

Fig. 6.47 Margin of a corpus luteum of a porcupine (*Erethizon dorsatum*) with a 16-cell tubal embryo. Note mitosis and the luteal cells differentiating from the stroma. *E. dorsatum*, 25L. × 400.

studies point to secretion of steroid hormones by the thecal gland, direct proof of this and of the exact nature of the secretion is lacking.

From the comparative standpoint, the only animals known to have unusually large, well-developed thecal glands are *Orycteropus afer* (aardvark) and the following genera of Geomyoidea: *Geomys* and *Thomomys* (smooth-toothed pocket gophers), *Dipodomys* (kangaroo rats), and *Perognathus* (pocket mice) (Figs. 6.48 and 6.49). Members of the following groups have thecal glands only a few cells thick and often interrupted, that is, absent over some areas of the follicle: Chiroptera (bats), Manidae (pangolins), Ochotonidae (pikas), Aplodontidae (sewellels), Sciuridae (squirrels), Erethizontidae (New World porcupines), Caviidae (guinea

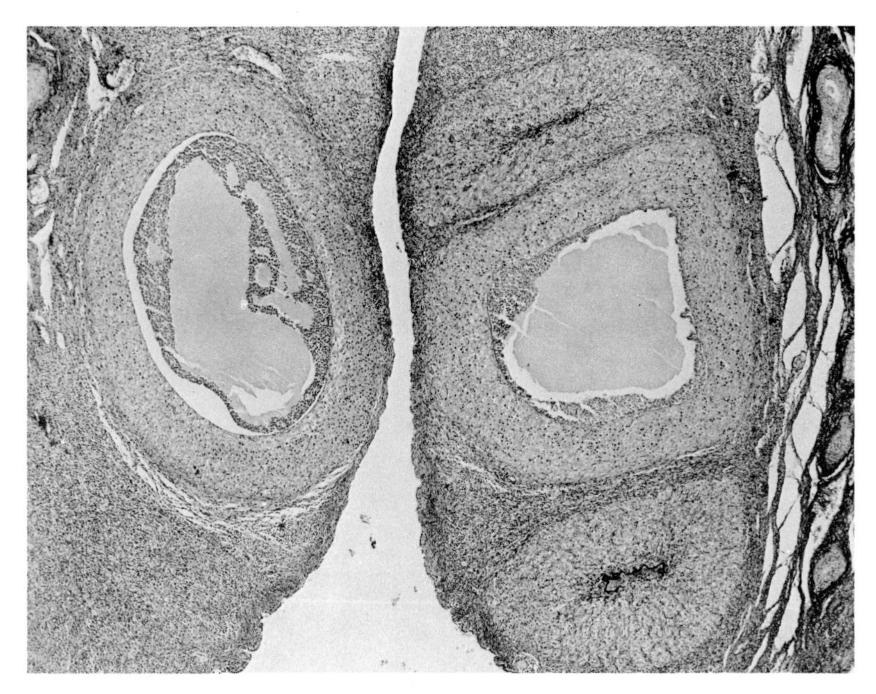

Fig. 6.48 Cortex on either side of a fissure in the lobate ovary of an aardvark (*Orycteropus afer*), showing two medium vesicular follicles (one on each side of the fissure) with thick developing thecal glands. On the right (above and below the vesicular follicle) are two corpora atretica with their thick zones of thecal type interstitial gland tissue. The thickness of the thecal glands at this stage indicates that, when fully differentiated, they must be comparable in relative size with those of the plains pocket gopher. *O. afer*, 3 .× 40.

pigs), Canidae (dogs), Mustelidae (mustelids), Felidae (cats), Tragulidae (chevrotains), and Cervidae (deer). Intermediate in this respect are most insectivore families, the Anthropoidea (monkeys, apes, and man), Dasypodidae (armadillos), Leporidae (rabbits and hares), Hyracoidea (hyraxes), Equidae (horses, donkeys, and zebras), Suidae (pigs), and Bovidae (bovids).

Knowledge of the thecal gland has been hampered by several factors. 1) It is a relatively inconspicuous feature of the ripening follicle in many species, and even when recognized as a theca interna its glandular nature is often discernible only by careful microscopy. 2) It has been commonly confused with the thecal type of interstitial gland tissue which develops from the theca interna when a follicle undergoes atresia. 3) It is a difficult element of the ovary to segregate for experimental analysis. 4) Because of the marked hypertrophy of its cells at the time of follicu-

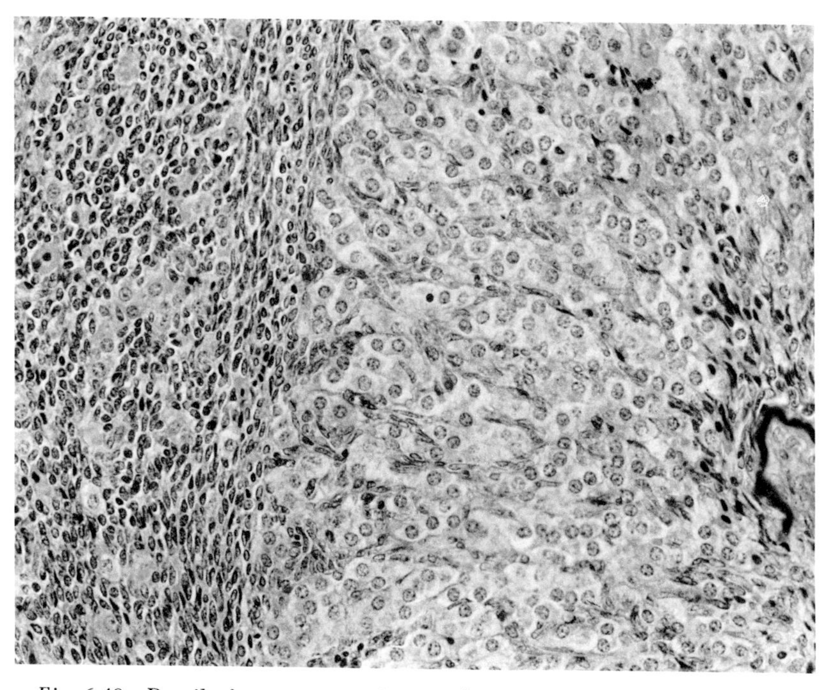

Fig. 6.49 Detail of a corpus atreticum and intervening stroma and stromal type interstitial gland cells of the ovary of the aardvark (*Orycteropus afer*) shown in Figure 6.48. *O. afer*, 3. × 200.

lar rupture and the collapse and consequent apparent disruption of the organization of the follicular wall at this time, it has been confused in some species with the lutealizing granulosa cells and with paraluteal cells. The second of these problems is discussed further in the section below on the interstitial gland (pp. 166–209); the fourth, under the luteal gland (pp. 209–20).

Follicular atresia

Because this subject has been well reviewed by Ingram (1962), only a few comments will be attempted here. (See Chapter 7 for a more detailed discussion of atresia.) Atresia is one of the least understood of ovarian phenomena, yet it is a prominent feature of the ovaries of all mammals. It often begins during fetal life.

Atresia of secondary and vesicular follicles, that is, of follicles having a theca interna, is closely tied to the production of interstitial gland tissue in all Eutheria, as we shall see in the following section. Since interstitial gland cells are without doubt steroid hormone secretors, it is immediately apparent that atresia of secondary and vesicular follicles has real physiological significance and is more than just a way of ridding the ovary of unfit or unnecessary eggs.

Ovarian interstitial gland tissues

Any cell of an endocrine gland type that occurs in or closely associated with the mammalian ovary and that is not part of a thecal gland or a luteal gland can be considered a form of interstitial gland cell. To group all such cells under this general class is convenient, although it is often illogical so far as the connotation of the word "interstitial" is concerned. Based on origin, location, time of appearance, resemblance to endocrine tissue elsewhere in the body, and relation to the reproductive cycle, there are at least seven fairly distinct types of ovarian interstitial gland tissue: fetal, thecal, stromal, medullary cord, rete, gonadal adrenal, and adneural. Once differentiated, some of these may be indistinguishable anatomically from one another and may, of course, have similar functions, whether or not they closely resemble one another structurally. Unlike thecal and luteal gland cells, interstitial gland cells ordinarily do not degenerate after a functional period; instead, they dedifferentiate, either to remain as indifferent stromal cells or to redifferentiate later into the same or some other cell type.

A few cases have been described in the literature of degeneration of interstitial gland cells, especially in Pinnipedia (seals, sea lions, and walruses) (Amoroso et al., 1965). In our experience, we have come to believe that, although occasional degeneration of interstitial gland cells does occur, these cells normally go through recurring cycles of glandular differentiation, followed by periods of dedifferentiation and glandular inactivity. We have little or no evidence for this in the case of the adneural and rete types. It should also be recalled that, in contrast, differentiated cells of the thecal gland do degenerate soon after follicular rupture and that luteal cells also eventually degenerate.

Fetal type

Fetal type interstitial gland cells occur in the medulla and to some extent in the cortex of the late fetal ovary of all mammals examined for their presence. They sometimes persist into the juvenile period, especially in altricial species (Fig. 6.50 and p. 70). They may be relatively scarce and scattered, or they may be so numerous and hypertrophied as to greatly enlarge the ovary and to occupy almost the entire medulla and much of the cortex, as they do in the horse and zebra (*Equus*) fetus and foal (Kohn, 1926; Cole et al., 1933; Amoroso and Rowlands, 1951) (Figs. 6.51, 6.52, and 7.1).

These fetal type interstitial gland cells are the developmental homologs of the interstitial gland cells of the testis. Like those of the testis, their formation is apparently initiated in late fetal life by gonadotropic hormones. (For information on fetomaternal hormone relations, see Diczfalusy, 1964.) In man, they are best developed in the testis from the end of the second month to about the middle of the sixth month; after this they slowly revert to small, undifferentiated cells difficult or impossible to distinguish from stromal cells, and they do not differentiate again until puberty. In the human ovary, fetal type interstitial cells appear during the sixth month of gestation and persist to about the time of birth, when interstitial gland cells of the thecal type begin to appear around atretic medullary follicles. Probably there is a cytological difference between these two types, but no one has yet studied them carefully enough to prove this; hence, because of the abundance of thecal type cells, we cannot be certain how long fetal type cells persist in the postnatal ovary. In our experience, fetal type interstitial gland cells have fine cytoplasmic granules (see Fig. 6.52), while the thecal type

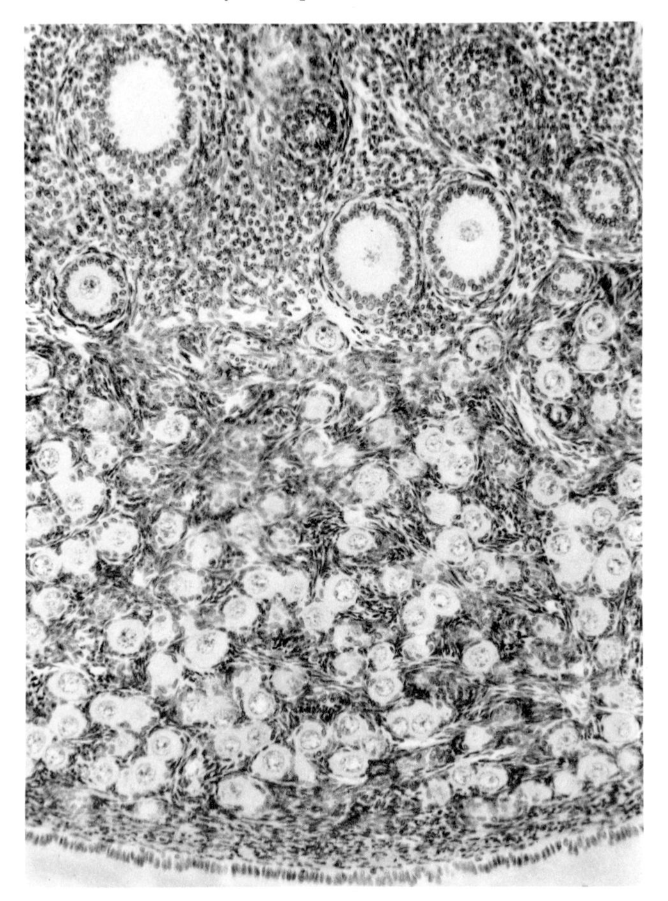

Fig. 6.50 Cortex and adjacent medulla of a juvenile sewellel (*Aplodontia rufa*). The large primary and secondary medullary follicles and fetal type interstitial gland tissue are chiefly confined to the medulla. *A. rufa*, 9. × 130.

usually, but not always, has coarsely vacuolar cytoplasm (Fig. 6.53). In species where this difference holds, the two should be easily distinguishable in well-fixed material by either light or electron microscopy.

Thecal type

Thecal type interstitial gland cells occur in all eutherian mammals so far studied. Since they develop from the cells of the theca interna of

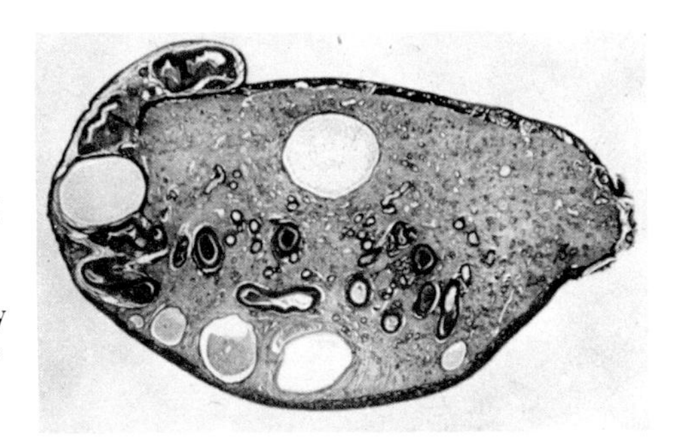

Fig. 6.51 Ovary of a late fetus of a Burchell's zebra (*Equus burchelli*), showing the lightly stained matrix of fetal type interstitial gland tissue which fills the ovary and surrounds the follicles and blood vessels. *E. burchelli*, 1. × 2.8.

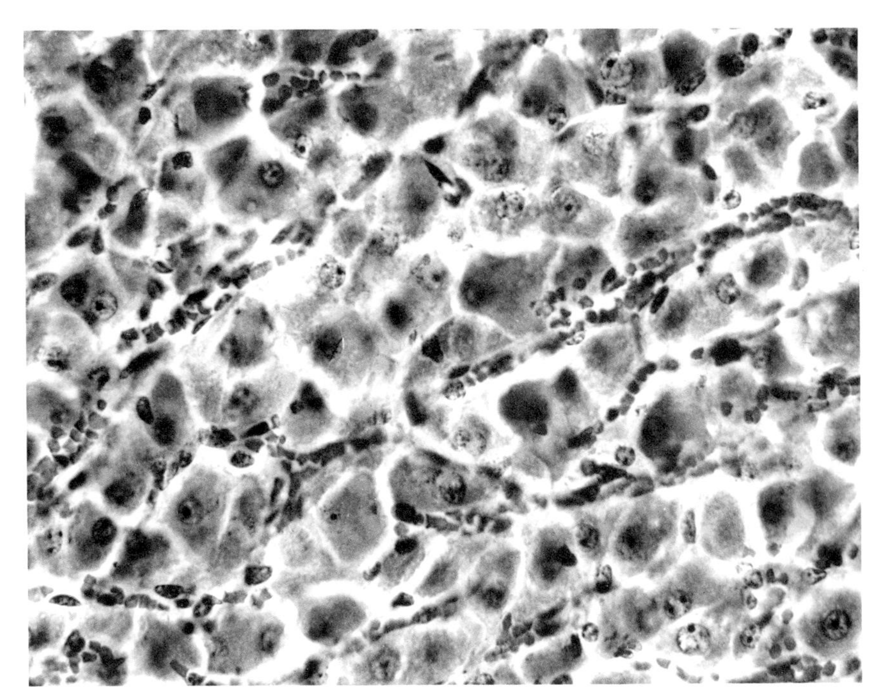

Fig. 6.52 Fetal type interstitial gland cells of a newborn horse (*Equus caballus*), showing the rich capillary supply. *E. caballus*, 2. × 400.

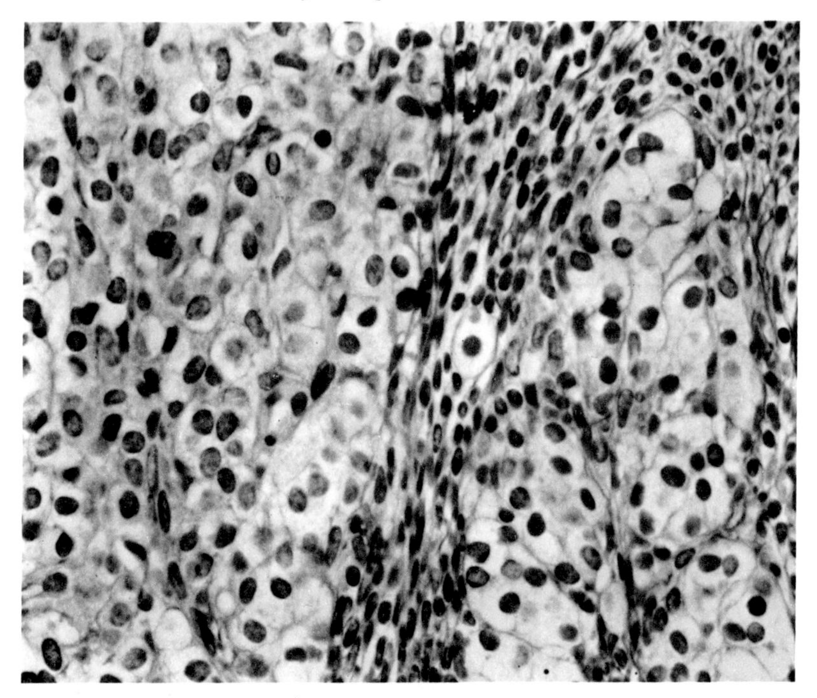

Fig. 6.53 Thecal type interstitial gland cells of a red squirrel (*Tamiasciurus hudsonicus*). This type of vacuolar cytoplasm is typical of thecal type interstitial gland cells of many mammals. *T. hudsonicus*, 100. × 800.

atretic secondary and vesicular follicles, they are present as soon as follicular atresia begins, which is during the late fetal or infantile period in most mammals; and they continue to appear as long as follicular atresia occurs, which means even in senile females of such species as the higher primates and man. However, these gland cells are usually fully differentiated (Fig. 6.54) only at certain periods. Much of the time they are small, cytoplasm-poor, rounded cells arranged about the degenerating follicle (Fig. 6.55). So far as we can determine, these do not degenerate after they have become fully differentiated and have passed through a functional period. Instead, they usually simply dedifferentiate and remain mingled with the stroma and other ovarian elements, apparently being capable of again differentiating if they happen to be included in the area of the theca interna of another follicle undergoing atresia. This has never been proven experimentally; as will become ap-

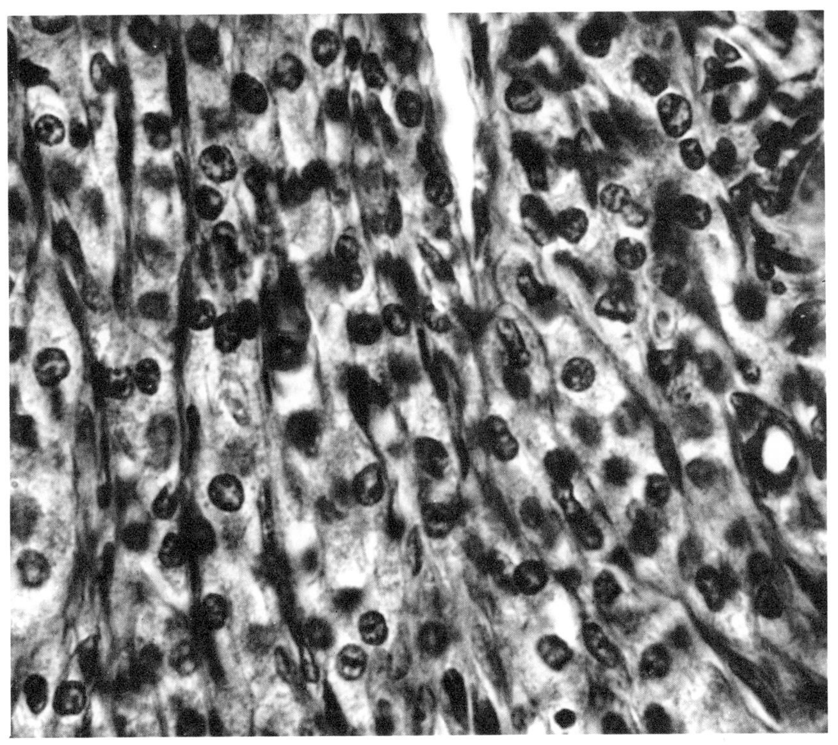

Fig. 6.54 Fully differentiated thecal type interstitial gland cells of a human ovary at 8½ months gestation. Human, 104B. × 695.

parent below, if this should be true, these interstitial cells must revert to an essentially pluripotential condition, for they might often be in a position to redifferentiate as thecal gland cells, or as paraluteal and eventually as luteal gland cells.

In the guinea pig, at least some of the differentiated cells of this type seem to be gradually crowded into the medulla where they appear as even larger, more vacuolar cells than they were in the cortex (Stafford, Collins, and Mossman, 1942). We have also seen evidence of this apparent migration toward the medulla in the plains pocket gopher, the gray and fox squirrels (*Sciurus carolinensis* and *S. niger*), and the red squirrel (*Tamiasciurus hudsonicus*). Apparently, these medullary cells do eventually degenerate. Whether this migration toward the medulla is confined to a particular class of cells or is purely fortuitous is un-

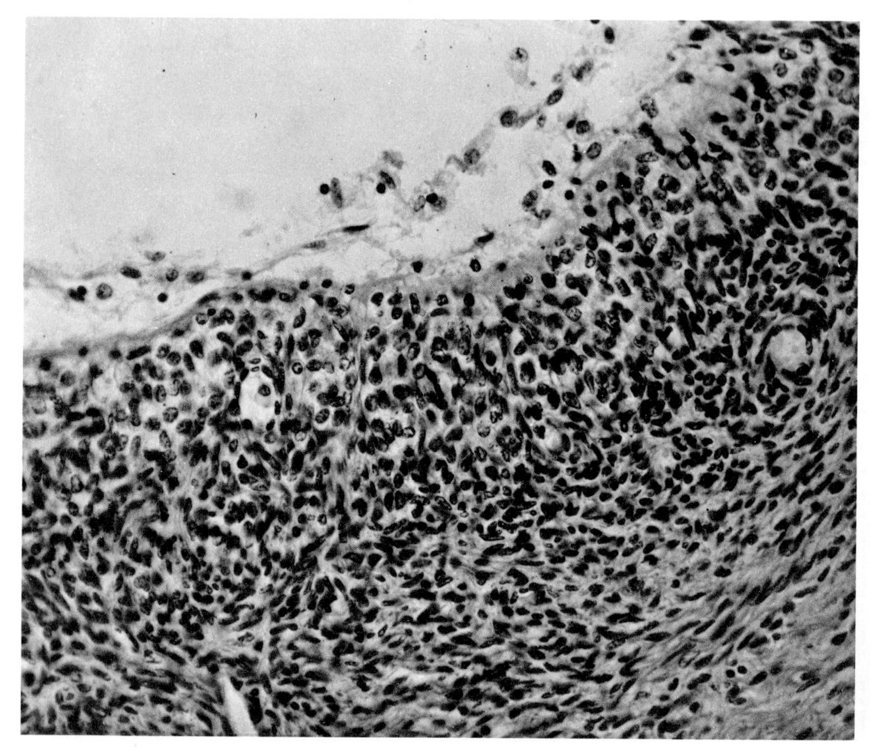

Fig. 6.55 Relatively undifferentiated thecal type interstitial gland cells of the ovary of a 7-year-old girl. Human, 121. × 340.

known. (See also Chapter 7, pp. 272–73, for a discussion of the movement of cortical elements toward the medulla.)

Any follicle having undifferentiated epithelioid theca interna cells usually gives rise to interstitial gland cells as it undergoes atresia, but these are not necessarily differentiated functionally. These cells are usually arranged in somewhat irregular, short columns radiating outward from the basement membrane of the follicular epithelium. The columns are separated by delicate trabeculae of connective tissue containing capillaries which, however, are never as plentiful or as dilated as those of an active thecal or luteal gland. The cells at the central ends of the columns are most differentiated; the outer, the least. In fact, at the periphery there is almost always an abrupt but obvious gradation into the small undifferentiated stromal cells. The radial arrangement of these cords of cells is particularly noticeable in man, apes, monkeys, and

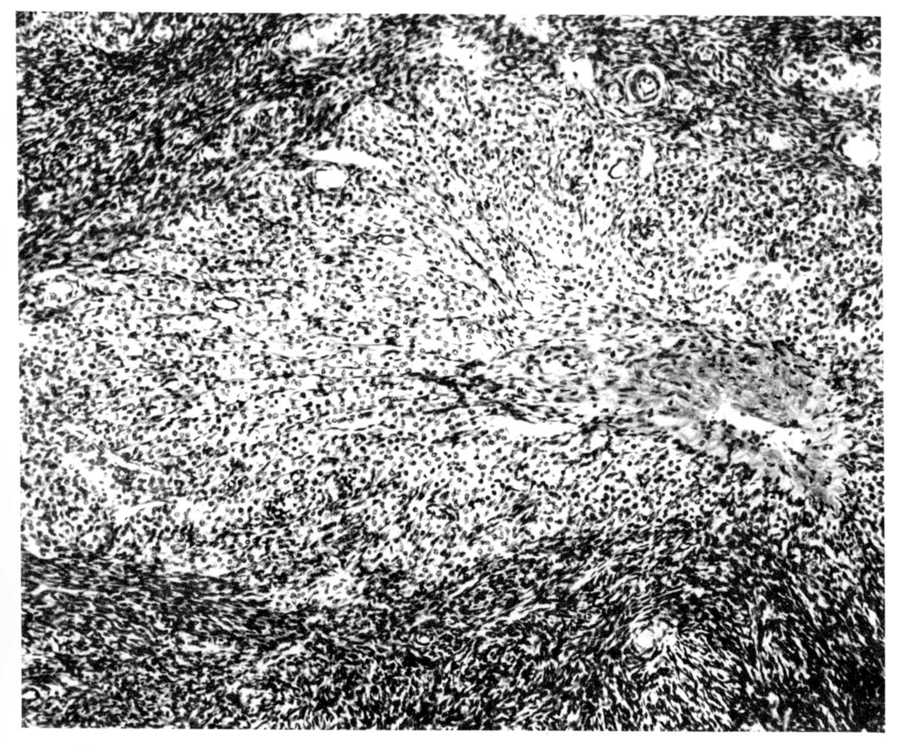

Fig. 6.56 Corpus atreticum of a woman at 8½ months gestation, showing the radial arrangement of the thecal type interstitial gland cells. Human, 104B. × 100.

bovids (Figs. 6.56 and 6.57). In bats, rabbits and hares, squirrels, the hamster, bears (Ursidae), weasels (Mustelidae), and cats (*Felis catus* and *Lynx rufus*), they simply form compact zones around the follicular remnants (Figs. 6.58–6.67).

In the case of atresia of nearly ripe follicles that have a well-differentiated zone of thecal gland cells, the differentiated cells degenerate and do not give rise to interstitial cells. However, the outer, less-differentiated cells of such a thecal gland zone do change into thecal type interstitial gland cells, just as the whole theca interna of younger atretic follicles does. This is easily observed in the plains pocket gopher. In other words, these outer cells were still embryonic or pluripotential.

Cytologically fully differentiated thecal type interstitial gland cells of most species contain very large vacuoles of lipoid material which

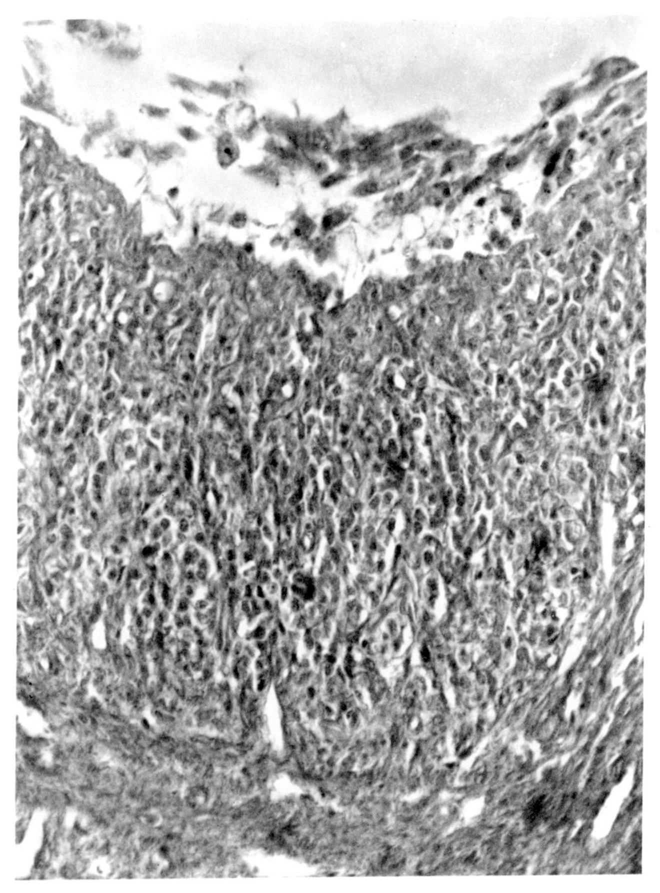

Fig. 6.57 Corpus atreticum of a bison (*Bison bison*) with a 205-mm CR fetus. Note the radial arrangement of the cords of thecal type interstitial gland cells. *B. bison*, 1. × 250.

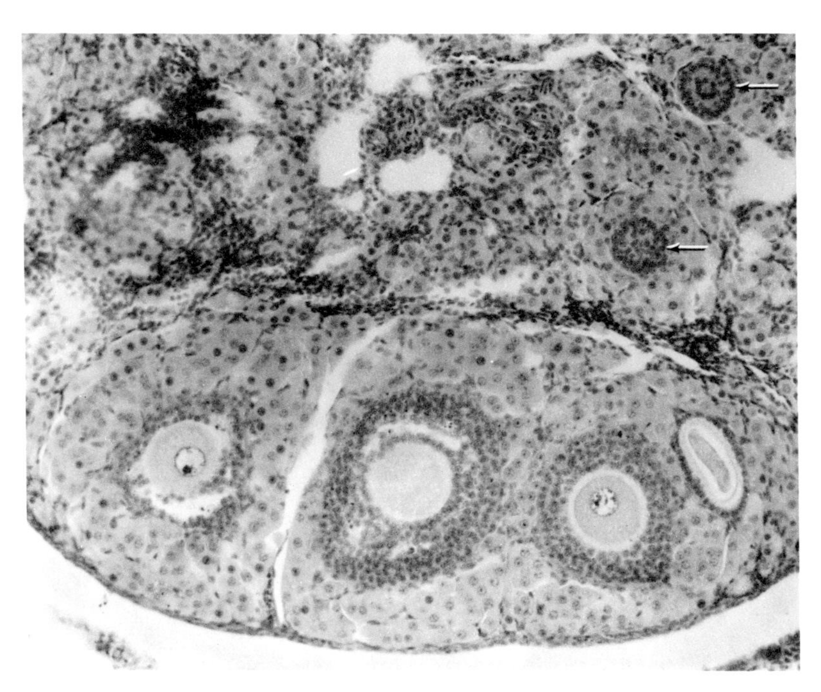

Fig. 6.58 Very early atresia of one primary and three secondary follicles of a nonpregnant California mouse-eared bat (*Myotis californicus*) in late July. The unusually early differentiation of thecal type interstitial gland tissue is typical of this genus. Two medullary cord remnants appear at the upper right (*arrows*). *M. californicus*, 10. × 140.

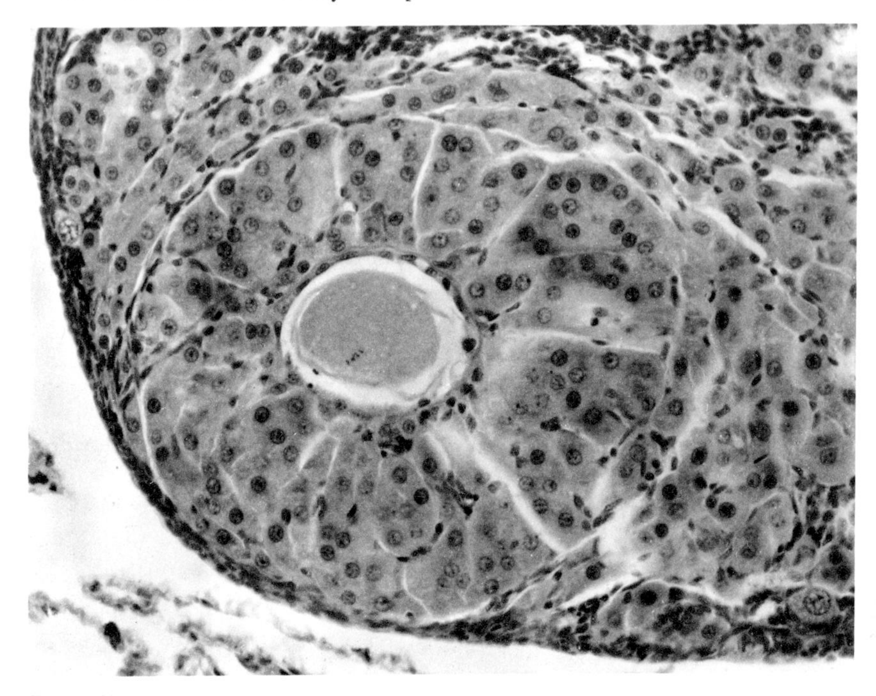

Fig. 6.59

Fig. 6.59 Atretic follicle with a maturation spindle from the ovary of the California mouse-eared bat (*Myotis californicus*) shown in Figure 6.58. *M. californicus*, 10. × 250.

Fig. 6.60 Very late atresia of three follicles of a nonpregnant black mouse-eared bat (*Myotis nigricans*). Their interstitial gland tissue is becoming vacuolated and mingled with the general mass of this tissue. *M. nigricans*, 16. × 250.

Fig. 6.61 Cortex of a snowshoe hare (*Lepus americanus*). Four corpora atretica with relatively undifferentiated thecal type interstitial gland tissue are embedded in the differentiated interstitial gland tissue, which is presumably of both thecal and stromal origin. *L. americanus*, 272RF6604C. × 100.

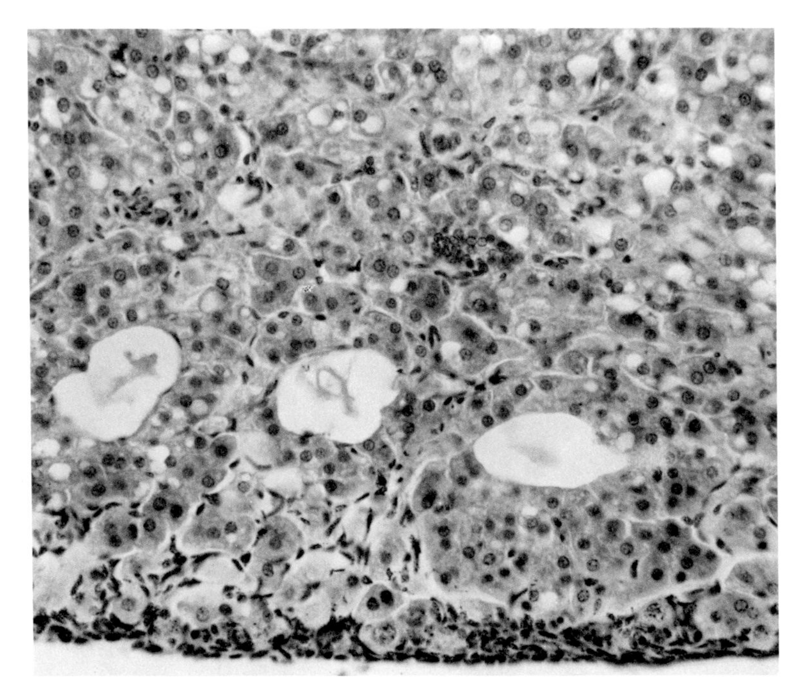

Fig. 6.60

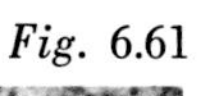

Fig. 6.61

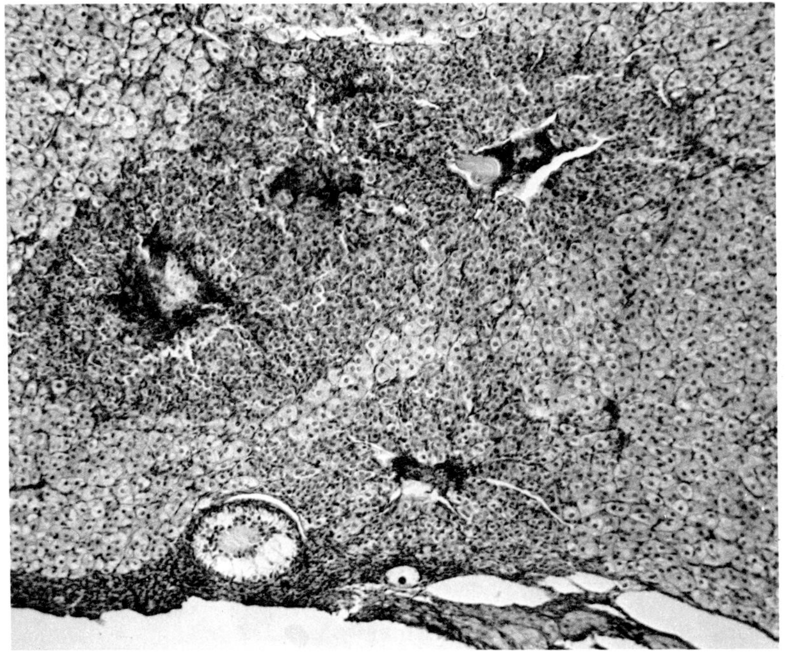

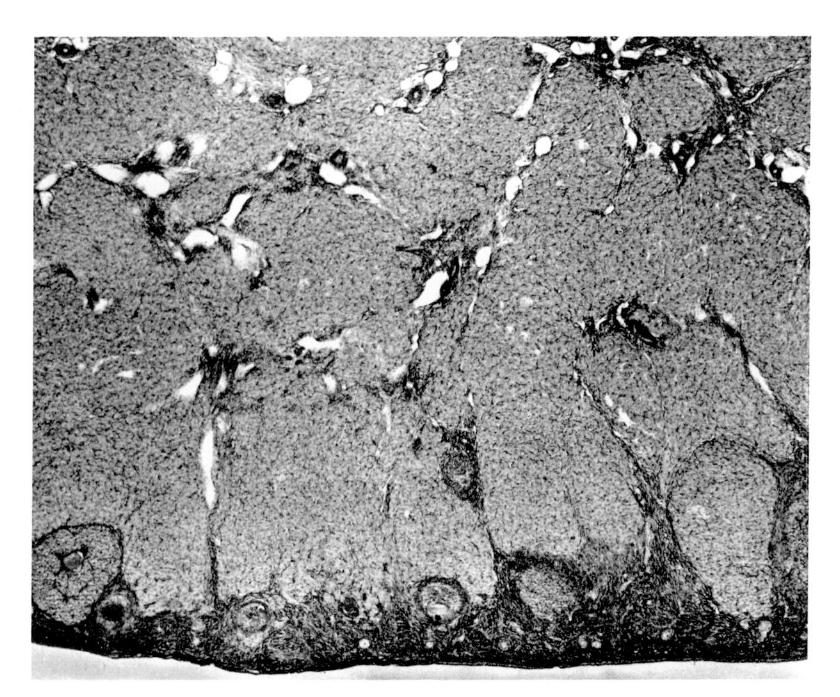

Fig. 6.62 Cortex of a New England cottontail (*Sylvilagus transitionalis*), showing masses of interstitial gland tissue. Most of them are certainly of the thecal type, but some are probably of the stromal type. *S. transitionalis.* × 40. (Courtesy of Clinton H. Conaway.)

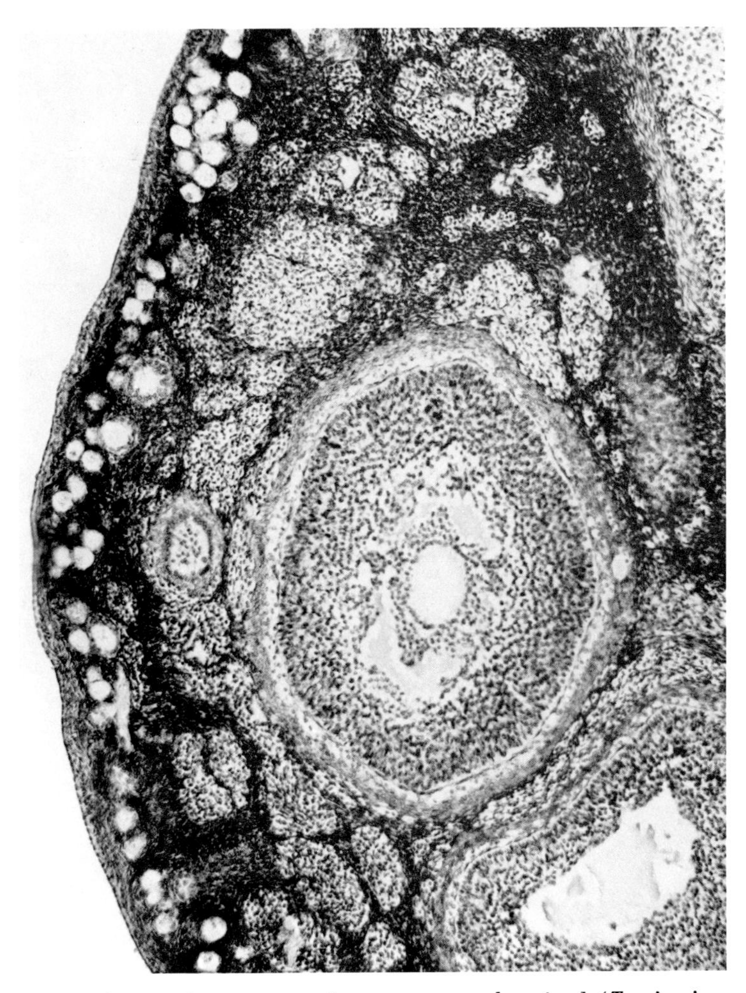

Fig. 6.63 Cortex of a parous, early proestrous red squirrel (*Tamiasciurus hudsonicus*) in early January. Note the numerous corpora atretica with the thecal type interstitial gland tissue and the growing follicle with small antral spaces. *T. hudsonicus*, 32. × 100.

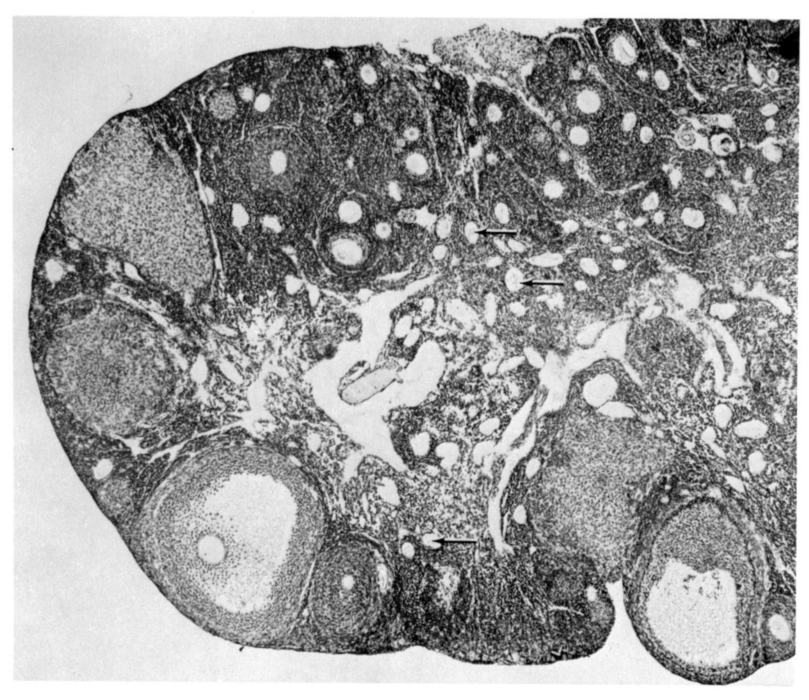

Fig. 6.64

Fig. 6.64 Cortex and medulla of a nonparous golden hamster (*Mesocricetus auratus*), showing numerous corpora atretica with remnants of ova and zonae pellucidae (*arrows*). *M. auratus*, 19. × 40. (Courtesy of Margaret W. Orsini.)

Fig. 6.65 Corpus atreticum of a nonpregnant grizzly bear (*Ursus horribilis*). Part of this corpus completely blends with the general stromal type interstitial gland tissue, which fills most of the ovary. *U. horribilis*, 1. × 40.

Fig. 6.66 Outer cortex of the ovary of an ermine (*Mustela erminea*) with tubal embryos. From left to right, medullary cord type interstitial gland tissue, corpus atreticum with thecal type interstitial gland tissue, more medullary cord type interstitial gland tissue, and a corpus luteum. *M. cicognanii* (= *M. erminea*), 1. × 198.

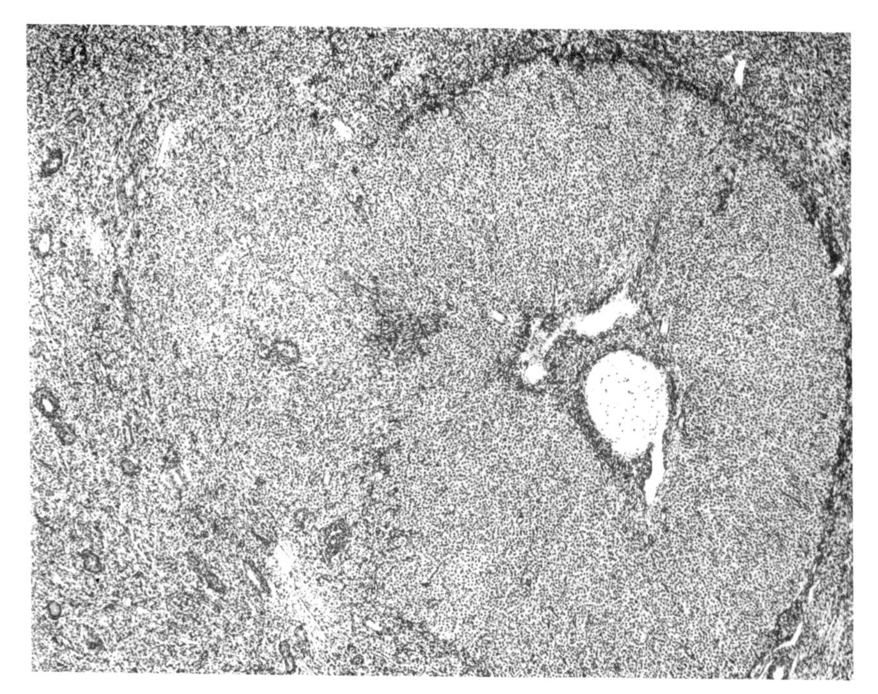

Fig. 6.65

Fig. 6.66

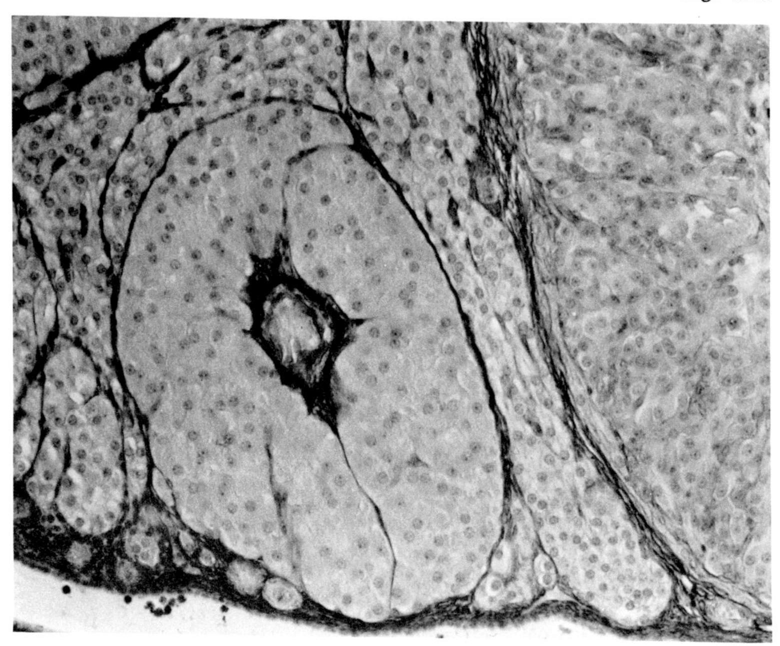

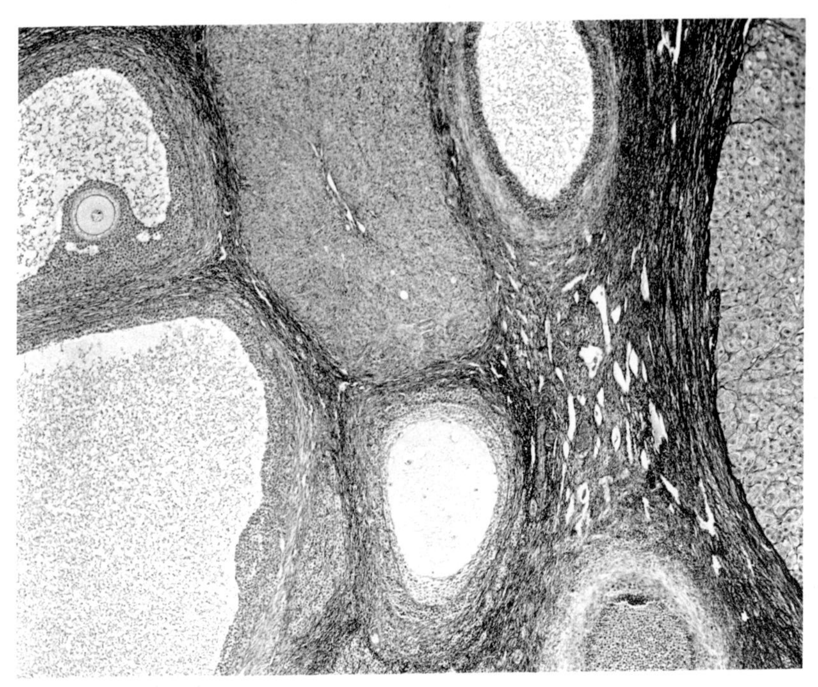

Fig. 6.67 Ovary from a cat (*Felis catus*) in the late limb bud stage of pregnancy. Note the two normal vesicular follicles (*left*), one large corpus atreticum and one atretic follicle with some thecal type interstitial gland tissue and portions of two other follicles (*center*), and the edge of a corpus luteum (*right*). *F. catus*, 7. × 40.

usually dissolve in tissue-processing, leaving so much clear space in the cytoplasm that, in sections, masses of these cells are readily recognized by the lightness of their stain. Usually, this clearly distinguishes them from thecal gland or luteal gland cells. Unfortunately, in some mammals, notably the laboratory rat, they rather closely resemble both thecal and luteal gland cells. This has led to much confusion, and to the use of the term "luteinization" of thecal cells, instead of the perfectly descriptive term "hypertrophy." There is no proof that in mammals in general these cells secrete progestins or contain lutein pigment, which would seem to be the only valid reasons for considering them "luteinized." Even in species such as the bears, where large corpora atretica composed of these cells exist, they have neither the structure nor color of corpora lutea (see Fig. 6.65).

Ultrastructurally, thecal type interstitial cells are typical steroid secretors with many osmiophilic vacuoles (Muta, 1958; Davies and Broadus, 1968). Histochemical studies have shown that their cytoplasm contains phospholipids, triglycerides, and cholesterol and its esters (Deane, 1952; McKay et al., 1961; Ben-or, 1963; Guraya and Greenwald, 1964*a*, 1964*b*; Deane and Rubin, 1965; Guraya, 1966; Motta and Takeva, 1971). Recent histochemical work has shown the presence in these cells of enzyme systems necessary for steroid hormone synthesis (Rubin, Deane, and Balogh, 1969). Experiments designed to determine by ovarian vein blood analysis and other methods the nature of the functional secretion of these cells have yielded conflicting results (Hilliard, Archibald, and Sawyer, 1963; Nalbandov, 1964; Keyes and Nalbandov, 1968). A little evidence has been produced for the secretion of androgens, estrogens, and progestins. Since these cells are present and apparently fully differentiated, at least periodically from birth to old age, it is a logical hypothesis that their secretion may be responsible for the development and maintenance of secondary female sex characters, just as testicular interstitial cell secretions relate to secondary male sex characters. Presumably, this would involve an estrogen or an estrogen analog. A feminizing ovarian tumor made up of cells resembling thecal type interstitial gland cells was described in a child of 18 months by Walczak and Pieńkowska-Mikotajczyk (1960). Hayward, Hilliard, and Sawyer (1963) found progestin in preovulatory ovarian vein blood of the rhesus monkey (*Macaca mulatta*). This could have come from a corpus luteum of the previous cycle, from thecal gland, interstitial gland, or from follicular epithelium prematurely beginning its progestin secretion. Hilliard, Archibald, and Sawyer (1963) activated preovulatory synthesis and release of progestin in the rabbit, but whether it came from the unusually abundant interstitial tissue of that species is an open question. (For a review of the older literature, see Zuckerman, 1962.)

It is apparent in many species that the thecae internae of some degenerating secondary and vesicular follicles have not produced functionally differentiated interstitial gland cells. One wonders whether they will eventually do so, or if perhaps they are follicles which began to degenerate at a time when there was no stimulus for the differentiation of interstitial gland cells. Such follicles sometimes occur along with others which had apparently been in the same stage of development and which do have interstitial gland cells. However, these two types do not usually occur in the same individual. It is probable, therefore, that the cause of follicular degeneration is not always directly associated with the factors that induce interstitial gland development from the theca interna.

Thecal type interstitial gland cells in an apparently fully differentiated state also commonly persist long after all other elements of the follicle have disappeared, at the same time that there are many growing follicles and that little or no atresia is taking place. In other words, the glandular function is sustained by some mechanism other than that which causes follicular growth or atresia itself.

Unsicker (1970) demonstrated sympathetic nerve fibers ending in large boutons embedded in the cytoplasm of interstitial gland cells of the laboratory mouse. Since these were presumably thecal type interstitial cells, it would be interesting to know when this innervation takes place. Does it happen while they are undifferentiated theca interna cells? If so, then one would expect thecal gland cells to be innervated. Thus, direct neural control could be a factor in the secretion of both cell types and possibly even in follicular growth and atresia.

Stromal type

The stromal type of interstitial gland tissue is well illustrated by the ovaries of rabbits and hares and of *Manis*. Adult ovaries of rabbits and hares consist of an almost solid mass of interstitial gland tissue occupying both cortex and medulla (Fig. 6.68 and 6.69). Except for a few large follicles or corpora lutea, the cortex proper appears to be a very thin interrupted zone of scattered primary and secondary follicles between which the interstitial gland tissue almost reaches the tunica albuginea. There is not even the usual relatively vague and irregular corticomedullary boundary. A few large vessels enter along the cephalic portion of the long attachment to the mesovarium and ramify through the central part of the ovary. The only obvious collagenic connective tissue in the medulla seems to be the delicate adventitia of these vessels and their branches. There is none of the loose medullary stroma and heavy collagenic strands so characteristic of this region in most other mammalian ovaries. Leporid interstitial gland cells are also nearly as large as fully differentiated luteal cells, and they remain relatively large and distinct throughout the year. There is, however, some decrease in the size of these cells in the snowshoe hare (*Lepus americanus*) and the European hare (*L. europaeus*) during the nonbreeding season (Fig. 6.70). The cells also tend to become less polyhedral and more spheroidal. Deep in the ovary of certain snowshoe hares, we have seen small groups of interstitial cells which contained large irregular vacuoles indicative of actual cell degeneration, but we have insufficient data to correlate this condition with any phase of the reproductive cycle.

The interstitial gland tissue of leporids originates from two different sources, but once formed it looks and behaves alike. Ovaries of juvenile and puberal rabbits and hares show very clearly that most of their interstitial gland tissue is derived from thecae internae of atretic follicles, that is, it is of the thecal type. Ovaries of adult animals also show that interstitial gland tissue is still being formed from atretic follicles, but because the ovaries are packed with gland tissue the individual corpora atretica are often hard to distinguish. This origin of interstitial gland tissue in the rabbit was first described by Lane-Clayton in 1905.

There are two reasons why leporid interstitial gland tissue is considered to be mainly of the stromal rather than thecal type. First, as already mentioned, it tends to fill the entire ovary and to remain relatively well differentiated throughout the year. Second, much of the interstitial gland tissue of leporids appears to arise from stroma without regard to its topographical relation to follicles. This stromal type fills the irregular spaces between the masses of thecal origin, but toward the central portion of the ovary the two become completely indistinguishable from one another. The area of origin seems to be mainly a thin zone of differentiating stromal cells adjacent to the slightly fibrous tunica albuginea. This "germinal" zone is interrupted frequently by primordial and primary follicles and by an occasional larger follicle or corpus luteum (see Figs. 6.68 and 6.70). Except for their location, we think that these germinal cells and those of the thecae internae of atretic follicles are alike.

Although, regardless of their dual origin, the interstitial gland cells of leporids appear to be exactly alike once they differentiate, they do differ somewhat from thecal type interstitial gland cells of most other mammals. We have already mentioned their less marked cyclic changes. The thecal type interstitial gland cells of other mammalian groups are also usually much smaller relative to other ovarian cells. In leporids, their ratio of cytoplasm to nucleus is unusually high. Instead of the few very large irregular-sized vacuoles characteristic of most thecal type interstitial gland cells, their cytoplasm is packed, depending upon fixation methods, with small uniform granules or with vacuoles. In fact, in this respect the interstitial gland cells of leporids resemble fascicular zone cells of the adrenal cortex, yet there is no evidence that they are ho-

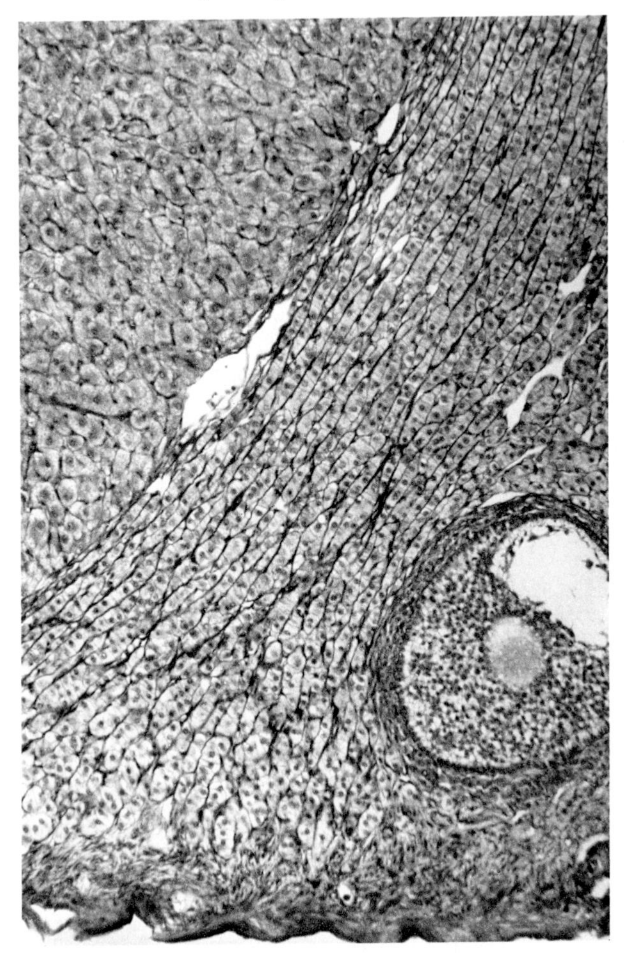

Fig. 6.68 Outer cortex of the ovary of a snowshoe hare (*Lepus americanus*) in early limb bud stage of pregnancy, showing the stromal type interstitial gland tissue, a small vesicular follicle in early atresia, and small sector of a corpus luteum. *L. americanus*, 3. × 130.

mologous to the gonadal adrenal cells of this or any other mammalian group.

The allegedly most primitive group of living lagomorphs, the pikas, have a large amount of weakly differentiated interstitial gland tissue occupying the cortex of the ovary during the juvenile period. Apparently,

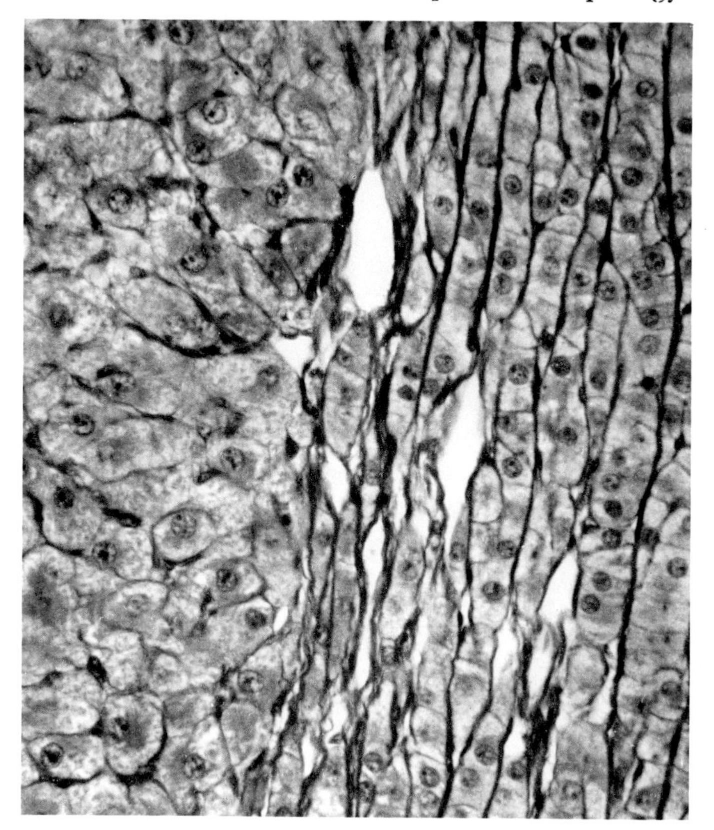

Fig. 6.69 Comparison of the interstitial gland cells and luteal cells of the ovary of the snowshoe hare (*Lepus americanus*) shown in Figure 6.68. *L. americanus*, 3. × 425.

this is not the thecal type, and it completely disappears in adults (Duke, 1952). It may correspond to the fetal type of other groups; certainly it is not comparable to the stromal type of the leporids. Eastern cottontails (*Sylvilagus floridanus*) have large amounts of persistent interstitial gland tissue, but almost all of it is in the form of encapsulated masses, indicating its origin from atretic follicles. This is especially true of younger breeding females. Domestic rabbits show more evidence of large amounts of tissue derived directly from the stroma. The hares (*Lepus*), especially the snowshoe hare, the white-tailed jack rabbit (*Lepus townsendii*), and the European hare, have relatively very large,

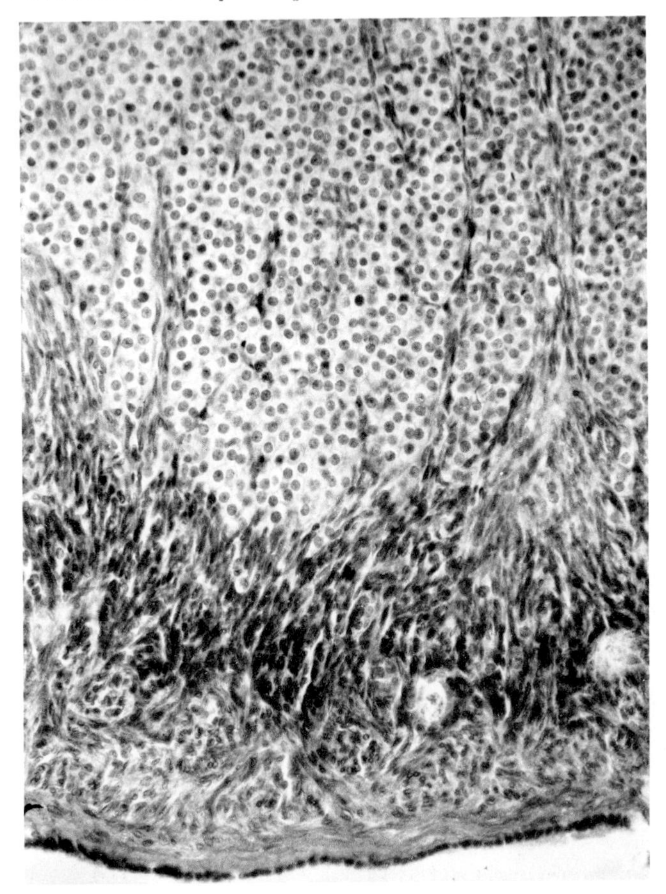

Fig. 6.70 Involuted stromal type interstitial gland tissue of a nonpregnant snowshoe hare (*Lepus americanus*) in late November. *L. americanus*, 5. × 200.

thick ovaries composed chiefly of interstitial gland tissue, with evidence that most of it is derived directly from stroma without any relation to atretic follicles. In the lagomorphs, then, there appears to be a progression from no interstitial gland tissue of direct stromal origin in pikas, to some in the cottontails (*Sylvilagus*), to an intermediate amount in the domestic rabbits, to an extreme amount in the hares.

Until more is known about the comparative morphogenesis and physiology of interstitial gland tissue, it seems justifiable to classify that of

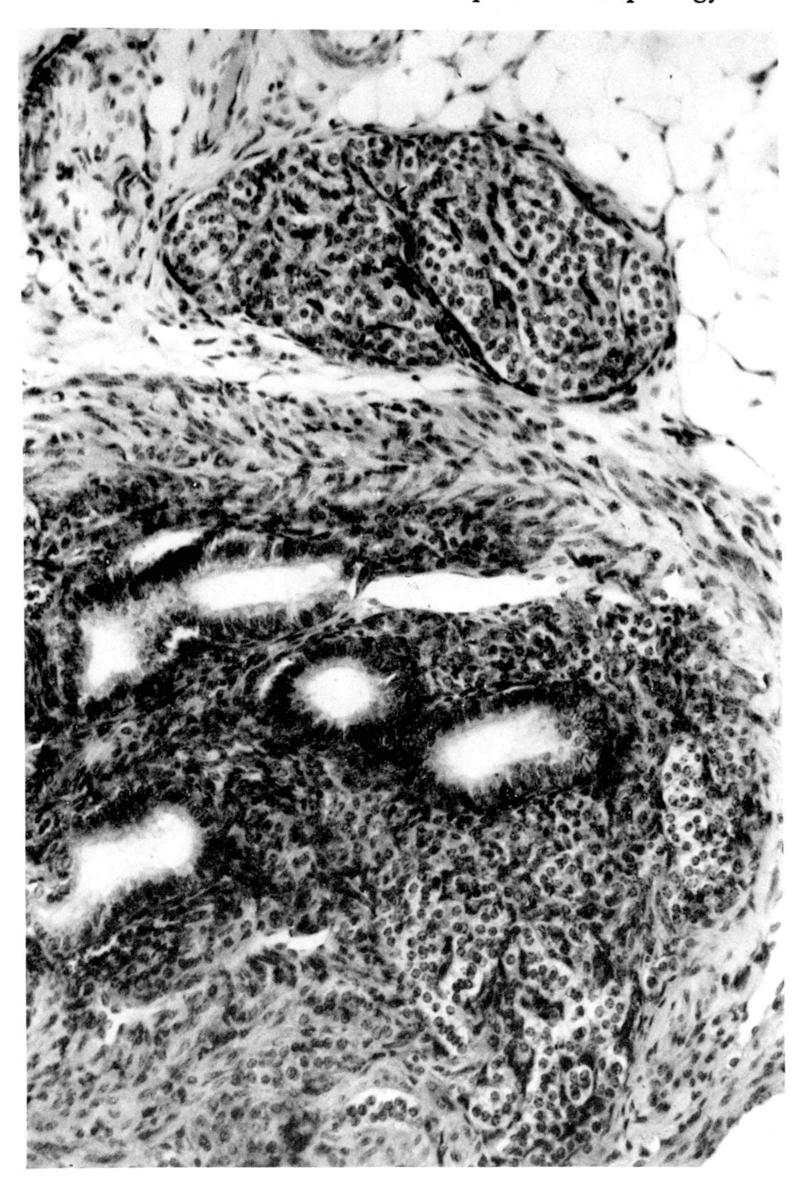

Fig. 6.71 Gonadal adrenal tissue in the epoophoron area of a gray squirrel (*Sciurus carolinensis*) with uterine morulae. This is the most common location and arrangement of this tissue in mammals. *S. carolinensis*, 17. × approx. 200.

the leporids as stromal, even though some of it does arise exactly like the thecal type of other groups.

Gonadal adrenal type

The gonadal adrenal type of interstititial gland tissue usually occurs just outside the ovary in close association with the epoophoron, commonly as small scattered cell groups adjacent to the tubule basement membrane (Fig. 6.71). Rarely these small cell groups are nearer the rete (Fig. 6.72). In some mammalian groups (lagomorphs, horses, and Manidae), the cells are partially or entirely aggregated into definite, small, encapsulated nodules (Figs. 6.73, 6.74, and 6.75). In others, the nine-banded armadillo (*Dasypus novemcinctus*) and the Asiatic elephant (*Elephas maximus*), they form masses comparable in size to corpora lutea, and have frequently been recognized as accessory adrenals (Fig. 6.76) because they have all the characteristics of an adrenal gland, except that they always lack medullary tissue. Accessory adrenals of this

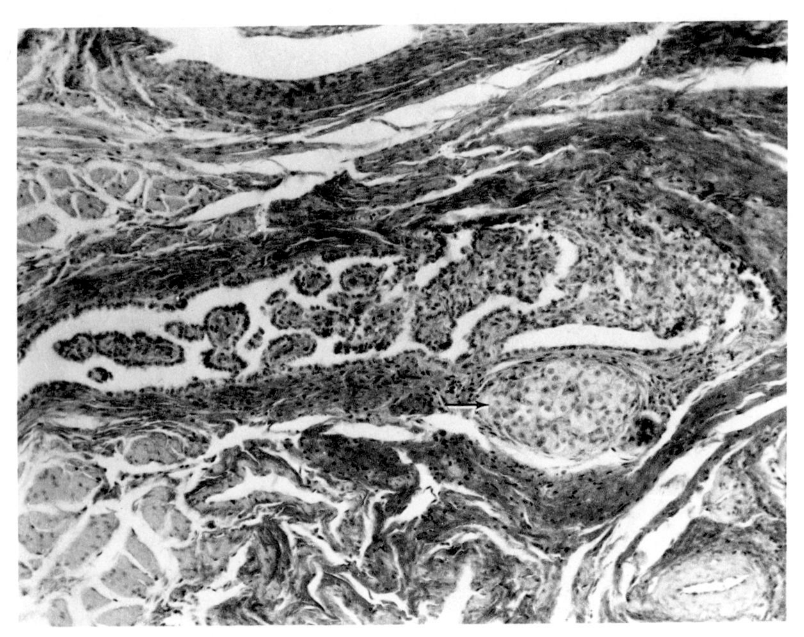

Fig. 6.72 Gonadal adrenal tissue (*arrow*) associated with the rete of one of the rock hyraxes (*Heterohyrax brucei*) containing 135-mm fetuses. *H. brucei*, 3. × 110.

sort were described in the broad ligament of the human fetus and newborn and in other mammals by Aichel (1900), but have been considered to be inconstant minor anomalies, except when they become pathologically active tumors. In the pangolin (*Manis temminckii*) there is a large mass of gonadal adrenal tissue in the mesovarium near the hilus, as well as scattered cells and small groups adjacent to the rete and epoophoron (Fig. 6.77). However, the large mass does not have the arrangement of cell cords typical of the adrenal cortex.

Groat (1943, 1944) showed that after adrenalectomy of the 13-lined ground squirrel (*Spermophilus tridecemlineatus*) large numbers of cells resembling those of the adrenal cortex appeared throughout the ovaries, including the areas of the hilus and epoophoron. He produced considerable evidence that they functioned and compensated for the loss of the adrenals. (See Chester Jones and Henderson, 1963, for further studies on the female ground squirrel, and Seliger, Blair, and Mossman, 1966,

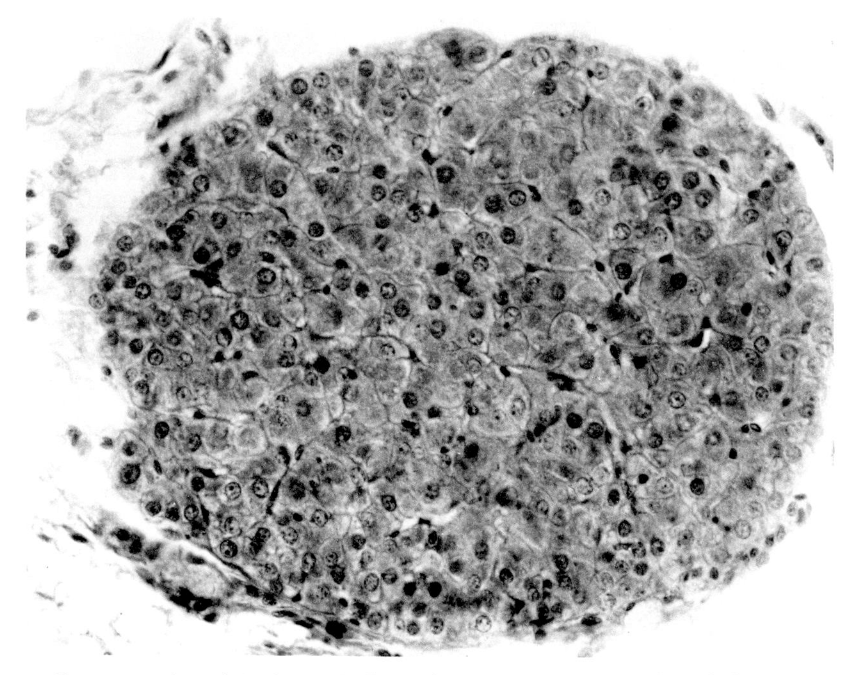

Fig. 6.73 Gonadal adrenal body in the mesovarium of a pika (*Ochotona princeps*) in early uterine gestation. *O. princeps*, 9. × 300.

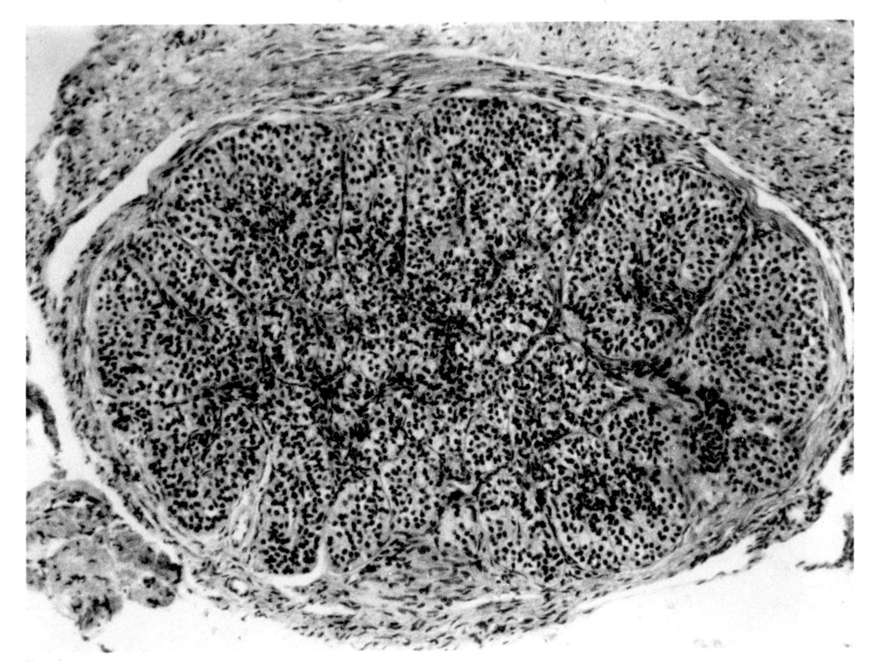

Fig. 6.74

Fig. 6.74 Gonadal adrenal nodule from the mesovarium of a mare (*Equus caballus*). *E. caballus*, 69–2896. × 140. (Courtesy of O. J. Ginther.)

Fig. 6.75 Gonadal adrenal tissue from the ovary of a mare (*Equus caballus*), showing the typical zona glomerulosa. *E. caballus*, 69–3028. × 250. (Courtesy of O. J. Ginther.)

Fig. 6.76 Accessory adrenal from the mesovarium of a parous nine-banded armadillo (*Dasypus novemcinctus*). *D. novemcinctus*, 66–56. × 63.

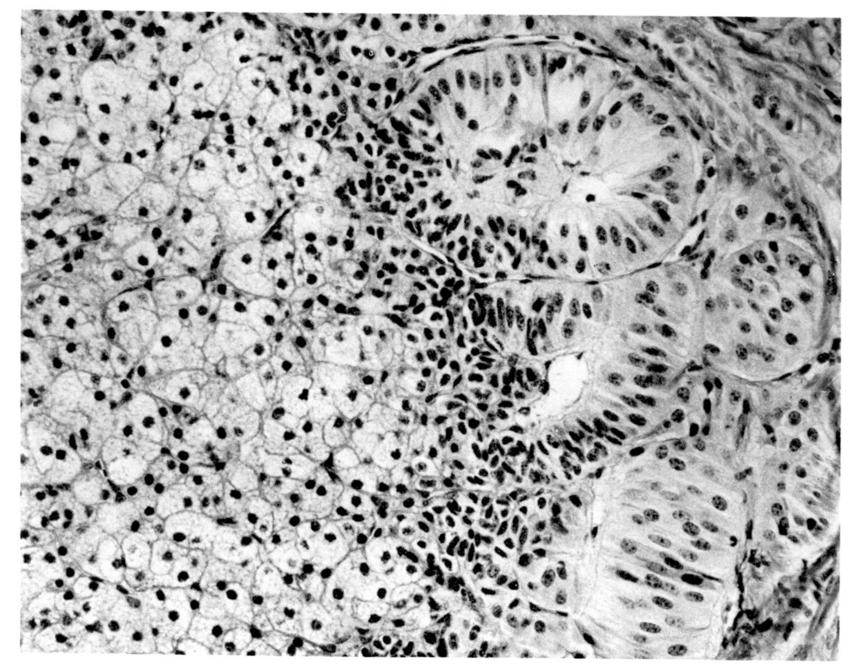

Fig. 6.75

Fig. 6.76

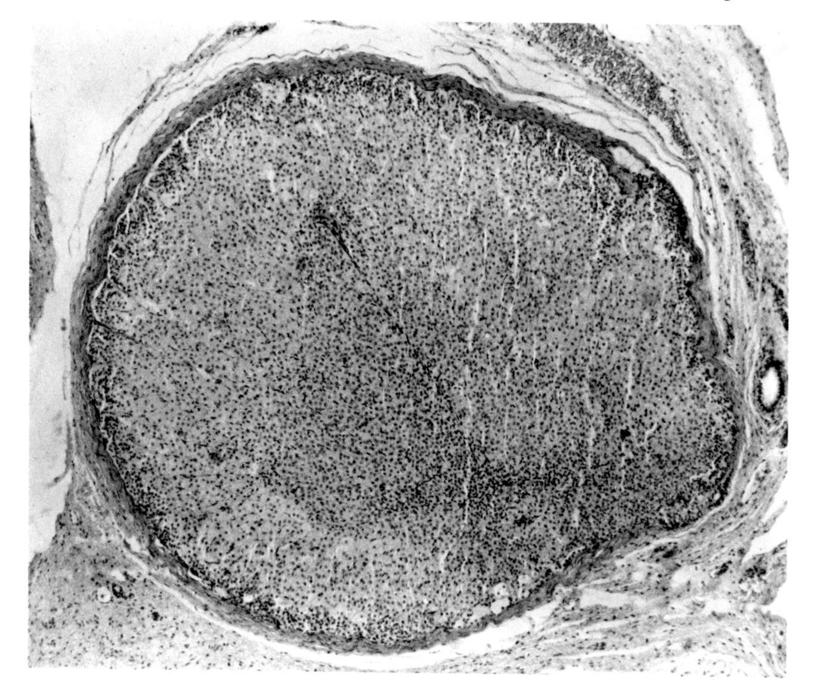

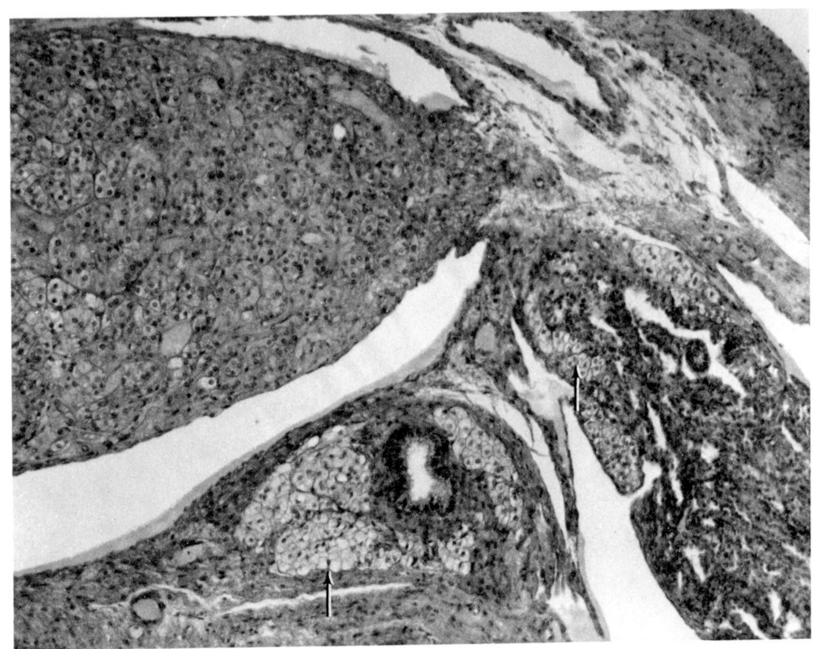

Fig. 6.77 Mesovarium of a pangolin (*Manis temminckii*) in mid-gestation, showing scattered groups of gonadal adrenal cells (*arrows*) around the epoophoron and rete and a portion of a much larger gonadal adrenal body (*upper left*). *M. temminckii*, 1. × 95.

for similar studies on the male ground squirrel.) Seth and Prasad (1967) also demonstrated the development of nodules of gonadal adrenal tissue in the region of the epoophoron of the Indian striped palm squirrel (*Funambulus pennanti*) after adrenalectomy. However, their so-called "follicular bundles" of "adrenocortical like" tissue are apparently partially lutealized atretic follicles, and their "syncytial nests" of "adrenocortical like" tissue are cortical cords, which are common in sciurid ovaries. Groat's papers appear to have been the first to correlate a specific cell type with ovarian cortico-adrenal function, although a close relation between the reproductive cycle or ovary and the adrenal cortex had been indicated several times previously (Guieysse, 1901; Zalesky, 1934; Emery and Schwabe, 1936; Nahm and McKenzie, 1937; Fekete, Wooley, and Little, 1941). We have seen well-differentiated adrenal cortexlike cells in the ovaries of several normal females of various genera of sciurids (Figs. 6.71 and 6.78), but never in the quantity or

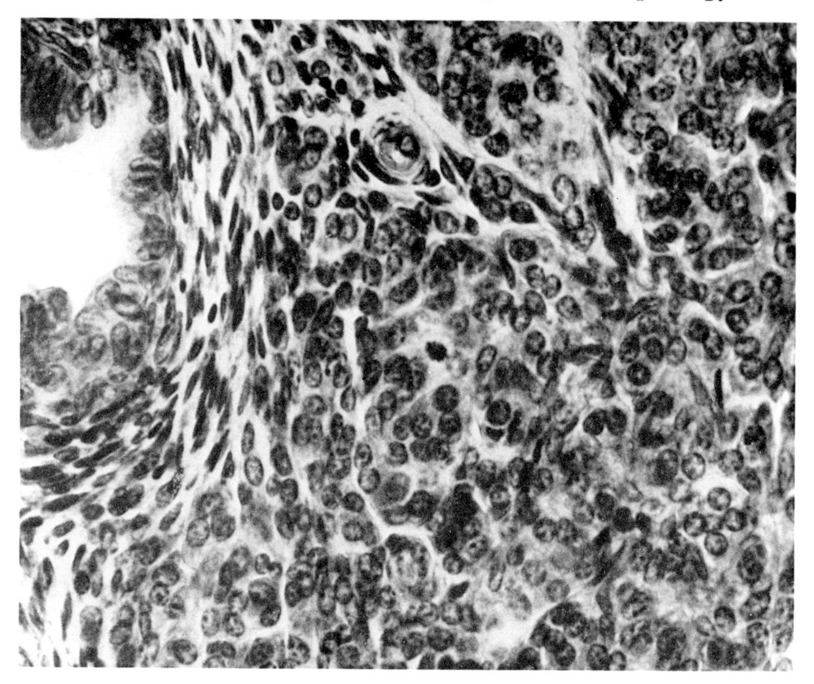

Fig. 6.78 Relatively undifferentiated gonadal adrenal cells in the epoophoron area of a parous 13-lined ground squirrel (*Spermophilus tridecemlineatus*) in August. *S. tridecemlineatus*, 49. × 520.

degree of differentiation that was characteristic of Groat's adrenalectomized animals. No one has ever followed closely enough the history of these cells in the sciurids during the various phases of the reproductive cycle to be able to say whether their state of differentiation bears any relation to the reproductive cycle, except that they are always in an undifferentiated state in prepuberal animals and during anestrus.

As with other types of interstitial gland cells there is often a problem of recognition and identification. When they occur in the form of definite accessory adrenals there is no question, and even when they occur in relation to the epoophoron only as scattered cells and groups of cells, there is reason to accept their identity simply because there is no other known type of glandular cell in such a position, and this part of the mesovarium is the known location of the clear-cut accessory cortical masses and of transitional conditions between these two. However, when gonadal adrenal type cells occur in the ovary proper, as they do in the

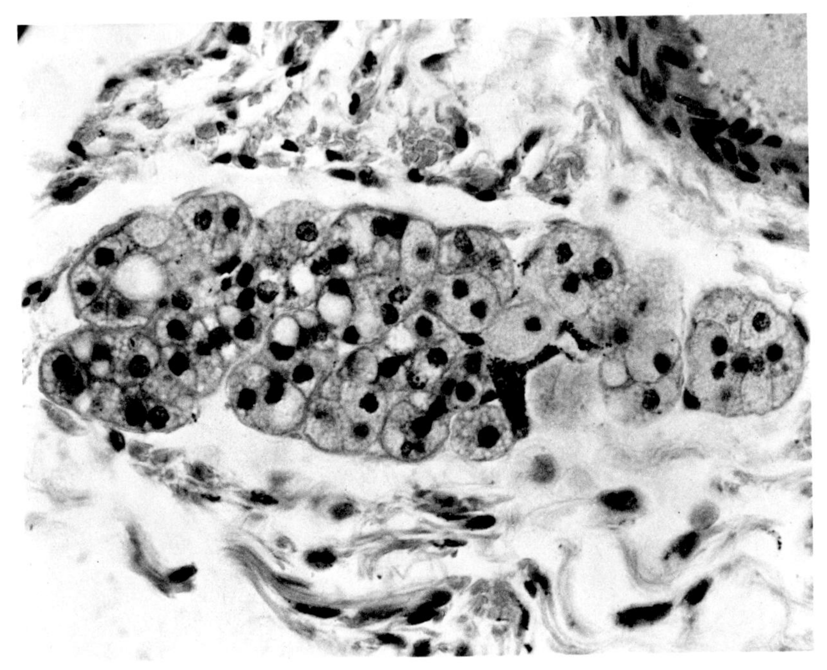

Fig. 6.79 Gonadal adrenal cells in the epoophoron area of a porcupine (*Erethizon dorsatum*) with a 180-mm fetus. Note the "foamy" cytoplasm and the melanocyte, of which there were several associated with this mass. *E. dorsatum*, 166R. ×400.

squirrels and many other groups, they must be distinguished by their cytological characteristics. Fortunately, Groat (1943 and 1944) and Chester Jones and Henderson (1963), by correlating physiology with morphology, have established with reasonable certainty the identity of these intra-ovarian cells. We know that, when fully differentiated, they closely resemble typical zona fasciculata cells of a normal adrenal, their most prominent feature being a large amount of cytoplasm, all of which is packed with small vacuoles of very uniform size (Figs. 6.75 and 6.79). (When fixed with steroid-hardening chemicals, such as OsO_4, these vacuoles are seen as small, uniformly sized black granules.) Cells having this general appearance at times do appear in other parts of the ovary, such as in the corpus luteum, and occasionally in other types of interstitial cell masses, but because of their different topographical positions these are not considered to be gonadal adrenal cells, although no one can say whether their similar cytological features indicate similarity of

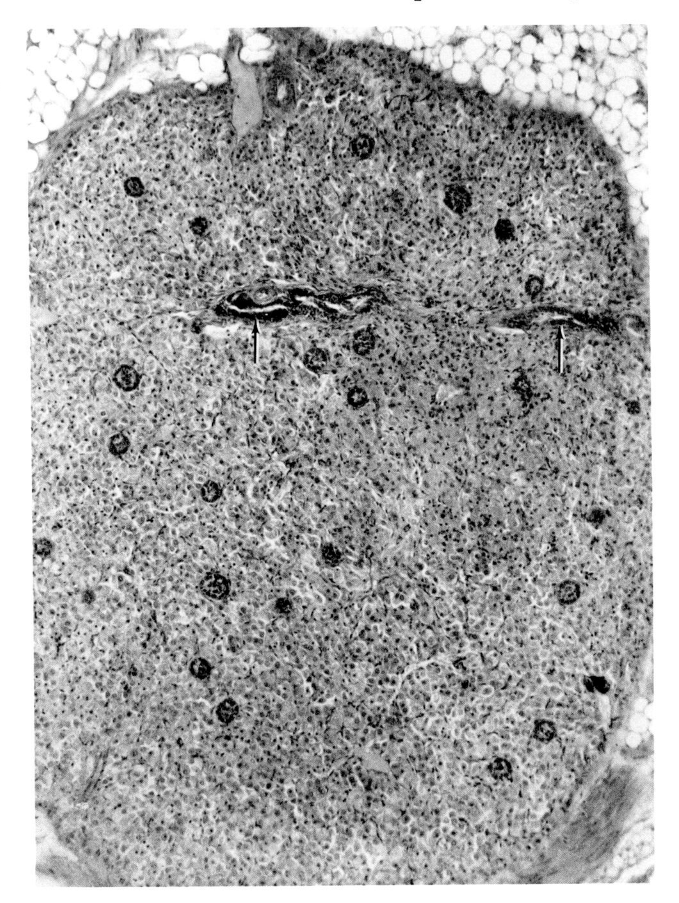

Fig. 6.80 Ovary of a star-nosed mole (*Condylura cristata*) in estrus, showing the medulla with rete (*arrows*), medullary cord remnants, and gonadal adrenal type interstitial gland tissue. *C. cristata*, 1. × 63.

function. Obviously, there is need for cytochemical and physiological studies on all of the glandular cells of the ovary in order to determine what their functions are and how their functions are specifically correlated with a particular cell type. Until this has been done, their classification must rest mainly on comparative studies based on similarities in origin, location, and cytology.

A good example of the difficulties involved in using these comparative criteria is provided by the gonadal adrenals of the lagomorphs where definite small nodules of tissue with very large cells occur in association

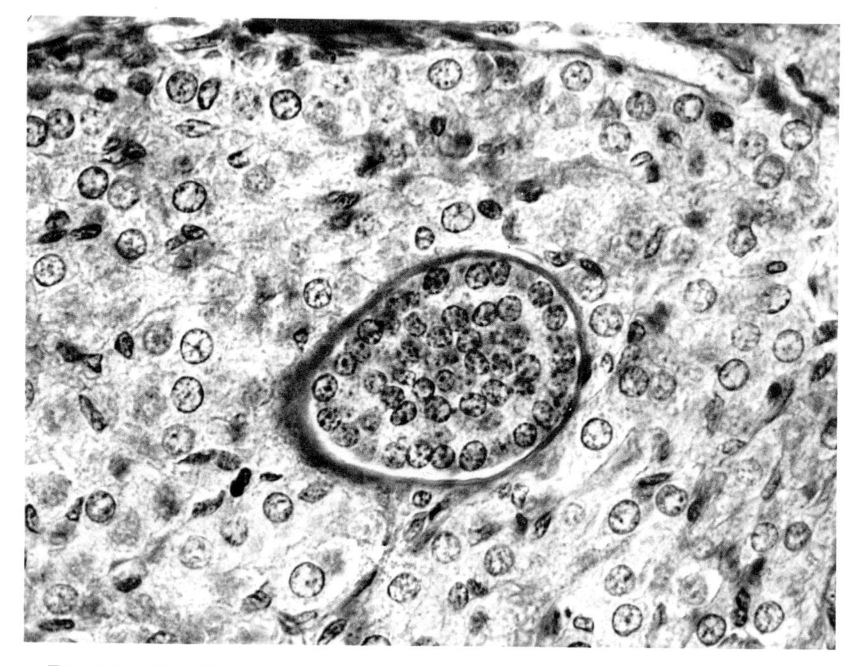

Fig. 6.81 Detail of the ovary of a star-nosed mole (*Condylura cristata*), showing a medullary cord remnant with its basement membrane and surrounding gonadal adrenal type interstitial gland cells. *C. cristata*, 13. × 640.

with the epoophoron, but where the characteristic small, closely packed vacuoles or granules just described are seldom seen (Fig. 6.73). Tentatively, we consider these as gonadal adrenals, mainly on the basis of their topographical location.

The value of comparative studies can be demonstrated in the case of the massive so-called interstitial gland of some of the talpids, the common European mole (*Talpa europaea*) and the star-nosed mole (*Condylura cristata*) (Figs. 6.80 and 6.81). One of the more primitive members of this group, the shrew-mole (*Neurotrichus gibbsi*), has a much more conventional-appearing ovary (Figs. 6.82 and 6.83). It does have large groups of gonadal adrenal type cells associated with its epoophoron, but these extend past the rete into the ovarian medulla where they are intermingled with persistent medullary cords. In the common European mole and the star-nosed mole, this tissue has been excessively developed and concentrated in the medulla and hilus, and

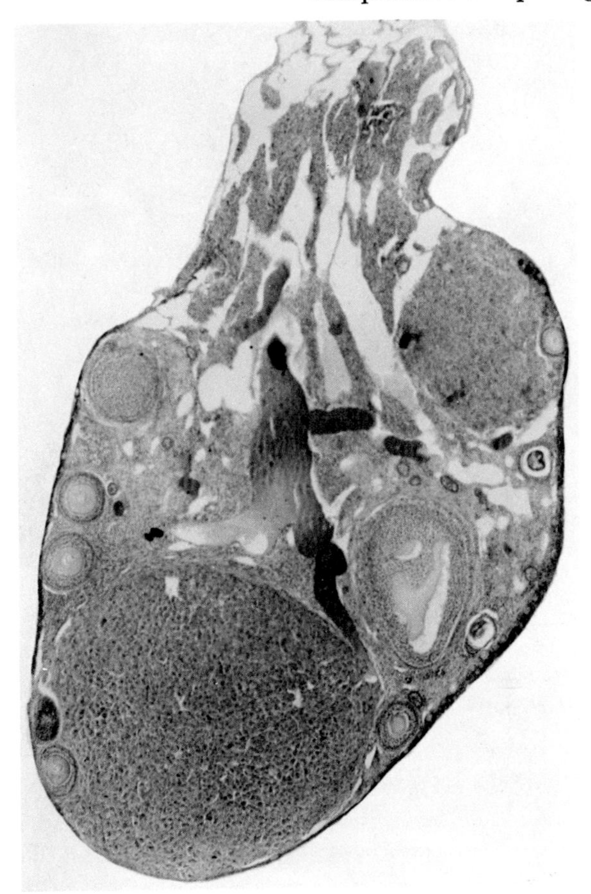

Fig. 6.82 Ovary of a shrew-mole (*Neurotrichus gibbsi*) with embryonic shield stage embryos. The gonadal adrenal type interstitial gland tissue surrounds the rete at the hilus and extends to the cortex. Lymphatic vessels and veins are greatly distended. *N. gibbsi*, 2. × 47.

the medullary cords have broken up into isolated balls of epithelium.

It is our belief, then, that in many but not all mammalian groups adrenal-cortexlike cells differentiate from mesenchyme between the cephalic end of the gonad and the mesonephros, that is, in the area of the epoophoron, and are not incorporated into the definitive adrenal cortex. Whatever the inductive mechanism is that accounts for their presence, it may in some groups also be effective in the ovarian hilus

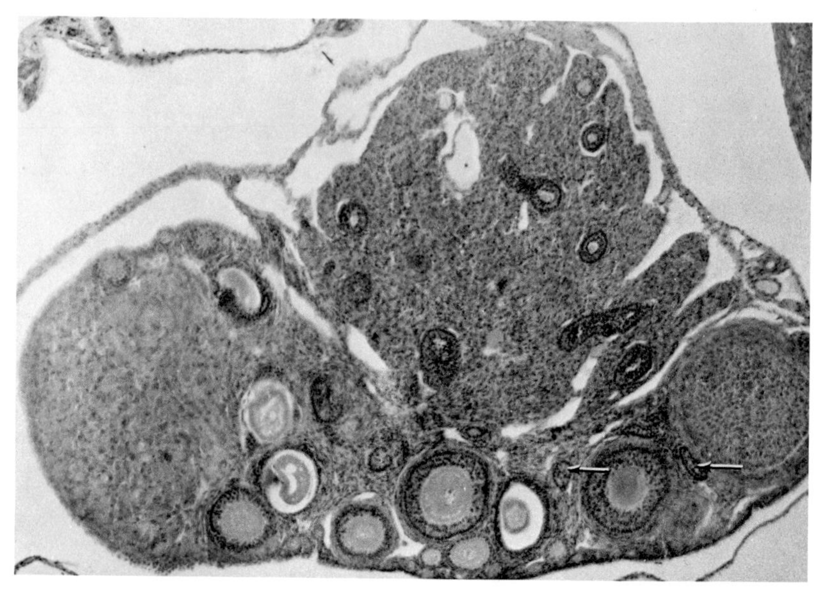

Fig. 6.83 Ovary of a shrew-mole (*Neurotrichus gibbsi*) with uterine blastocysts. The remnants of medullary cords and tubules are surrounded by gonadal adrenal type interstitial gland tissue. Note that some medullary cords (*arrows*) apparently lie in the cortex. *N. gibbsi*, 3. × 80.

and even within the ovary proper. This tissue is usually organized in the form of scattered or small groups of cells, but may appear as grossly visible accessory cortical adrenals. In some groups of mammals, this gonadal adrenal tissue is constant, in others inconstant. In the 13-lined ground squirrel, it no doubt functions and can even substitute for the normal adrenal cortex. Whether, when normally present, it has a function different from the adrenal cortex and is more closely tied to reproductive physiology is an open question, but it is known to be best developed during the nonbreeding season in the common European mole (Deanesly, 1966).

Gonadal adrenal tissue is constant in all sciurids examined, except the red squirrel, is voluminous and constant in some talpids (shrew-mole, star-nosed mole, and common European mole), is common but possibly not constant in the form of grossly visible accessory adrenals in the nine-banded armadillo (A. C. Enders and Buchanan, 1959), and is gross and constant in the patas monkey (*Erythrocebus patas*) (Conaway,

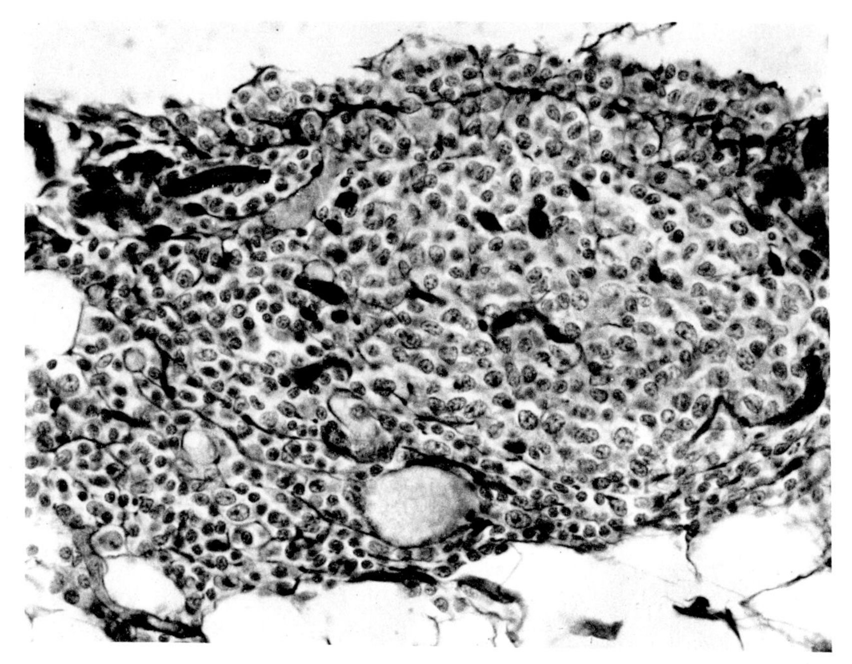

Fig. 6.84 Nodule of lymphoid cells from the epoophoron region of one of the bush babies (*Galago demidovii*). Such lymphoid nodules may occupy a position very similar to that of the gonadal adrenal masses, but should not be confused with the latter. (See also Fig. 1.33.) *G. demidovii*, 67. × 400.

1969). We have seen a gonadal accessory adrenal of large size (14-mm diameter) in a two-year-old Asiatic elephant.

Gonadal adrenal tissue is probably far more common and more constant than now recognized. In preparing ovaries for sectioning, most investigators trim off the mesovarium and so fail to see this tissue, unless it occurs in large enough masses to be grossly visible. Even then it is easily confused with grossly visible lymph nodes, which often occur in this area (Fig. 6.84). Also, when it is plentiful, as in some talpids, it has usually not been identified as adrenal type tissue.

Adneural type

Adneural type interstitial gland cells were described by Berger (1922, 1923) in the mesovarium and hilus of human ovaries and testes. He called them "sympathicotroph cells" because of their close association with involuntary nerves. Campenhout and Demuylder (1946) described them in the tunica albuginea of the adult human testis.

Since there is no evidence of any nerve-related function, we consider the term "sympathicotroph" misleading. It is also cumbersome. Because the distinctive feature of these cells is their presence near or within nerve sheaths, the term "adneural" is appropriate. Except for the obvious disadvantage of eponyms, we would prefer "Berger cells," for he was apparently the first to describe them.

Hilus cell tumors derived from these cells are fairly common in women. They are almost always strongly masculinizing. Cytologically, the cells resemble interstitial gland cells of the testis, even in their cytoplasmic crystalloids, so there seems to be little doubt that they are androgen producers. It is uncertain whether they have any other function and whether even their androgen function is ordinarily of any significance. So far as known, they normally occur in very small numbers — only a few hundred cells scattered along the course of small nerves in the cephalic portion of the mesentery of the gonad or at its hilus (Fig. 6.85). It is somewhat of a mystery why they have not been reported in mammals other than the higher primates. We have looked

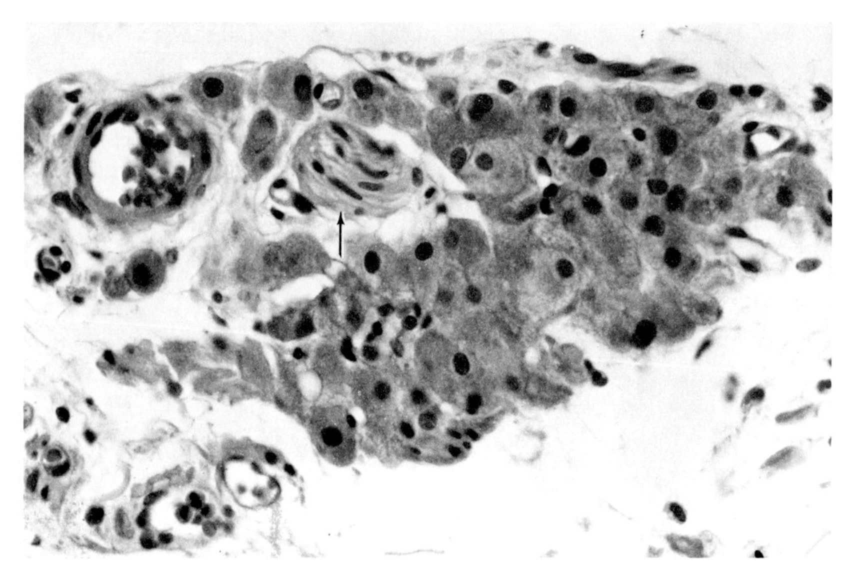

Fig. 6.85 Adneural gland cells surrounding a small nerve twig (*arrow*) in the mesovarium of a woman carrying a 51-mm fetus. Human, 51. × 400.

for them as we have searched mesovaria for gonadal adrenal tissue, but have never seen them except in anthropoids and man.

While hilus cell tumors are commonly comprised of adneural cells, it is probable that some are derived from gonadal adrenal tissue. We have seen small groups of cells in a few human mesovaria which appeared to be gonadal adrenal in character. Since adrenal tumors are also often masculinizing, this syndrome is not trustworthy as a differential feature.

Medullary cord type

Medullary cord type interstitial gland tissue is best developed only in certain groups of Carnivora, especially the Mustelidae, but also to some extent in Procyonidae (raccoons, coatimundis, etc.), Ursidae, Canidae, Viverridae (civets, genets, and mongooses), and Felidae. It seems to be derived by hyperplasia and glandular hypertrophy of the medullary cords, and often extends into the cortex where it may connect with cortical cords (Figs. 6.86–6.88). In the long-tailed weasel (*Mustela frenata*) and ermine (*M. erminea*), these epithelial cells are differentiated as active gland cells from proestrus through preimplantation pregnancy (Fig. 6.89). They begin to dedifferentiate just before implantation and are in their least-differentiated state during late pregnancy and lactation (Fig. 6.90). They begin redifferentiation during anestrus.

In many carnivores, it is by no means clear that these interstitial gland cells are actually derivatives of medullary cords, but in the mink (*Mustela vison*), the long-tailed weasel, and the ermine the derivation is fairly obvious. In serial sections, the network of glandular cords can be traced to their junction with the rete (Fig. 6.88). Also, some infantile and juvenile ovaries of weasels have shown the extensive network of embryonic medullary cord epithelium. In most of the other carnivores, the homologous tissue tends to be confined chiefly to the outer medullary and the inner cortical zones. Its cordlike arrangement is usually evident, but its connection with the relatively scarce and fragmented deeper medullary cords and, through these, with the rete is not easy to demonstrate. In fact, in most cases these connections probably do not exist, at least in the adult ovary.

Another problem in interpreting the morphological relations of this tissue is that, even in the mink and weasels, where its relation to medullary cords is clear, the picture is confused by the fact that similar cell types develop from the theca interna of atretic follicles (see Fig. 6.87).

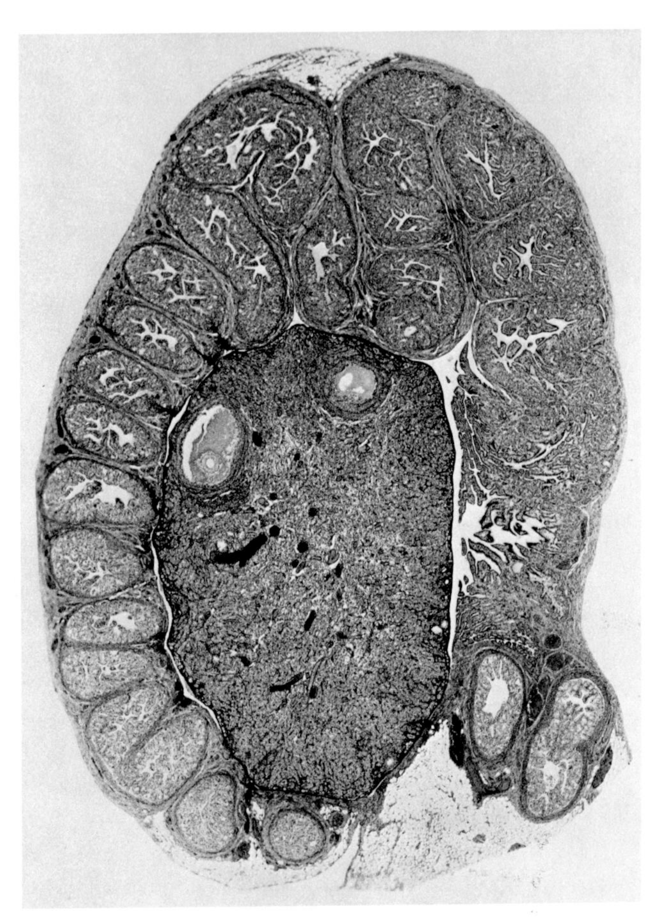

Fig. 6.86

Fig. 6.86 Interstitial gland tissue occupies most of the ovary of a non-pregnant mink (*Mustela vison*) in March with nearly ripe follicles. The ovarian bursa was cut parallel to the encircling oviduct. Note the inwardly directed funnel. (Compare with Fig. 1.18.) *M. vison*, 9. × 13.

Fig. 6.87 Cortical zone of a mink (*Mustela vison*) ovary in late March, showing the medullary cord type of interstitial gland tissue. At the right is an atretic follicle whose granulosa is partially lutealized (accessory corpus luteum) and whose theca interna is developing into interstitial gland cells of the thecal type. These cells intergrade with the medullary cord type. The zonae pellucidae (*arrows*) of atretic eggs have gland cells grouped around them which may be of thecal origin, but are indistinguishable from medullary cord type interstitial cells. (See text pp. 203–7.) *M. vison*, 7. × 120.

Fig. 6.88 A mink (*Mustela vison*) ovary, showing medullary cord type interstitial gland tissue continuous with the rete (*r*). *M. vison*, 4. × approx. 400.

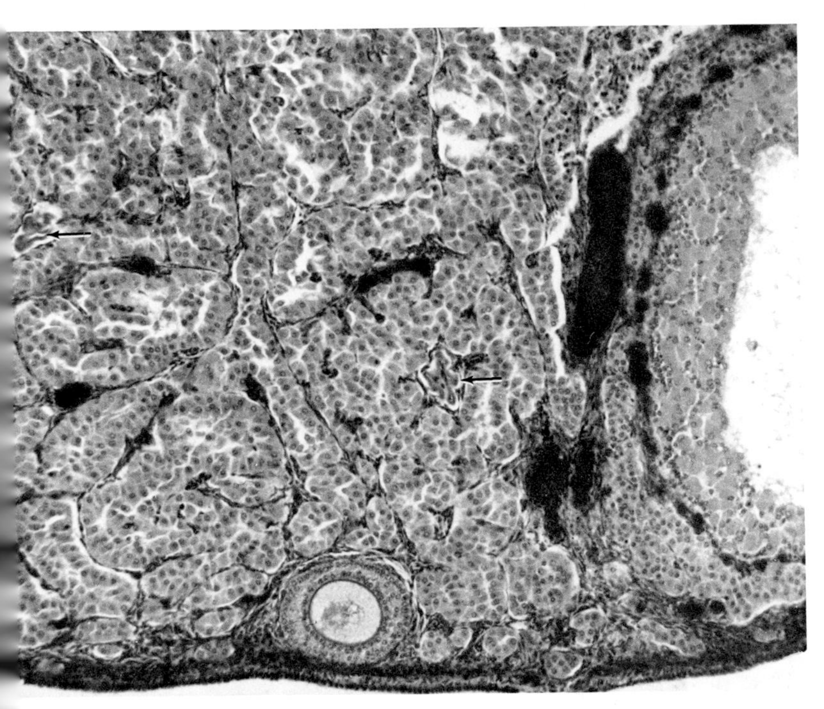

ig. 6.87

Fig. 6.88

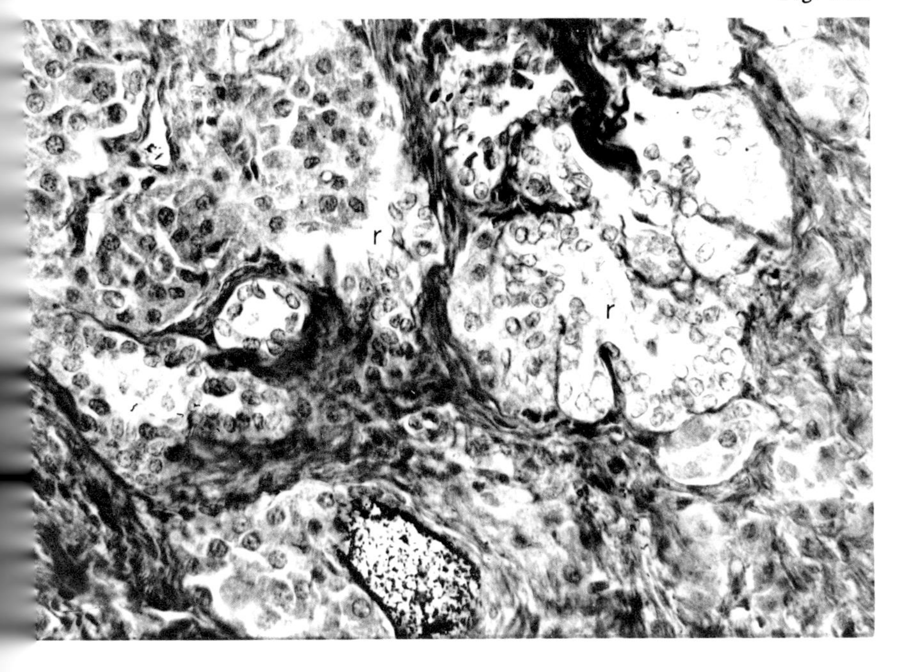

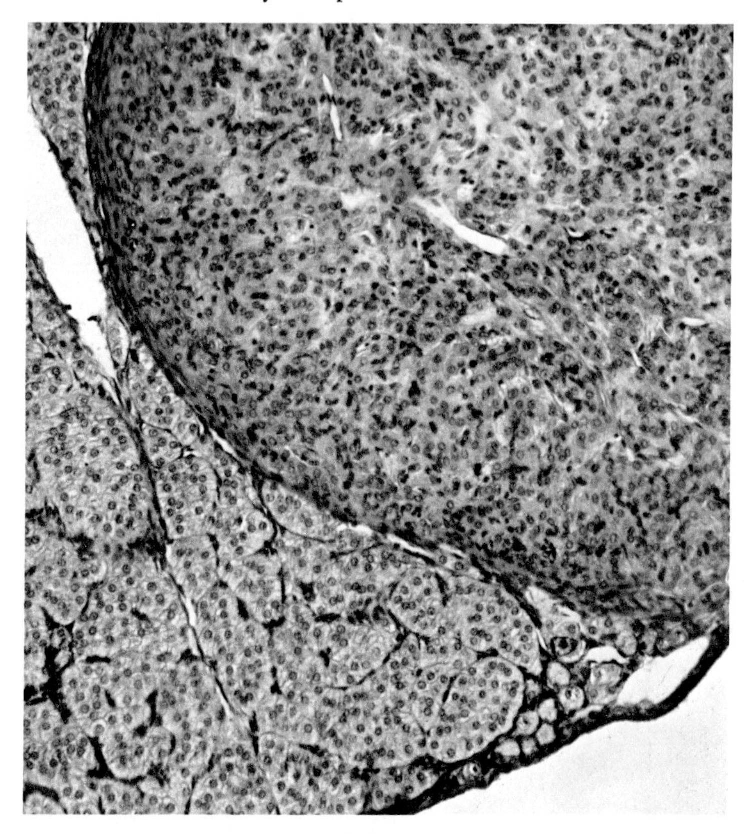

Fig. 6.89 Corpus luteum and medullary cord type interstitial gland tissue of a long-tailed weasel (*Mustela frenata*) in September, with "delayed" uterine blastocysts. *M. frenata*, 81. × approx. 150.

This thecal type of interstitial gland tissue does not take the form of cords, at least at first, but as it differentiates it eventually becomes so intimately intermingled with the medullary cord type that it is indistinguishable from it by cursory examination. In fact, one gets the impression that it probably becomes rearranged and incorporated as part of the medullary cord net. This is another example of the pluripotential nature of ovarian cells: epithelial cells give rise to the medullary cord type of interstitial gland cells; stromal cells of the theca interna of atretic follicles give rise to thecal type interstitial gland cells; the two gland cell types appear identical by any criterion so far used, and there is no

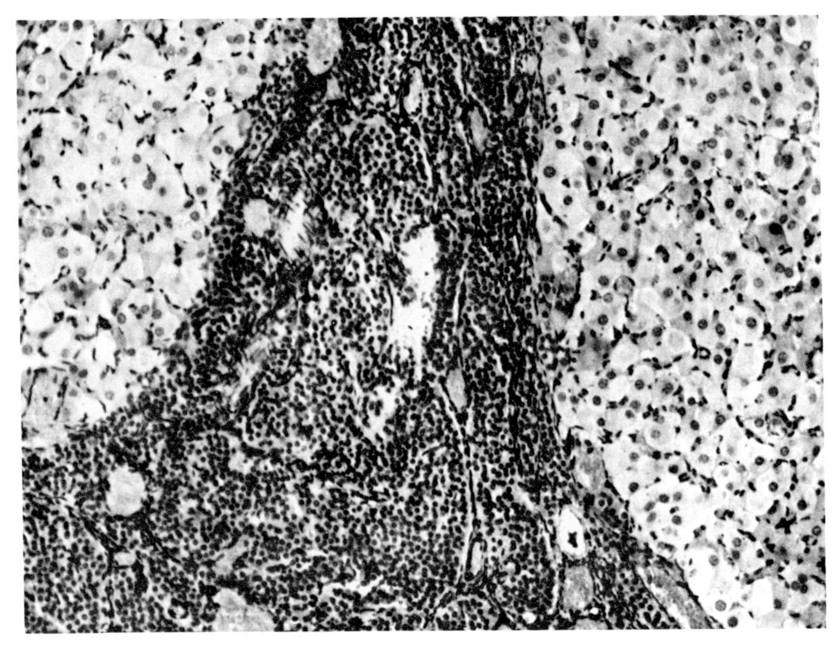

Fig. 6.90 Corpora lutea and medullary cord type interstitial gland tissue of a long-tailed weasel (*Mustela frenata*) in March with 21-mm embryos. Note the differentiated luteal cells and the involuted interstitial cells. (Compare with Fig. 6.89.) *M. frenata*, 143. × approx. 150.

reason to believe that they are not identical in function, for they also show the same cyclic changes in relation to the reproductive cycle.

A less conspicuous glandular differentiation of medullary cord epithelium occurs in some moles. (See the synoptic tables and their supplementary notes.)

Rete type

Rete type interstitial gland tissue has been seen in only one genus of bats — *Uroderma* (the tent-building bat). In *Tadarida* (free-tailed bats), *Myotis* (mouse-eared bats), *Eptesicus* (big brown bats), and *Eumops* (mastiff bats), the rete is large and complex, and many of its epithelioid cells appear not to border its lumen (Fig. 6.91). However, in *Uroderma* the rete area is only recognizable as such by its relation to the tubules of the epoophoron (Fig. 6.92). Its epithelial cells are typical large gland cells and show little evidence of their primitive arrangement as the lining of

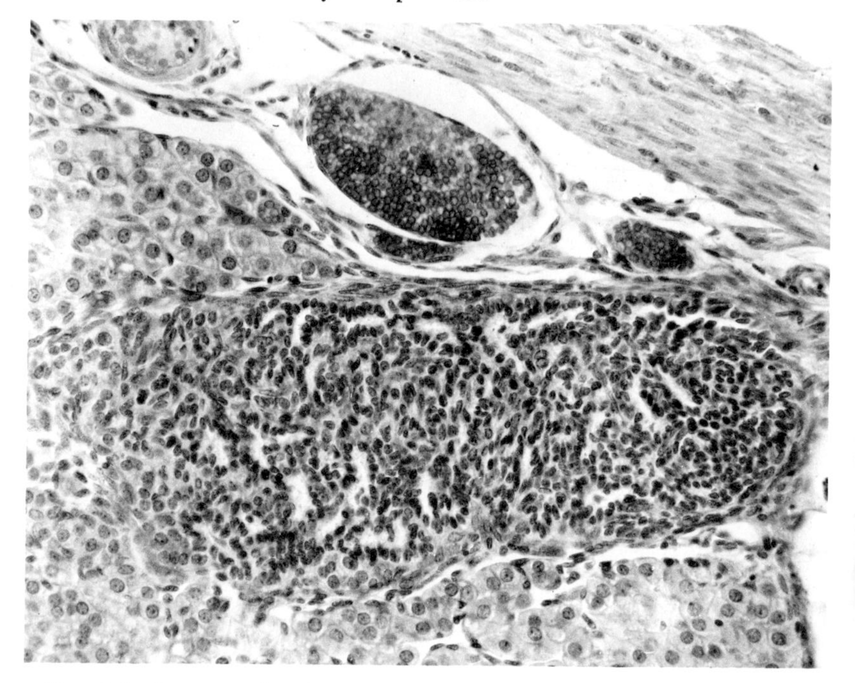

Fig. 6.91 Rete of a South American mouse-eared bat (*Myotis chiloensis*), showing the excessive number of epithelial cells characteristic of several genera of Microchiroptera. *M. chiloensis*, 18. × 250.

a labyrinthine lumen, or "rete." Only a study of the development of this epithelium can prove that it is actually derived from the rete. Too few specimens have been closely examined to give significant information on the relation of the degree of differentiation of this tissue to the breeding cycle; nor do we know anything of its histochemistry or ultrastructure. Highly developed ovarian retia are common in many groups of mammals, but in no group other than these few bats have we seen any evidence of its differentiation to resemble an endocrine gland. Obviously, no speculation as to the significance of rete type interstitial gland tissue is warranted at this time.

In a few species, for example, jumping mice (*Zapus*), guinea pig, and white-tailed deer (*Odocoileus virginianus*), the rete epithelium is columnar and the lumen wide, thus indicating exocrine secretion into the rete lumen. Archbald et al. (1971) saw indications of a holocrine type secretion during diestrus in what they considered the rete of the ovaries of

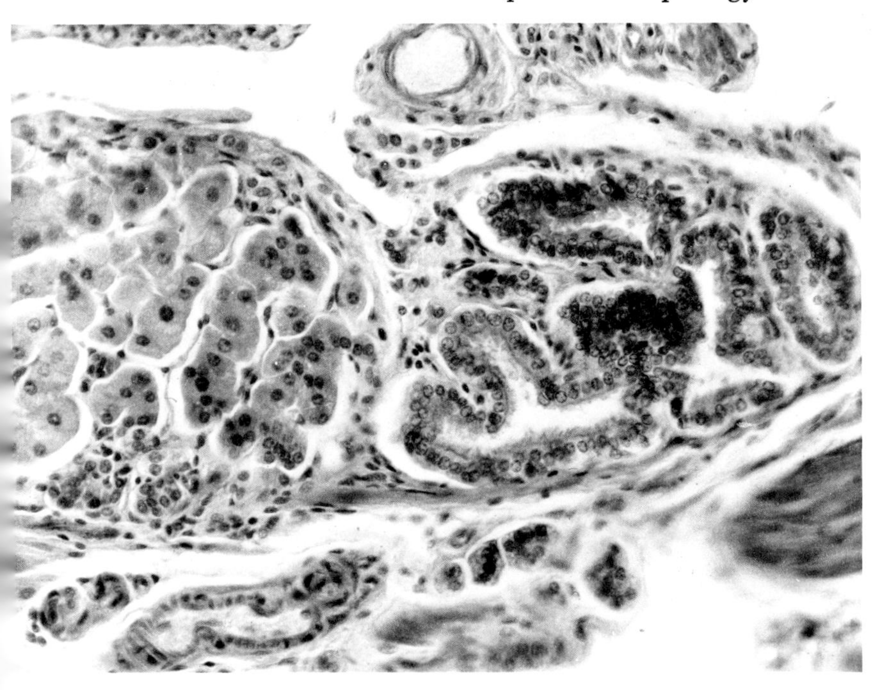

Fig. 6.92 Rete type interstitial gland tissue (*left*) and epoophoron tubules (*right*) from the hilus and mesovarium of a tent-building bat (*Uroderma bilobatum*) with an 8-mm embryo. *U. bilobatum*, 1. × 250.

heifers. However, their figures indicate that the structures they studied were medullary cords and tubules, not the rete.

Luteal gland

The luteal gland, or corpus luteum, is the best understood of all the glandular tissues of the mammalian ovary. With one exception, all recent workers agree that luteal cells are derived from the follicular epithelium (granulosa cells), supplemented in some cases by differentiation of surrounding thecal or stromal cells. P. N. O'Donoghue (1963) maintained that the luteal cells of one of the tree hyraxes (*Dendrohyrax arboreus*) are derived exclusively from hypertrophied cells of the theca interna. His field-collected material was poorly fixed, and hence is very difficult to interpret. We have several series of sections (also not well fixed) of the two other genera of hyraxes, but none of them has recently ruptured follicles, which are crucial for determining corpus luteum origin. It is true that their corpora lutea are unusual in that they show some indefinite zonation and contain a variety of cells other than the typical large luteal cells, but we see no evidence of origin from thecal cells. A final decision must await the study of well-preserved specimens during early stages of corpus luteum development.

It is commonly stated that theca interna cells give rise to luteal gland cells. If by "theca interna cells" undifferentiated stromal type cells are meant, we agree. If differentiated thecal gland cells are meant, we disagree. As mentioned above, where adequate material is available, differentiated thecal gland cells clearly degenerate immediately and rapidly after ovulation. However, the cells of the periphery of the theca interna are not always differentiated even in a ripe follicle, and these, along with other stromal cells, are likely to differentiate into paraluteal cells and to true luteal cells in the case of ovulated follicles, or into thecal type interstitial gland cells in the case of follicular atresia.

In the North American porcupine, paraluteal cells transform into luteal gland cells in great numbers, and a definite paraluteal zone persists during early pregnancy (see Fig. 6.46) (Mossman and Judas, 1949). However, instead of disappearing by full conversion to luteal cells and by failure of continued paraluteal differentiation of surrounding stroma, the paraluteal zone of this species ceases to transform into luteal cells and differentiates as a narrow zone of interstitial gland cells identical in appearance with the thecal type interstitial gland tissue in the same ovary (Fig. 6.93). Weir (1970) shows a similar zone of interstitial cells in the green acushi (*Myoprocta pratti*). These periluteal zones of interstitial gland tissue remain distinct from the rest of the corpus until it degenerates. This, of course, raises the question as to the possible functional similarity between paraluteal cells and interstitial cells. Could they both be producing an analog or a precursor of progesterone?

The corpora lutea of Perissodactyla (odd-toed hoofed mammals), Artiodactyla (even-toed hoofed mammals), and Cetacea (whales, dolphins, porpoises, and related forms) pose other problems. All have two distinct classes of differentiated luteal cells, the usual large polyhedral type and a small irregularly fusiform or stellate type. The latter lie along the terminal branches of the blood vessels throughout the lobules of luteal tissue, not just at the periphery or in the larger septa as is characteristic of the usual type of paraluteal cell (Figs. 6.94–6.99). These small luteal cells also differ from typical paraluteal cells in that they show little evi-

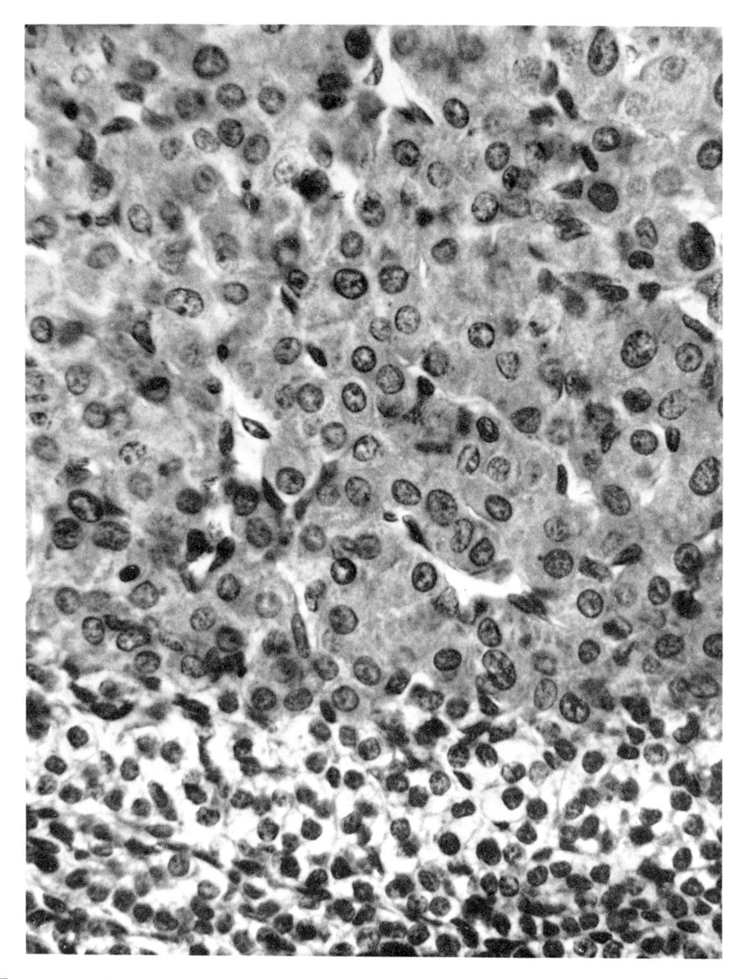

Fig. 6.93 Margin of the year-old corpus luteum from a previous pregnancy in a porcupine (*Erethizon dorsatum*) with an embryo in the embryonic disc stage. The corpus luteum of the current pregnancy is shown in Figure 6.46. The zone of small, lightly stained cells is derived from paraluteal cells and is identical with thecal type interstitial gland cells. *E. dorsatum*, 29R. × 600.

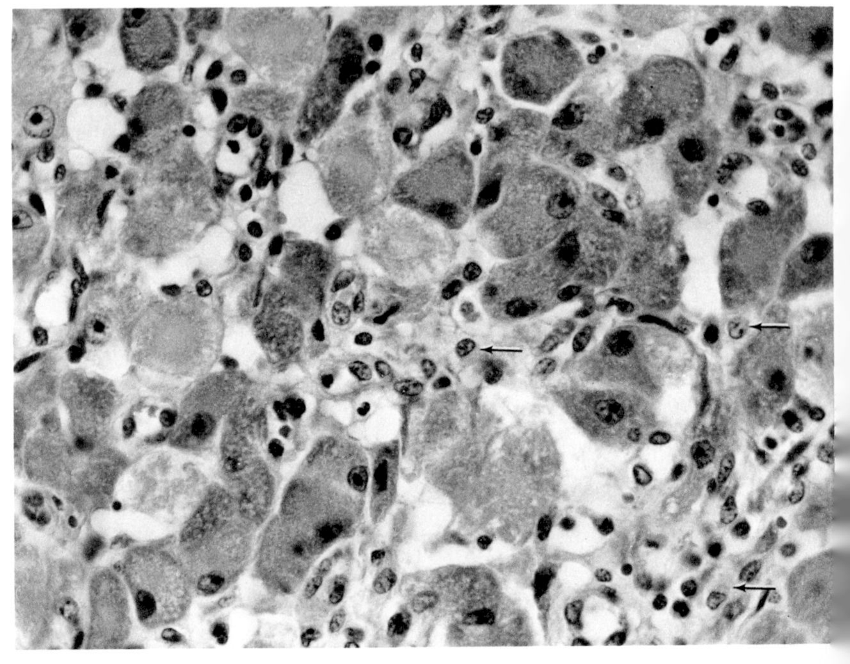

Fig. 6.94

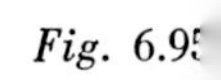

Fig. 6.95

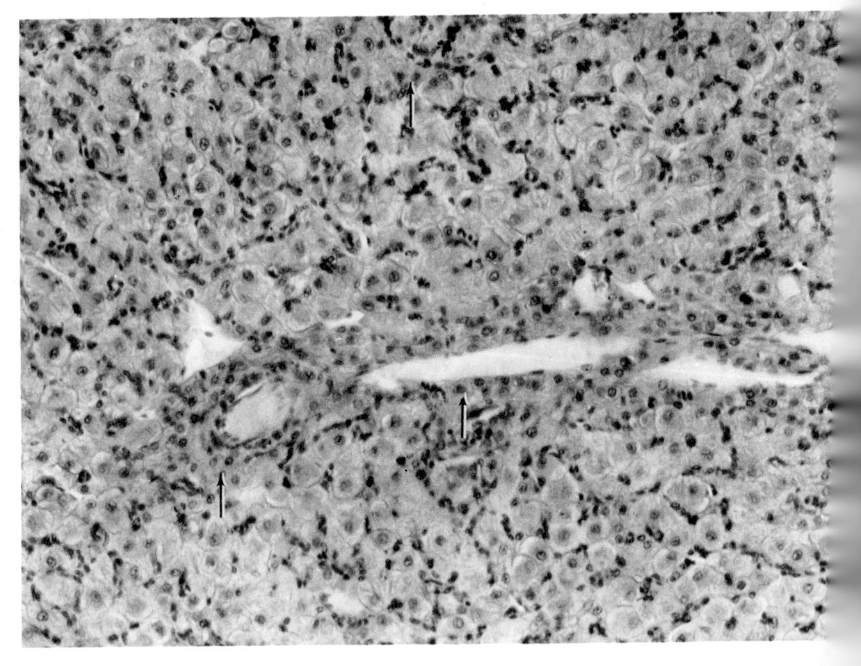

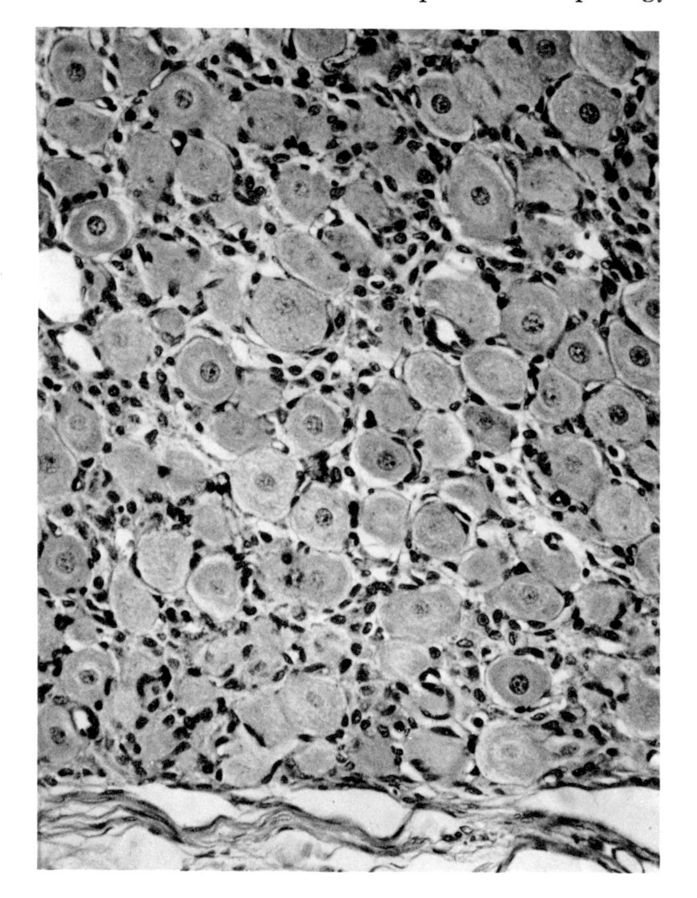

Fig. 6.96

Fig. 6.94 A corpus luteum of the 11th day of the estrous cycle of a mare (*Equus caballus*), showing the ordinary large luteal cells and the small ones (*arrows*) close to the capillaries. *E. caballus*, 69–2972. × 400. (Courtesy of O. J. Ginther.)

Fig. 6.95 Corpus luteum of a pig (*Sus scrofa*). The small luteal cells (*arrows*) are grouped along small blood vessels. × 155.

Fig. 6.96 Periphery of a corpus luteum of a mouse deer (*Tragulus javanicus*) in mid-gestation, showing uniform distribution of the large and small luteal gland cells. *T. javanicus*, 2. × 250.

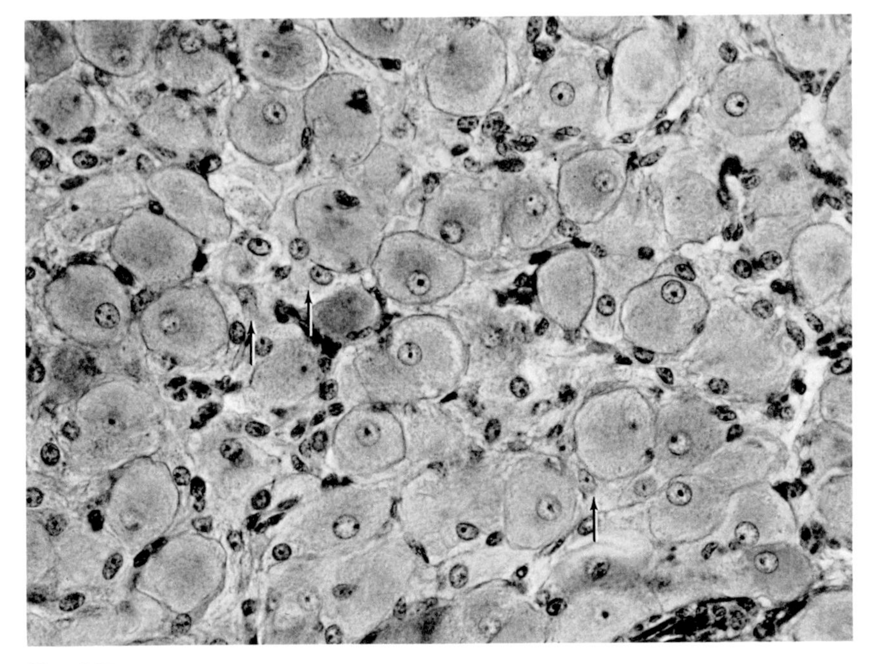

Fig. 6.97

Fig. 6.98

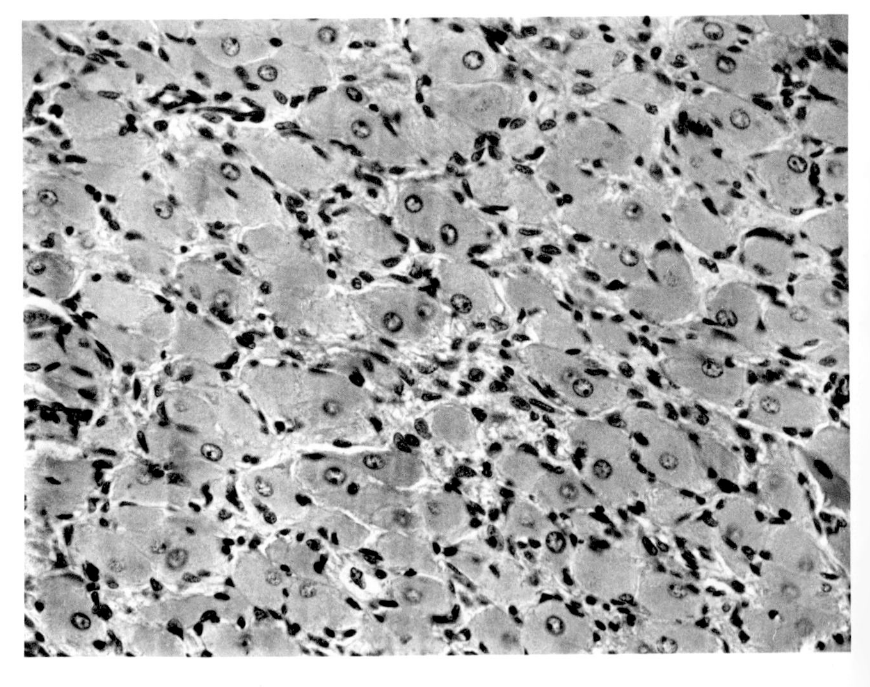

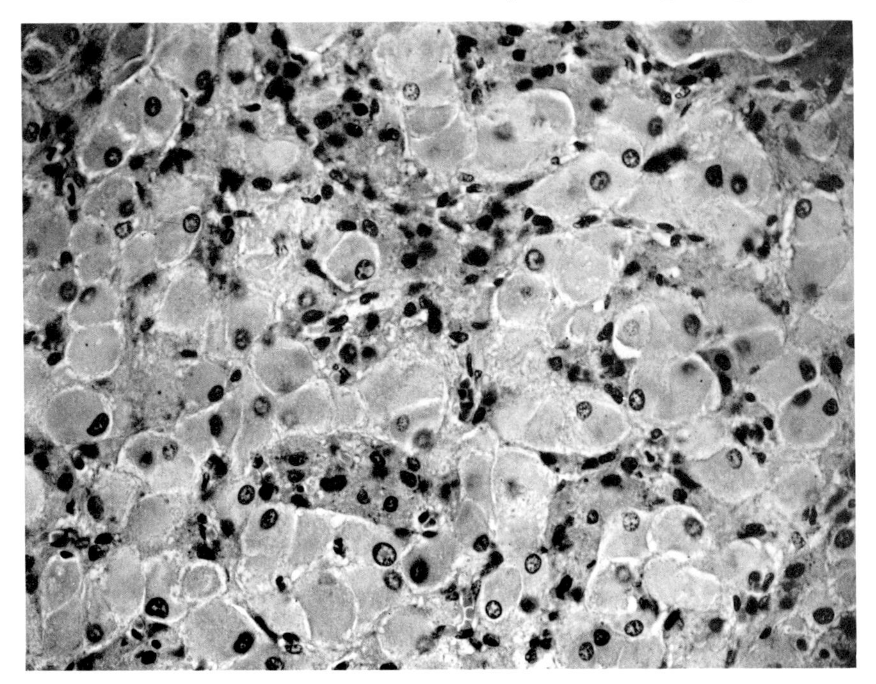

Fig. 6.99

Fig. 6.97 Corpus luteum of a white-tailed deer (*Odocoileus virginianus*). Note the typical large spheroidal luteal cells with distinct borders and the small irregularly shaped luteal gland cells (*arrows*). *O. virginianus*, 532. × 270.

Fig. 6.98 Corpus luteum of a pronghorn (*Antilocapra americana*), showing the two types of luteal gland cells. *A. americana*, 54–108. × 250.

Fig. 6.99 Corpus luteum of a white whale (*Delphinapterus leucas*) with a small embryo, showing the two types of luteal cells. *D. leucas*, 101. × 200.

dence of being transitional stages between undifferentiated stromal cells and the fully differentiated luteal cells of the larger polyhedral sort. We are uncertain of their origin, but they most likely arise from stromal cells which invade the early granulosa luteal mass along with the blood vessels. At this time, there is probably some transformation of these cells of stromal origin into the large granulosa type of luteal cell that occurs in many other mammals, but they certainly do not do this to any extent once the corpus is vascularized, for they remain as a distinctively different cell population until luteal involution takes place. Sinha, Seal, and Doe (1971*a*, 1971*b*), and others for many years, have called them "theca lutein cells" in contrast to "granulosa lutein cells." We prefer to speak of them as "secondary" luteal cells and of the others as "primary" luteal cells. They are secondary in time of appearance, in size and conspicuousness, and quite possibly in function. They are probably only in part derived from undifferentiated thecal cells. Certainly, in many mammals some of the primary ones are derived from stroma as well as from granulosa. Hence, *primary* and *secondary* are more appropriate and less-misleading terms.

The existence of two classes of luteal cells in artiodactyls was recognized by Corner (1915, 1919) in swine. He considered the larger ones to be of granulosa cell origin and the smaller ones to be modified "epithelioid" cells of the theca interna. He also correlated the various cytological appearances within these two classes with the stages of the reproductive cycle, and gave excellent reviews of the literature concerning luteal cell origin.

We have found these two classes of luteal cells in the horse, pig, llama (*Lama peruana*), mouse deer (*Tragulus javanicus*), wapiti (*Cervus canadensis*), white-tailed deer, pronghorn (*Antilocapra americana*), cow, bison (*Bison bison*), kob antelope (*Kobus kob*), goat, sheep, and in the white and pilot whales (*Delphinapterus leucas* and *Globicephala melaena*). Therefore, it seems likely that this is a characteristic feature of Perissodactyla, Artiodactyla, and at least of the toothed whales (Odontoceti). It may be further evidence of a phylogenetic relationship between these groups (cetaceans, perissodactyls, and artiodactyls), but it is remarkable that such a seemingly minor feature would have persisted so universally and so long in each group.

The corpora lutea of the European mole, beaver (*Castor canadensis*), and the African springhare (*Pedetes capensis*) have two types of luteal cells, but they are not as distinctly different in size and morphology as those of the hoofed mammals and whales. More study will be necessary to determine whether these are actually two distinct types or simply opposite ends of an intergrading series.

There is evidence that a domestic mare may ovulate during the second to fifth month of gestation and that these follicles form secondary corpora lutea, which supplement the degenerating primary corpus (Cole, Howell, and Hart, 1931; Amoroso et al., 1948; Amoroso and Rowlands, 1951). To our knowledge, no histological evidence as to the nature of these secondary corpora lutea has ever been published. Kupfer (1928) studied the fresh ovaries of over 260 donkeys and more than 100 horses. He made no histological sections, but sliced the ovaries grossly; his published fine color plates show the whole and sliced ovaries at all stages of the estrous and pregnancy cycles. Nowhere does he mention or illustrate secondary corpora lutea, but in all cases he does describe a small degenerating primary corpus luteum at term. We have microscopic sections of a pair of ovaries from a Burchell's zebra mare (*Equus burchelli*) at almost full term in which there is one small degenerate corpus luteum. This and Kupfer's work casts some doubt on the identity of the alleged secondary corpora of the mare. Are they actually developed from ovulated follicles, or are they accessory corpora lutea formed by lutealization of atretic follicles, or perhaps masses of thecal type interstitial gland tissue? A thorough histological study should be made of equine ovaries. It should be correlated with other data on the reproductive cycle, including particularly the fluctuations in circulating gonadotropins and steroid hormone output from the ovary.

Secondary corpora lutea may also be considered to occur in the mink, in which a second ovulation takes place after a short "delay" in implantation of the embryos from the first ovulation (R. K. Enders, 1952). However, here the second set of eggs usually contributes most of the embryos, whereas in the mare a second set of eggs, even if present, certainly does not. Because embryos are derived from both ovulations in the mink, both sets of corpora are usually considered primary. Naturally occurring true secondary corpora lutea derived from follicles which ovulated spontaneously during pregnancy are yet to be demonstrated.

At about the same time as the primary ones, accessory corpora lutea are formed from follicles which do not ovulate but are undergoing atresia. This is well illustrated in the porcupine (Figs. 6.100 and 6.101) (Mossman and Judas, 1949), but also occurs to some extent in several other groups including some other hystricomorph rodents, the mink, and

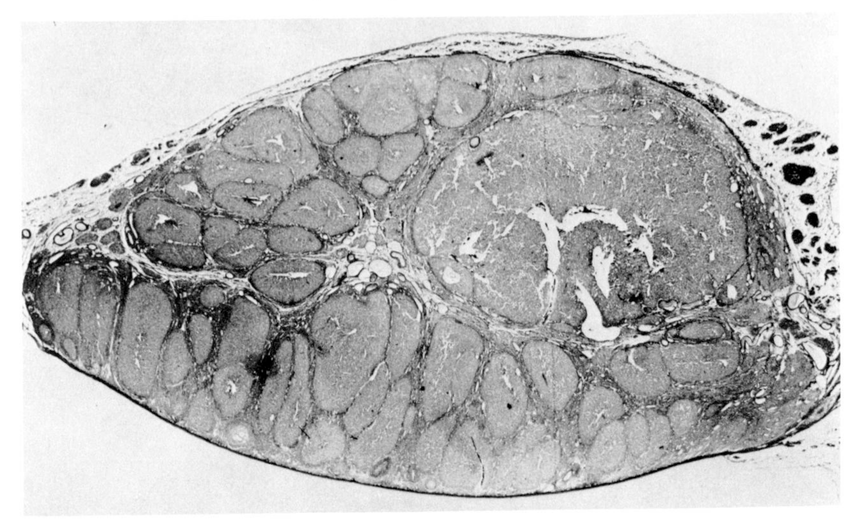

Fig. 6.100 The ovary of the pregnant side of a porcupine (*Erethizon dorsatum*) in March with a 135-mm fetus. Note the retention of accessory corpora lutea. *E. dorsatum*, 159. × 6.5.

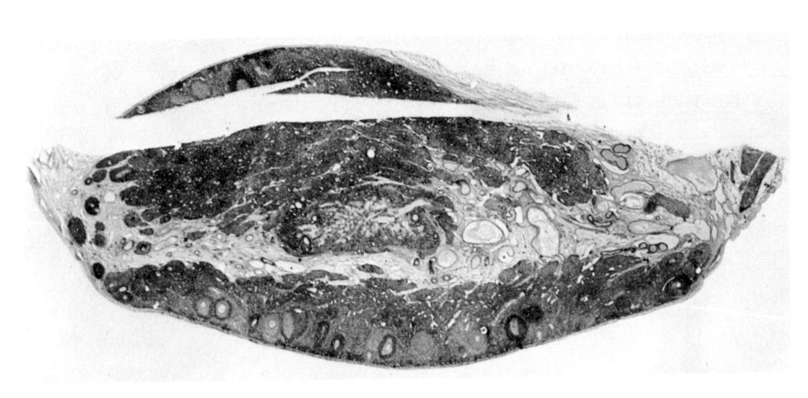

Fig. 6.101 The ovary of the nonpregnant side of the porcupine shown in Figure 6.100. Here the accessory corpora lutea are almost completely absent. Note, also, the centrally located corpus albicans and the numerous corpora atretica comprised of thecal type interstitial gland tissue. *E. dorsatum*, 159. × 6.5.

man (see Figs. 5.7, 5.8, and 6.87). Apparently, some follicles are so sensitive to the lutealizing factors that they respond before ovulation takes place. We have seen as many as three partially lutealized follicles in one human ovary. In one case, the accessory luteal mass had a diameter about one-third that of the primary corpus (see Figs. 5.7 and 5.8); in other words, it was clearly visible and could have been mistaken by gross examination for a primary corpus.

The behavior of the accessory corpora lutea in the porcupine is remarkable. They are very numerous in both ovaries from estrus until about implantation time. The porcupine ovulates from only one follicle, and implantation is always on the same side. The accessory corpora of the pregnant side behave exactly as the primary corpus, growing during early pregnancy partly by hyperplasia and partly by accretion from paraluteal cell zones and remaining as accessory corpora lutea for more than one year, that is, until about implantation time of the next pregnancy when, like the primary corpus, they rapidly involute and disappear completely. On the nonpregnant side, they involute completely at about the time of implantation, so that from early gestation until the next proestrous period, while this ovary may rarely contain one or two small accessory corpora lutea, there is ordinarily little but interstitial gland tissue in it, and it is distinctly smaller than the ovary of the pregnant side (see Figs. 6.100 and 6.101). This offers evidence of a local effect of the uterus on the ovary, but the nature and route of the uterine influence are still unknown. For discussions of this problem, see L. L. Anderson (1966) and Fischer (1967). Tam (1970), using various techniques of chemical analysis, produced evidence that the function of primary and accessory corpora is the same.

Another form of luteal tissue that may be called accessory occurs in pregnant juvenile eastern chipmunks. These animals breed when only about three months old, while their ovaries are infantile in character except for a few follicles which develop from medullary cords (Fig. 6.102). When these medullary follicles ovulate, they develop into corpora lutea in the usual manner, but luteal gland tissue also develops throughout the medulla, apparently from remaining epithelium of medullary cords. This accessory luteal tissue so crowds the ovary that the outlines of the primary corpora are obscured, and the whole medulla appears to be an irregularly lobulated mass of luteal tissue.

So-called aberrant corpora lutea were described in the rhesus monkey by Corner, Bartelmez, and Hartman (1936). Koering (1969), who also

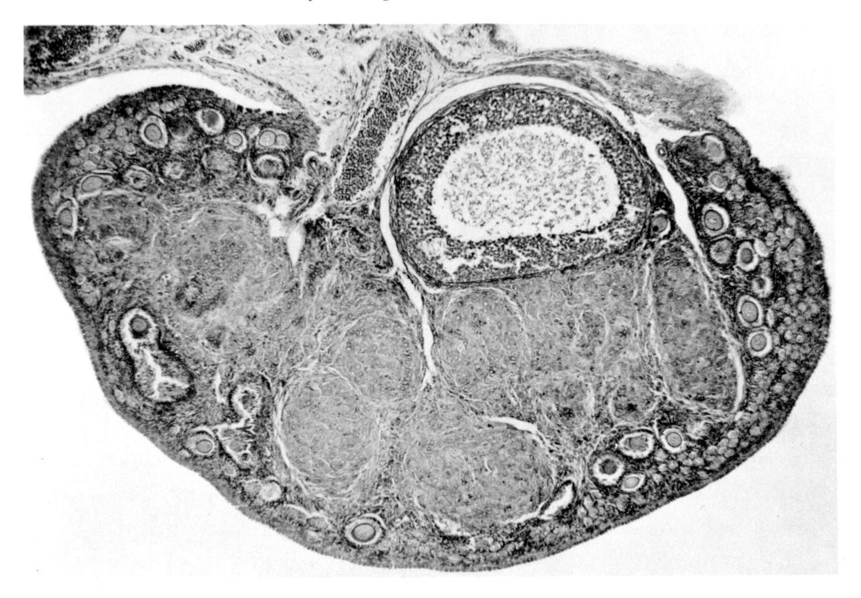

Fig. 6.102 Diffuse luteal tissue in the medulla of the ovary of a pregnant juvenile eastern chipmunk (*Tamias striatus*) in July. *T. striatus*, 93. × 60.

observed them in the rhesus monkey, had evidence that they are corpora which persist from one menstrual cycle into another and perhaps take a new "lease on life" in the second cycle. They are not constant, and their significance is a mystery.

Ultrastructure of the ovary

Weakley (1969*a*) described the differentiation of the surface epithelium of the hamster ovary from full term to one year. Lobulation of the nuclei, and the number of microvilli, pinocytotic vesicles, fluid-filled vesicles, and free ribosomes increase to maturity. Apparently, fluid transport is an increasingly important function of the epithelium.

Electron microscopy has revealed the presence of ciliated cells in unexpected places. Appley and Richter (1970) found isolated cilia on follicular cells of the gray mouse-eared bat (*Myotis grisescens*). Motta, Takeva, and Palermo (1971) described cilia on cells of the follicular epithelium and stroma, and of the theca interna of both normal and atretic follicles. Luteal and "interstitial" cells had only centrioles. They studied ovaries of the domestic rabbit, cat, guinea pig, laboratory mouse and rat, and fetal and adult humans. Since the theca interna of atretic follicles of all of these contains thecal type interstitial gland cells, their statement that cilia did not occur on "interstitial" cells leaves some doubt as to the cells they included in this category.

Evidence for the presence of contractile elements has been strengthened by O'Shea (1970), who found bundles of fibrils with associated dense bodies, such as are characteristic of smooth muscle, in cells of the theca externa and of the capsule of corpora lutea of the rat. He believed that these cells develop in situ from undifferentiated stroma, but admitted that they may exist throughout the stroma and simply become passively arranged around the follicles during their expansion.

Several key studies have been made of the ultrastructure of the oocytes and developing follicles in the domestic rabbit (Hashimoto et al., 1960), pig (Bjersing, 1967*a*), hamster (Weakley, 1969*a*, 1969*b*), and mouse (Odor and Blandau, 1969*a*, 1969*b*). Norrevang (1968) discussed these matters from the comparative aspect. Besides the classical questions concerned with oogenesis, these studies have dealt with the origin of the zona pellucida and follicular fluid, and have indicated that follicular cells are the primary factors in the latter two processes. Papers dealing more strictly with the oocytes themselves have been written by Adams and Hertig (1964) on the guinea pig; Zamboni et al. (1966) on a human pronuclear stage; Stegner (1967), who studied stages of fertilization, cleavage, and the blastocyst; T. G. Baker and Franchi (1967), who chiefly discussed nuclear structure; Hertig and Adams (1967) and Hertig (1968), who were concerned with oocyte organelles, particularly Balbiani's body; and Weakley (1968), who described nonmembranous lamellae in the oocytes of several species.

As would be expected, the only glandular tissue of the ovary that has received much attention from electron microscopists is the corpus luteum. The differentiation and structure of the luteal cells of the domestic rabbit were described by Blanchette (1966*a*, 1966*b*). Green and Maqueo (1965) and Green, Garcilazo, and Maqueo (1967, 1968) showed that human luteal cells are cytologically practically unimpaired at term. They also demonstrated the intercellular and intracellular canaliculi. Incidentally, Sinha, Conaway, and Kenyon (1966) described an elaborate labyrinth of fluid-filled spaces of variable diameter ranging from microscopic to several millimeters in corpora lutea of the sea otter (*Enhydra lutris*). If these gross spaces of the sea otter are exaggerations of the canaliculi of other species, then aspiration and analysis of their contents are in order.

Bjersing (1967*b*) showed that, in the luteal cells of sows, whorls of agranular endoplasmic reticulum are most numerous during the period of highest progesterone secretion, and suggested that luteal cell secretory activity could be estimated at this time. Breinl, Andrzejewski, and Tonutti (1967) provided evidence of the cytological secretory competence of the rat luteal cell during lactation.

There are excellent papers on the fine structure of luteal cells by Priedkalns and Weber (1968*a*, 1968*b*) on the cow; Sinha, Seal, and Doe (1971*a*, 1971*b*) on the white-tailed deer and raccoon; Goodman et al. (1968), Cavazos et al. (1969), and Belt et al. (1970) on the sow; Motta (1969), Gillim, Christensen, and McLennan (1969), and Crisp, Dessouky, and Denys (1970) on the human; Bjersing (1970*a*, 1970*b*) on the sheep; and Crombie, Burton, and Ackland (1971) on the guinea pig.

Sinha has kindly sent us electron micrographs of luteal cells of the nine-banded armadillo (Fig. 6.103), beaver (*Castor canadensis*), black bear (Fig. 6.104), raccoon (Fig. 6.105), dog (Fig. 6.106), crab-eater seal (*Lobodon carcinophagus*), white-tailed deer (Figs. 6.107 and 6.108), pig (Fig. 6.109), and sheep (Fig. 6.110). These indicate that luteal cells in general have the characteristics of steroid secretors. Complexly plicate or microvillous surfaces facing intercellular spaces or tubules also appear to be characteristic. The most unique cells in this group are those of a black bear in early pregnancy, probably taken during early hibernation; these have very large numbers of conspicuous crystals often 4 μ or more in length (Fig. 6.104).

We know of only one paper describing the ultrastructure of theca interna cells (Santoro, 1965). Santoro studied the domestic rabbit, but his figures give little indication that he was dealing with differentiated thecal gland of ripe follicles.

There are only three papers known to us which specifically consider the ultrastructure of interstitial gland cells. Muta (1958), studying mouse ovarian interstitial tissue, pointed out smooth endoplasmic reticulum and modified mitochondria similar to those of the adrenal cortex and luteal cells. Davies and Broadus (1968) found that ovarian interstitial cells of the domestic rabbit are of the steroid secretor type, but they could find no changes in them associated with pregnancy. Merker and Diaz-Encinas (1969) described the ultrastructure of the interstitial gland cells of the juvenile rat and rabbit after stimulation with gonadotropic hormones.

Obviously, much more work on all luteal cells must be done before many clear generalizations can be made as to their cytological changes in correlation with breeding cycles and hormone output. While a beginning has been made on this one gland, little or nothing is known of the ultrastructure of any of the others.

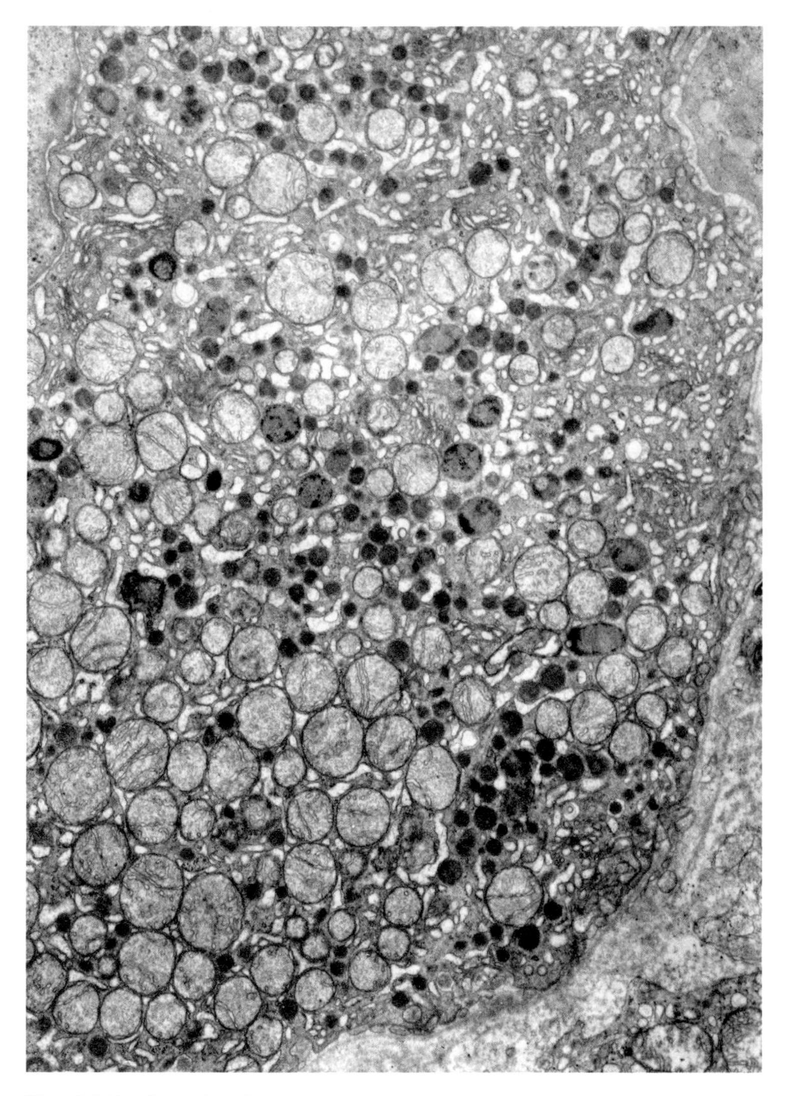

Fig. 6.103 Luteal cells of a pregnant nine-banded armadillo (*Dasypus novemcinctus*) with 105-mm fetuses. *D. novemcinctus*, 4–19–68. × 11,385. (Courtesy of Akhouri Sinha.)

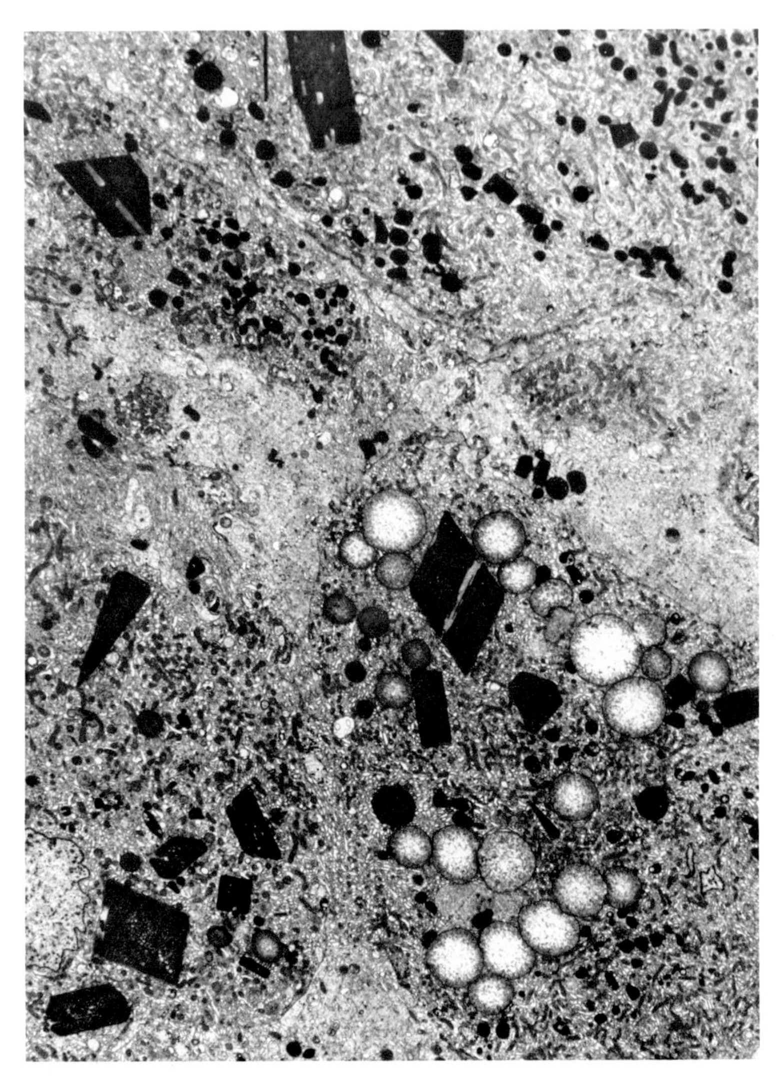

Fig. 6.104 Luteal cells of a pregnant black bear (*Ursus americanus*) with 190-mm fetuses. Prominent crystalloids apparently are characteristic of this species. They can be readily seen with the light microscope. *U. americanus*, 1–9–69. × 2,600. (Courtesy of Akhouri Sinha.)

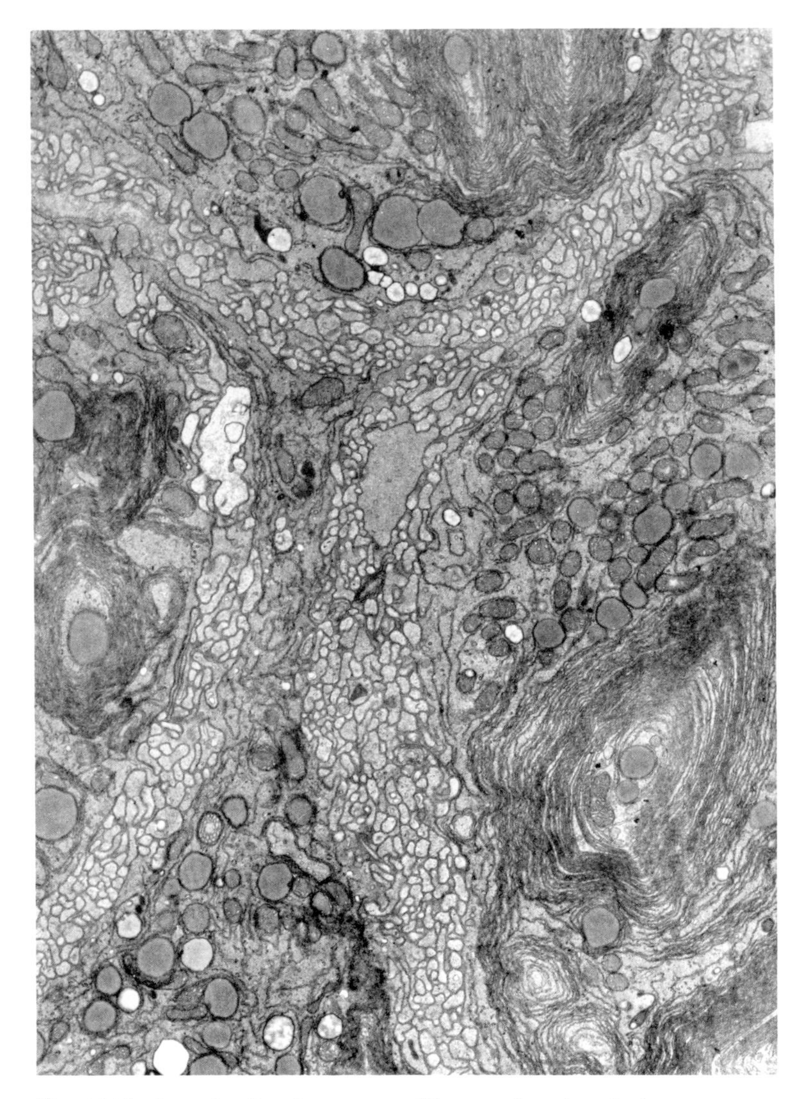

Fig. 6.105 Luteal cells of a raccoon (*Procyon lotor*) early in pregnancy. This shows, in exaggerated degree, two things that are common in corpora lutea: 1) complex interdigitation of adjacent cell membranes; 2) whorls of smooth endoplasmic reticulum. *P. lotor*, 4–9–68. × 7,500. (Courtesy of Akhouri Sinha.)

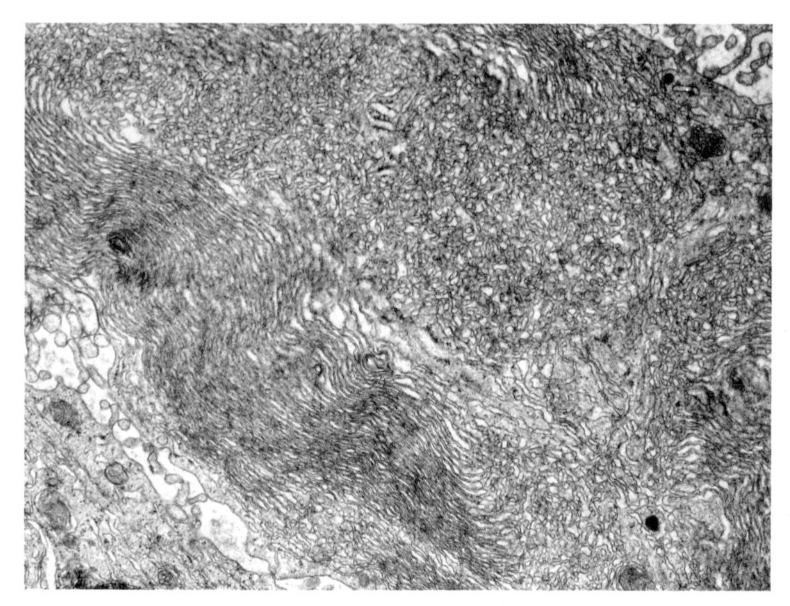

Fig. 6.106

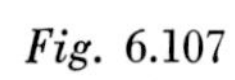

Fig. 6.107

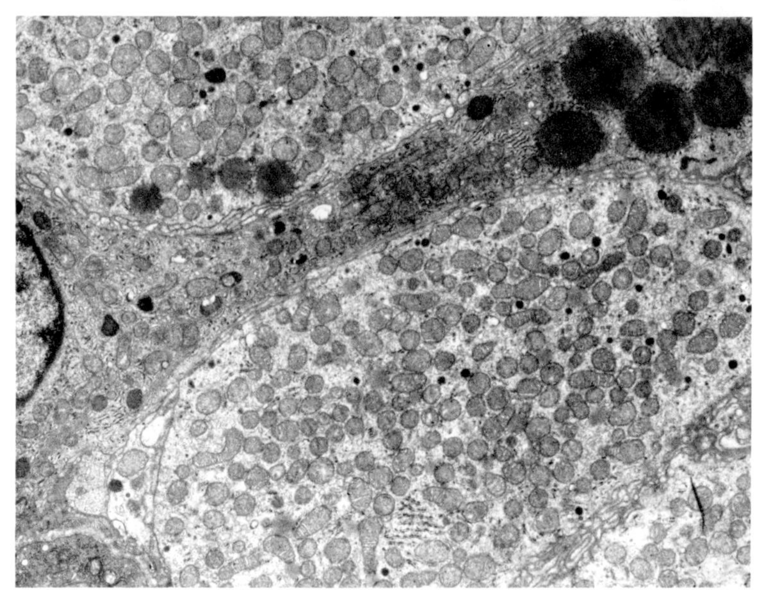

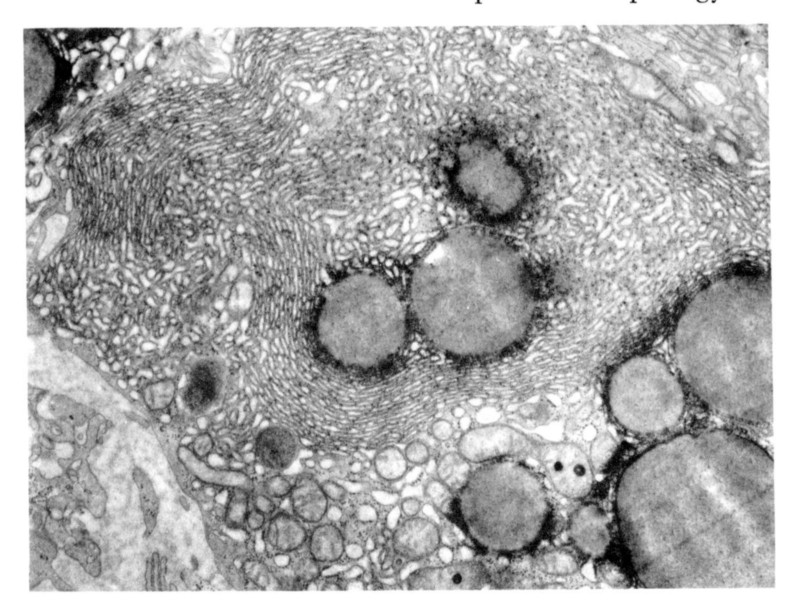

Fig. 6.108

Fig. 6.106 Luteal cell cytoplasm from a dog (*Canis familiaris*) in late gestation, showing large arrays of tubular and cisternal smooth endoplasmic reticulum. Note also the prominent intercellular space. *C. familiaris*, 4–2–68. × 10,500. (Courtesy of Akhouri Sinha.)

Fig. 6.107 Corpus luteum of a white-tailed deer (*Odocoileus virginianus*) with 320-mm fetuses. Part of one of the small luteal gland cells runs diagonally from lower left to upper right, separating two of the ordinary large luteal cells. *O. virginianus*, 3–21–68. × 5,210. (Courtesy of Akhouri Sinha.)

Fig. 6.108 Corpus luteum of a white-tailed deer (*Odocoileus virginianus*) with 65-mm fetuses, showing characteristic whorls of smooth endoplasmic reticulum and lipid vacuoles of a typical large luteal cell. *O. virginianus*, 1–16–68. × 13,510. (Courtesy of Akhouri Sinha.)

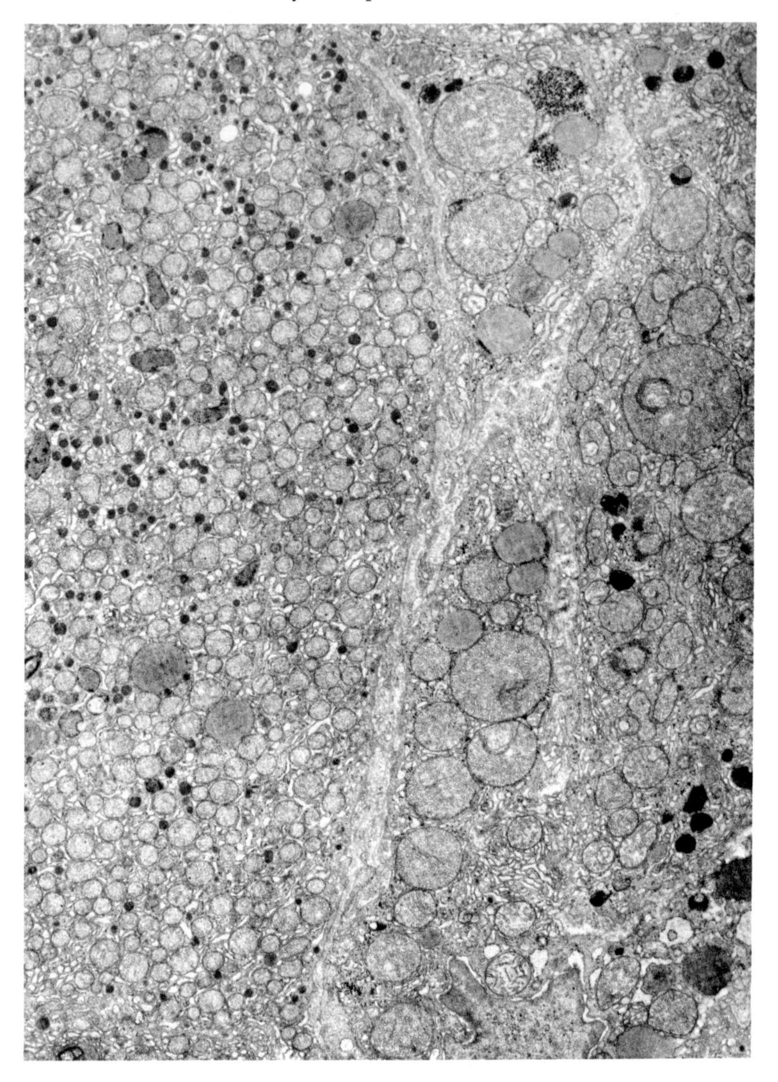

Fig. 6.109 Luteal cells of a sow (*Sus scrofa*) with near-term fetuses. On the left is one of the typical large luteal cells with spherical mitochondria and characteristically abundant smooth endoplasmic reticulum. At the right are portions of the small luteal cells with mitochondria of a wide size range. *S. scrofa*, 4–16–68. × 7,200. (Courtesy of Akhouri Sinha.)

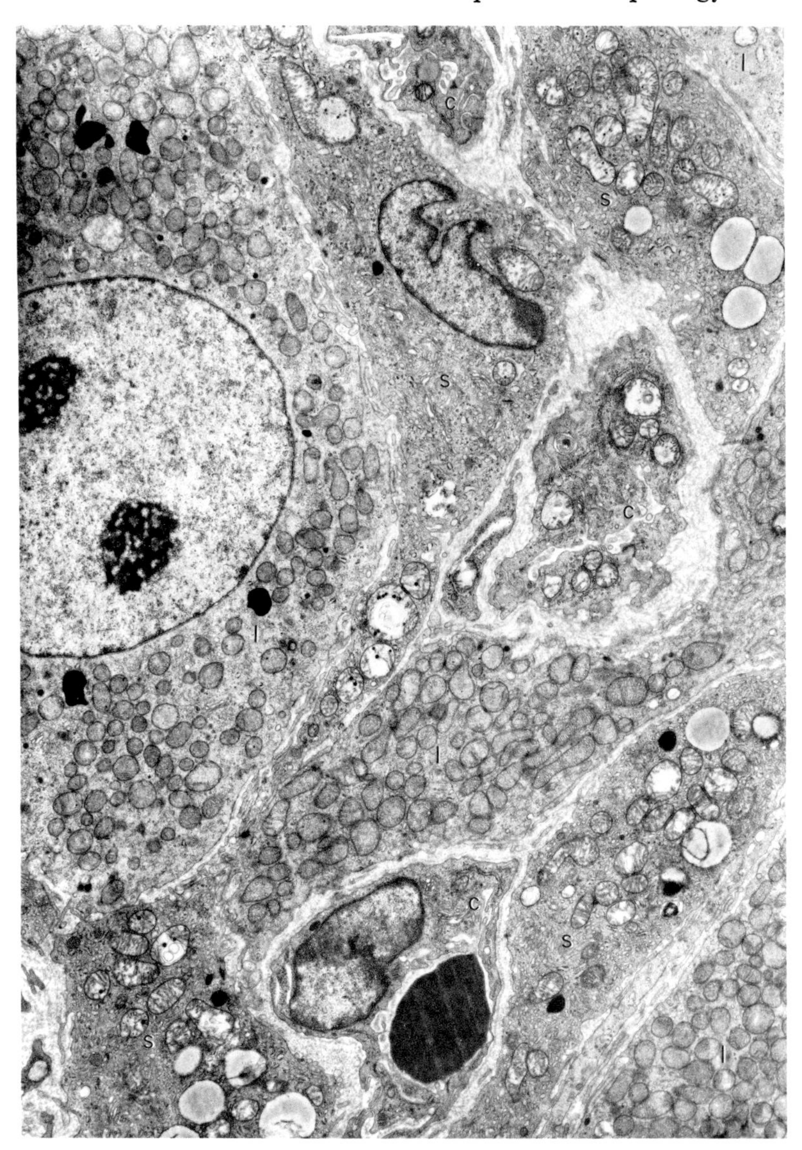

Fig. 6.110 Corpus luteum of a domestic ewe (*Ovis aries*) with 430-mm fetuses. Capillaries (*c*); portions of the ordinary large luteal cells (*l*); portions of the small luteal cells characteristic of artiodactyls (*s*). *O. aries*, 3–12–68. × 5,500. (Courtesy of Akhouri Sinha.)

Summary

The serosa of the ovary is continuous with that of the mesovarium and thus with the peritoneum in general. Its epithelium usually ranges from simple squamous to simple low columnar with a thin basement membrane and a relatively indistinct and cellular supporting layer of connective tissue. In a few cases the epithelium is pseudostratified, and in others it appears to consist of no more than stromal cells flattened on their exposed surfaces and with no basement membrane or fibrous supporting layer. The latter condition is essentially like that of the embryonic gonad. In most species, some more frequently than others, oogonia or oocytes may occur in the surface epithelium, which is therefore considered to be part of the germinal epithelium. Invaginations of the surface epithelium to form either subsurface sinuses, cortical (or "egg") tubules or cords, and shallow crypts, or simply localized groups of epithelial cells are common at some period in adult ovaries of most mammalian species. Oogonia or oocytes are usually present in the epithelium of these ingrowths.

The ovary consists of the cortex, medulla, and hilus. The classical concept is that the cortex is added late in the embryonic period to the more primitive medullary portion by a secondary proliferation of sex cords from the surface epithelium. This is fairly obvious in some species but very obscure in others. In the adult ovary, the boundary between cortex and medulla is usually only an irregular and indistinct transitional zone much distorted by large follicles and corpora lutea. The large ovarian vessels reach the ovary through the cephalic portion of the hilus. Here also are located the rete and the ovarian ends of the efferent ductules of the epoophoron; however, there is much variation in the location of the rete in different species. Typically, it is entirely within the ovary, but often it is partly or even almost entirely outside in the mesovarium.

The term "ovum," or "egg," is conveniently applied to oogonia and oocytes and to the ootid, or true ovum. Oogonia and oocytes frequently occur in surface epithelium and its derivatives and in medullary cords. They occur rarely in the epithelium of the rete and epoophoron. The mammalian egg is usually ovulated in the secondary oocyte phase and gives off the second polar body after spermatozoan penetration. Known exceptions are the red fox, which ovulates primary oocytes, and certain tenrecs, in which the eggs are fertilized before ovulation.

Four major stages of normal ovarian follicles are commonly recognized: primordial, primary, secondary, and vesicular. The final stage of the latter is the mature, ripe, or preovulatory follicle.

From the secondary stage onward, ovarian follicles are enclosed in a theca interna which, at the mature follicle stage, is an active endocrine layer, the thecal gland. An outer, probably somewhat contractile theca externa is usually present in larger species, but in many small mammals it is inconspicuous or absent.

The cells of the theca interna become fully differentiated as thecal gland cells only during proestrus and estrus. They degenerate rapidly after follicular rupture and, we believe, are not incorporated into the corpus luteum. The gland is present in all mammals that we have studied, but is very large in some and very inconspicuous in others. It is probably an estrogen secretor, but direct proof is lacking.

Atresia is the fate of about 95% of ova and follicles in any mammal where quantitative estimates have been made. In spite of the importance of atresia in the elimination of eggs and in the formation of interstitial gland tissue from the theca interna, little is known as to what initiates it or how long it takes for a follicle to disappear. A fair estimate of the latter appears to be at most a week or two for the smallest mammals and a month or two for the largest.

Interstitial gland tissue occurs in all mammalian ovaries, in massive amounts in some, in very small amounts in others. The term "interstitial" is very unsatisfactory as a name for these glandular tissues of the ovary, because they are usually not interstitial in the literal sense. In fact, they commonly occur as more or less discrete clumps or cords of cells, but sometimes as massive tissue in which other structures such as follicles and corpora lutea are embedded. However, since the term is in common usage, and since we know very little about the true nature or function of these tissues, it is a convenient name under which to assemble them. Until more is known about them, interstitial gland tissues of the mammalian ovary can tentatively be defined as any endocrine type of gland tissue found in the ovary or the mesovarium which is not thecal gland (glandularly differentiated theca interna of mature follicles) or luteal gland (corpus luteum, including its associated paraluteal cells, when present). We have provisionally divided interstitial gland tissues into seven types based on a variety of characters, including time of appearance, location, tissue of origin, relation to the reproductive cycle, and resemblance to glandular tissue elsewhere in the body. When more is known of their function and developmental origin, a more meaningful classification and nomenclature can be devised.

The seven types of ovarian interstitial gland tissue are:

1) Fetal. Present in the fetus and newborn, and sometimes in juveniles (see pp. 70 and 167); apparently derived from mesenchyme between the primary sex cords and therefore a homolog of the interstitial gland cells of the testis. This type has been found in all mammals examined for its presence.

2) Thecal. Present in the ovaries of all mammals as soon as follicular atresia begins and persisting as long as this occurs; derived from the theca interna of atretic follicles.

3) Stromal. Notably present in rabbits and hares, especially the hares; derived from a population of undifferentiated ovarian stromal cells without regard to their spatial relationship to other ovarian structures, except that the main zone of differentiation usually appears to be just deep to the tunica albuginea and associated with the zone of primordial and primary follicles; accumulates so that it is most conspicuous in the oldest individuals.

4) Gonadal adrenal. Known to be present more or less constantly in one to a few species of moles, rodents, lagomorphs, armadillos, primates, hyraxes, and horses, but is probably much more widespread, simply having gone unobserved because it commonly occurs in the seldom-examined mesovarium. It is usually associated with the epoophoron (and efferent ductules in the male); sometimes takes the form of a typical accessory cortical adrenal; and is known to have cyclic behavior only in the common European mole. In the 13-lined ground squirrel, adrenalectomy causes it to hypertrophy and so to compensate for adrenal loss in both male and female. It is massively developed in the common European mole and star-nosed mole, and occurs as accessory adrenals in the nine-banded armadillo and patas monkey.

5) Adneural. Demonstrated so far only in certain primates, including man. Cytologically, it closely resembles testicular interstitial gland cells, but occurs only within or closely adjacent to small involuntary nerves and nerve-fiber bundles near the ovarian hilus (also in the corresponding position in the male). It is a common source of hilus cell tumors, which are usually masculinizing.

6) Medullary cord. Clearly present only in Carnivora so far as known, and especially well developed in certain Mustelidae (mink, ermine, and long-tailed weasel). It is derived by hyperplasia and glandu-

lar hypertrophy of medullary cord epithelium and probably cortical cords as well; most highly differentiated from proestrus through the period of delayed implantation to just before implantation; most dedifferentiated during late pregnancy and through lactation. In the common European mole and in the shrew-mole the epithelium of the medullary cord remnants undergoes glandular hypertrophy along with the gonadal adrenal cells. However, these cord remnants are so scattered that the glandular tissue they form is much less conspicuous than that of mustelids.

7) Rete. Well differentiated in only one genus of bats, *Uroderma.* It is a glandularly hypertrophied rete epithelium. Nothing is known of its relation to the reproductive or life cycle of the species.

Primary corpora lutea, or luteal glands, are mainly derived from the follicular epithelium of ovulated follicles, supplemented in some species by luteal transformation of surrounding stromal cells. Accessory corpora lutea develop from atretic follicles, occasionally in many species, normally and in large numbers in several others. So-called secondary corpora lutea have been described from a second ovulation during a single pregnancy cycle in the mare. Aberrant corpora lutea, as reported in the rhesus monkey, appear to arise from originally primary corpora, some portion or all of which fails to atrophy and persists into one or more succeeding reproductive cycles.

Two or more types of luteal cells have been described for many species. Of these, the so-called granulosa luteal cells are of universal occurrence and make up the bulk of all corpora lutea. Paraluteal cells are stromal cells in the process of transformation into luteal cells; when fully differentiated, they are probably indistinguishable from those of granulosa origin. Therefore, we believe the definitive luteal gland cells of most mammals are of one type only, regardless of their origin. Artiodactyls, cetaceans, and possibly one or two minor groups do have two definitive luteal cell types. The secondary luteal cells of these groups are probably also derived from stromal cells, and, although some may transform into primary luteal cells, a large population remains as distinctly different definitive cells. We object to the use of "theca lutein" to designate either paraluteal or these secondary luteal cells, for this term implies that they originate from thecal gland cells and we believe this is not true.

All ovarian gland cells so far studied by histochemical or electron-microscopic methods are of the steroid secretor type. Unfortunately, these methods have been adequately applied only to luteal cells.

Seven

Features and problems associated with the mammalian ovary

Although some of the subjects discussed below have been mentioned or even treated at length in previous chapters, it seemed expedient to comment on them again as separate topics where there could be no problem of interfering with the thread of a more general discourse. Others are relatively isolated problems which simply did not fit elsewhere. In each case, our aims have been 1) to present the essential facts and the problems which they pose, 2) to outline the significant interpretations of other investigators, 3) to give our own analyses of them and in some cases to point out the direction we think future studies of these matters should take.

Follicular liquor

Robinson (1918) distinguished two types of follicular liquor: primary, formed first and responsible for the development of the antrum; and secondary, formed at the preovulatory period and responsible for the rapid enlargement of the ripening follicle. A third type appears just after ovulation. Since this often contains a few red blood corpuscles and forms a clot filling the residual follicular cavity and plugging the stigma, it is probably an exudate of blood plasma.

Primary follicular liquor begins to accumulate in small intercellular pockets in the secondary or multilaminar follicles. Continued accumulation of the fluid results in enlargement and confluence of scattered pools, finally terminating in the formation of a single large vesicle, or antrum. Although this appears to be the pattern for most mammals, there are some that deviate markedly. The ovaries of some insectivores have no visible antrum in the preovulatory follicles — *Setifer* (Strauss, 1938); *Echinosorex* and *Hylomys* (Duke, unpublished). There is a very small antrum in the mature follicles of some other insectivores — *Blarina* (O. P. Pearson, 1944); *Sorex* (see Fig. 6.31); *Condylura* (see Fig. 6.32) — and of some bats — *Myotis* (Wimsatt, 1944); *Plecotus* (O. P. Pearson, Koford, and Pearson, 1952); *Eptesicus* (see Figs. 6.34 and 6.35).

The follicular cells have been cited as the most likely source of the follicular liquor, although the manner in which the cells elaborate it is not clear. Some investigators thought that the follicular cells were holocrine cells and that the liquor was formed by the sloughing and dissolution of the cells. A more acceptable view, and the current one, was introduced by Honoré (1900) in his study of the domestic rabbit ovary. He also implicated the follicular cells, but thought they functioned as merocrine cells. Robinson (1918), in his study of ovarian follicles of the European ferret (*Mustela putorius*), accepted Honoré's conclusions in these words:

> . . . Honoré holds that it is an intercellular secretion, and he shows that the so-called bodies of Call and Exner are not . . . vacuoles in cells but, in reality, intercellular spaces. My own observations entirely support Honoré's point of view, and I believe, moreover, that the central parts of the bodies are merely isolated portions of the general antrum. (p. 321)

An electron-microscopic study made by Hadek (1963) on the mode of secretion of the primary follicular liquor in the mouse ovary indicated that follicular cells were active in secreting the fluid. The Golgi complex underwent a series of changes resulting in the accumulation of material in the Golgi vesicles. These vesicles then expanded into larger smooth-walled cisternae, and finally appeared to release the secretion into the intercellular spaces. According to Hadek, Honoré believed that the Call–Exner bodies (small amorphous masses surrounded by rosettes of follicular cells) were extracellular. He says, "Robinson has shown them to be cytoplasmic in their location, and the present studies bear this out" (p. 458). This, of course, is contrary to Robinson's own statement quoted above. Hadek considered intracellular droplets only about 1 μ in diameter to be Call–Exner bodies.

The function of the follicular fluid is still not well understood. A common opinion has been that the accumulation of fluid in the large antrum of preovulatory follicles might well increase the intrafollicular pressure and thus be an agent in the process of ovulation. Zachariae and Jensen (1958) concluded that there was an increase in colloidal osmotic pressure as the result of enzymatic depolymerization and mucopolysaccharides in the fluid. This osmotic pressure increase, they thought, would be an important factor in the ovulatory mechanism. Espey and Lipner (1963) used direct cannulation by micropipettes to determine intrafollicular pressure in the domestic rabbit ovary. The average pressure in the antrum of mature follicles was slightly over 17 mm Hg. A direct proportionality was observed between blood pressure and follicular pressure, suggesting that colloidal osmotic pressure contributed little to the total hydrostatic pressure in the follicle.

Then, too, what of the nonantral follicles of some species such as the large Madagascar "hedgehog" (*Setifer setosus*)? No doubt small amounts of follicular fluid exist in the intercellular spaces of their follicular epithelium, but it is unlikely that this would be a significant pressure factor in ovulation. Perhaps the fluid serves just as any other tissue fluid, as an accessory medium to transmit substances between the oocyte and the follicular cells. An appreciable amount of liquor could aid in flushing the egg out of the follicle at ovulation, but eggs obviously escape readily in those species having no reservoir of liquor folliculi.

The so-called Call–Exner bodies occur sporadically in a few species, but they are absent from most. The best evidence indicates that they are nothing but tiny pools of intercellular follicular fluid, surrounded by rosettes of follicular cells. Motta (1965) showed by electron microscopy that the fluid body had a smooth boundary, but that the surrounding cells were separated from it by a zone of microvilli. Because these cells were rich in organelles, they probably were actively secreting into the center. Plasma membranes of cells that happened to be adjacent to the central mass were sometimes interrupted at the area of contact, but this could be the result of tangential sectioning. Call and Exner (1875) were much more intrigued with the idea that the bodies might represent an additional source of oocytes! So far there is no evidence to indicate that Call–Exner bodies are of any functional significance.

Clues to the possible function, other than mechanical, of the follicular liquor may lie in its chemistry. We are not equipped to discuss this angle, but for those who are interested we suggest reference to Lutwak-

Mann (1954), who discussed the effects of its content of glucose and of lactic and ascorbic acid on the fructolytic activity of spermatozoa, and to Richardson (1967). Rondell (1964*a*), in studies directed toward the mechanisms of ovulation, gave the concentrations of Na and K in the follicular fluid of domestic rabbit follicles at various phases of the reproductive cycle.

Follicular atresia

We still know far too little about the process of degeneration of ovarian follicles. What initiates the process? It begins on a massive scale in late fetal life in many mammals, and in early postnatal life in others. Arai (1920), using the rat, calculated a decrease in the total number of ova (oocytes) from 35,000 at birth to 11,000 20 days later. This is an attrition rate of about 1,000 oocytes per day. If this rate continued, the ovaries would be depleted of all oocytes before sexual maturity (about 45 days of age) was attained. Deanesly (1970) stated that 60%–80% of the original oocytes had disappeared in the ovary of the European ferret by the end of the meiotic prophase (21 days postpartum). What factors slow down or control the process? It has been conjectured that the attainment by the oocytes of a layer of follicular cells determines whether they degenerate or enter into the "storage phase" (Ohno and Smith, 1964).

Once the ovary is mature and its cyclic pattern is initiated, numerous "selected" primary follicles are in some way stimulated to enter the growth phase. How are they selected? How are those in storage kept from degenerating? Of the crop of follicles that begins growth during a cycle, what determines which ones will ovulate and which ones will degenerate?

We know very little of the time required for follicles of various sizes to complete the degenerative process. It is probably rapid in small mammals, for Engle (1927*b*) was able to show in the mouse significantly more follicles in atresia the day following ovulation than at any other time in the estrous cycle; the fewest were present in diestrus. These "waves" of atresia that occur during the prepuberal period and in mature animals near each estrous period are an indication of hormonal involvement.

Estimates of the number of follicles in a human ovary at birth range from 370,000 to 500,000 and at puberty about 190,000. The number of eggs ovulated by a woman who had had a 30-year reproductive period would certainly not be over 400. This means that at least 99.8% of her eggs present at puberty would degenerate, most of them before the end of her reproductive period, if we assume that no postnatal oogenesis occurs and that relatively few oocytes remain after menopause. If postnatal oogenesis does take place, then the percentage of atresia is even higher. The ratio of degenerate eggs to ovulated ones is probably somewhat lower in animals that ovulate several eggs at each estrous period. It has been estimated that the laboratory rat is born with about 35,000 ova, and some postnatal oogenesis would add slightly to this figure. If we assume a reproductive life of three years, ovulation of 10 eggs every five days, and no pregnancies, about 2,200 eggs would be the maximum ovulated. This would still indicate a loss by degeneration of 94% of the ova formed — undoubtedly a much lower loss than would actually occur in this species. The magnitude of egg loss by atresia and the prominence of atretic follicles in the morphology of the mammalian ovary are seldom appreciated, even by students of mammalian reproduction.

There are no reliable data on the length of time needed for a follicle at a given stage of development to disappear by atresia. Almost no attempt has been made to determine this, except for follicles already having antra. Sturgis (1949) seemed to think it would take about five weeks in the rhesus monkey (*Macaca mulatta*). Ingram (1962), after reviewing the literature, suggested that it may take several years. It appears almost impossible to devise a method whereby individual follicles may be followed directly from the beginning of degenerative changes until they disappear. Even indirect methods are difficult. For one thing, end points are subjective and controversial. We believe that the first indication of atresia of a vesicular follicle is pyknosis and karyorrhexis in granulosa cells next to the antrum; others see the first evidence in changes in the oocyte, or in cessation of mitosis of granulosa cells, a process which we know continues in the outer granulosa for some time after the inner cells are dying. Still more difficult is the definition of the end of degeneration. Should it be the time of the disappearance of the oocyte, or of the antrum, even though the "scar" or an interstitial cell mass developed from the theca interna may still be visible? We consider atresia incomplete as long as an area identifiable as the former center of the follicle persists. But what should one assume as the termination in such species as the guinea pig and plains pocket gopher (*Geomys bursarius*) where remnants of the zona pellucida are identifiable in the cortical stroma and even in the medullary connective tissue after all other signs of a follicle have vanished? When one considers atresia of primordial, primary, and secondary follicles, still other problems appear. Once the antrum has disappeared how is one to know it was ever there and whether he is dealing with a former vesicular or nonvesicular follicle? What are the first signs of degeneration in the oocyte of a primordial follicle where one has only cytological criteria and little evidence from follicular structure by which to judge?

Since there is no direct means of following the course of a primary follicle through its life history to its ultimate fate of degeneration or ovulation, indirect methods have been used. To initiate degeneration of follicles, X-irradiation has been employed, because dosage can be easily calibrated and controlled. Vermande-Van Eck (1956) found all follicles larger than primary follicles were degenerate within seven days following irradiation. This method has been criticized as being unnatural, severe, and abusive of living tissues. Hypophysectomy of immature rats resulted in a wave of follicular degeneration about seven days after the operation (Young, 1961: 463). After ligation of the ovarian blood supply in adult rats, most follicles degenerated within ten days (Butcher, 1932). All of these procedures are unnatural, but it may well be that once an oocyte or follicle becomes moribund, regardless of the cause, the time sequence for completing the degenerative process may be the same.

The length of time necessary for atresia of follicles is not just an academic question. It involves an understanding of endocrine cycles as they affect reproductive capacity in both prepuberal and mature females. Without a knowledge of the length of time an atretic oocyte remains identifiable and without adequate criteria for deciding whether an oocyte is in initial atresia, data such as those of Zuckerman and Mandl on neo-oogenesis are basically unsound. (For a good review of this work, see Zuckerman, 1962.) Since the rate of atresia is unknown, the gradual decrease in oocyte numbers with increasing age of the female can be explained either by gradual loss from an original fixed supply or by a loss which is somewhat greater than the rate of replacement of oocytes by continuing oogenesis. We have made no counts, but the numbers of oocytes and follicles, especially atretic ones, that we have seen in ovaries during proestrus, estrus, and early uterine pregnancy are in many species strikingly greater than during late pregnancy and lactation. This is especially obvious in the squirrels (Sciuridae), weasels

(Mustelidae), and mink (*Mustela vison*). If there is no new formation of oocytes in the adults of these species, then it must be assumed that large numbers of oocytes exist in unrecognizable form throughout much of pregnancy and lactation.

In the absence of direct evidence on the duration of atresia, one is justified in forming a hypothesis based on indirect evidence. First, there is no indication whatsoever that atresia lasts more than a couple of months in any animal, if we consider it to end with the conversion of the area formerly occupied by the granulosa and the antrum into connective tissue similar to that of neighboring stroma, and if we disregard the persistence or absence of interstitial gland tissue formed from the theca interna, or of a resistant zona pellucida such as that of the guinea pig, or of accessory fibrous corpora albicantia such as occur in man and bovids. Second, the larger the follicle the longer the time one would expect the process to take. This means that follicles in later stages and those of the larger mammals would take longer to regress than follicles in earlier stages and those of smaller mammals. A ripe follicle of a rat-sized animal has a diameter of about 1 mm; that of a woman, approximately 15 mm; that of a cow, 20 mm. Fibroblasts migrate across a Clark–Sandison transparent chamber in a rabbit's ear at the rate of about 0.2 mm per day (Stearns, 1940), and macrophages and leukocytes probably move faster. On this basis, it would take about 2½ days for fibroblasts to reach the center of a 1-mm follicle and 37 days to reach the center of a 15-mm follicle. The latter is probably an excessive estimate, for it is believed that follicular fluid is rather rapidly absorbed, allowing atretic follicles to collapse. If one thinks of a decadent follicle as a small sterile wound in the relatively embryonic ovarian stroma, then it seems reasonable that it would be obliterated or "healed" at about the rate of a wound of the same size in any cellular connective tissue such as that of the uterine mucosa. On this basis, we would estimate that complete atresia would take a week or two at the most for follicles with a 1-mm diameter, and a month or two for 15-mm follicles: a matter of years, or even of several months, seems to us most unlikely.

For a recent review of factors influencing the number of oocytes in the ovary and the causes of atresia, see Krohn (1967). (See also pp. 47–49, 101, 103–4, 110, and 166.)

Anovular follicles

Anovular follicles have frequently been mentioned in the literature in connection with studies of the following mammals: domestic rabbit (Regaud and Lacassagne, 1913); a species of bush baby (*Galago*) (Gérard, 1919–1920); opossum (*Didelphis marsupialis*) (Hartman and League, 1924); opossum, rhesus monkey, and nine-banded armadillo (*Dasypus novemcinctus*) (League and Hartman, 1925); dog (Jonckheere, 1930; Barton, 1945); mouse (Brambell, Parkes, and Fielding, 1927); vagrant shrew (*Sorex vagrans*) (Wilcox and Mossman, 1945); and rat (Davis and Hall, 1950). A perusal of these papers makes it quite clear that the term "anovular follicle" means different things to different people. Clusters of epithelial cells in the peripheral part of the cortex, appearing in sections as nodules or cords or tubules surrounded by a basement membrane, have been called anovular follicles. Larger epithelial complexes often associated with the medullary cords and rete have also been so designated.

An ovarian follicle basically consists of an oocyte, first surrounded by an epithelial layer and later by one or two thecal layers. Growth and cytological changes take place in all of these elements as follicular development proceeds. By definition, then, an anovular follicle would consist of all of these elements except the oocyte. It is very doubtful if such a morphological unit exists. If an oocyte were never present in such a cellular complex, it is probable that the other layers, especially the thecae, would not differentiate. And any follicle containing a moribund oocyte would be nothing more than an atretic follicle.

The small, more peripherally placed epithelial complexes could be descriptively designated as epithelial nodules, cords, or tubules. However, we believe that they are homologs and vestiges of cortical cords and tubules, which not only occur in embryos, but commonly in adult ovaries such as those of the grizzly bear (*Ursus horribilis*) (see Fig. 6.7) and the raccoon (*Procyon lotor*) (see Fig. 6.6) and in the fetal and newborn horse (Fig. 7.1). The larger complexes, which League and Hartman (1925) called "mature anovular follicles," and which are often associated with the rete tubules, we believe are homologs and vestiges of medullary cords or tubules. These may be highly developed in a few species, and have been called "testis cords" by Wilcox and Mossman (1945) in the vagrant shrew (see Fig. 6.16) and by Davis and Hall (1950) in the rat, and "spermatic cords" by Brambell, Parkes, and Fielding (1927) in X-rayed ovaries of mice. Figure 7.2 shows "testis cords," i.e., medullary cords, in the ovary of an acushi (*Myoprocta*).

Theoretically, medullary cords differentiate deep in the embryonic ovary as epithelioid cords, whereas cortical cords are invaginations of surface epithelium into the superficial zone of the embryonic ovary. As

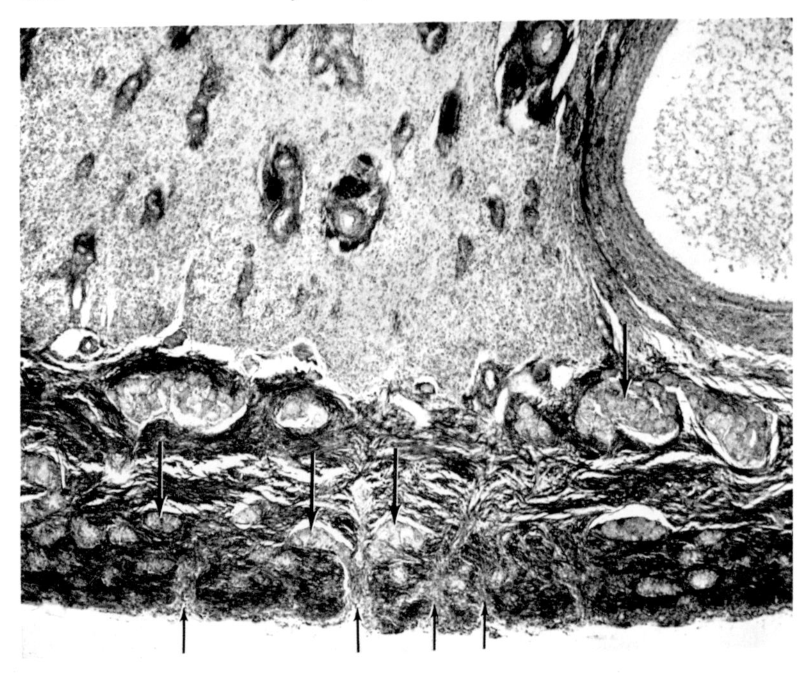

Fig. 7.1 A portion of the cortex and corticomedullary junction of the ovary of a near-term fetus of a Burchell's zebra (*Equus burchelli*) to show the numerous cortical cords (*short arrows*) which penetrate the tunica albuginea and end in egg nests (*long arrows*). Note the fetal type interstitial gland tissue at the left of the vesicular follicle; compare with Figure 6.51. *E. burchelli*, 1. × 60.

was pointed out in Chapter 3, page 59, and by Wilcox and Mossman (1945), cords of each type are sometimes continuous with one another, making it obvious that there is not always a sharp distinction between the two. It may well be, then, that the term "anovular follicle" is both inaccurate and inappropriate, and that other more descriptive and meaningful terms should be employed, as suggested above. Seitz's dictum, as quoted by League and Hartman (1925), is apparently sound: "Ohne Ei, kein Follikel."

Polyovular follicles

Polyovular follicles were reported in 1827 by Baer in the same communication in which he announced the discovery of the mammalian ovum. They have been described in the ovaries of several mammalian spe-

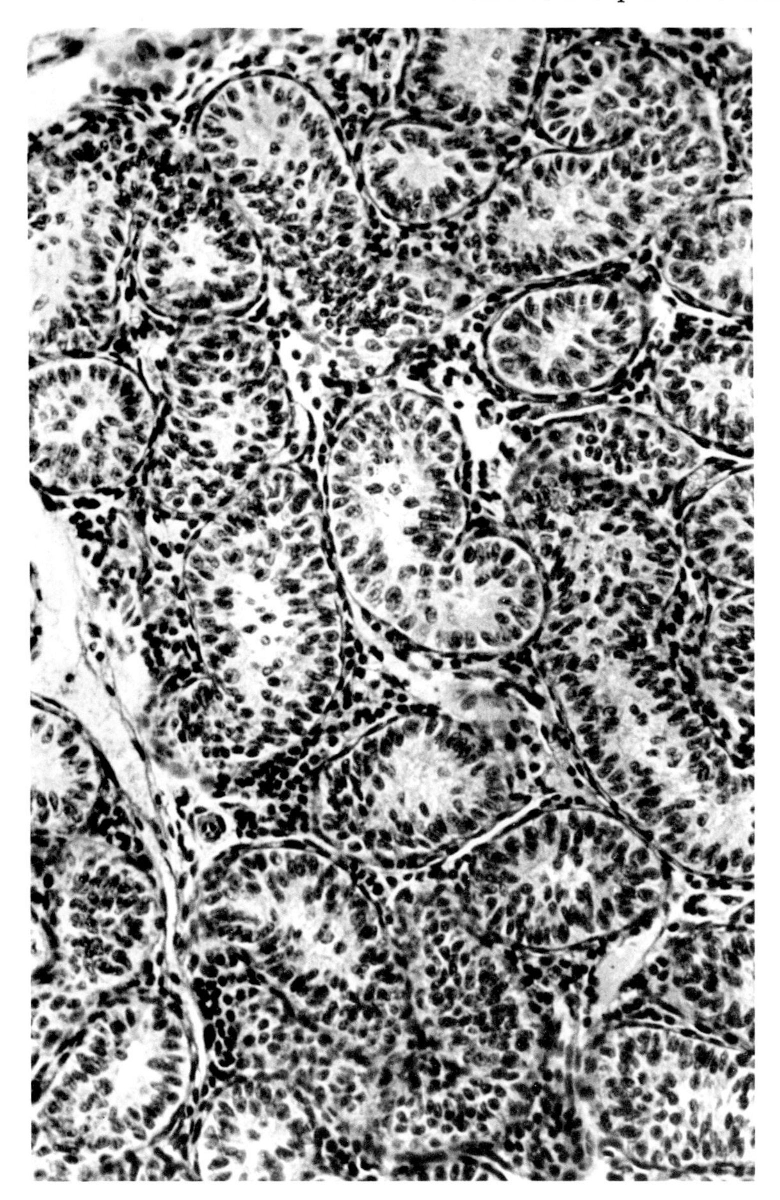

Fig. 7.2 Medullary cords, or "testis cords," in the ovary of an acushi (*Myoprocta*) in late gestation. These were found in 8 of 36 specimens of agoutis and acushis. *Myoprocta*, 1. × 250. (Courtesy of Barbara Weir.)

cies: Hartman (1926) listed them in 13 eutherian mammals, and we can add at least 24 additional genera to the list (unpublished data). They have been reported to occur more frequently in the ovaries of fetal and juvenile females than in adults. Their origin, significance, and fate have never been adequately studied. For reviews of this subject the reader is referred to the articles by Hartman (1926) and Mainland (1928).

Lane (1938) encountered a few polyovular follicles in the ovaries of immature rats. Of 28,522 follicles counted in over 100 ovaries from 60 rats, ranging in age from 15 to 64 days, only 13 biovular and three triovular follicles were counted. Lloyd and Rubenstein (1941) studied the effects of gonadotropic substances on the ovaries of immature rhesus monkeys. At varying intervals after treatment, unilateral ovariectomy was performed. The investigators were impressed with the number of polyovular follicles seen, although no actual counts were published, and they attributed the multiplicity of ova to mitotic division of the original oocyte, which had been stimulated by the injected gonadotropins. They apparently did not know that normal juvenile rhesus monkeys, weighing about 2.5 kg, have numerous polyovular follicles (Fig. 7.3). The idea of mitosis of the oocyte in a multilaminar follicle is interesting, but contrary to all we know about the primary oocyte. Engle (1927*a*) suggested this same process to account for a biovular follicle in the mouse ovary.

Bodemer and Warnick (1961*a*) studied the immature hamster ovary, and observed not only a large number of polyovular follicles but a quantitative fluctuation in the number of such follicles in animals of different ages. Ovaries of animals 21 days old showed 7% of the follicular population to be of the polyovular variety, while in 25-day-old females the percentage was 17. In 33-day-old females there was a decrease to about 6%. What is the cause of this apparent fluctuation in the number of polyovular follicles during this period? The most commonly accepted explanation for the origin of polyovular follicles is the failure of adjacent oocytes in the early sex cords to become separated by intervening epithelial cells or connective tissue fibers. Thus, two or more oocytes may be enclosed in a common follicular envelope. The immature hamsters used by Bodemer and Warnick were far beyond the developmental stage for the above explanation to hold.

It is known that the number and the size of ova in polyovular follicles vary. Dawson (1951:190) concluded that such follicles in immature rats

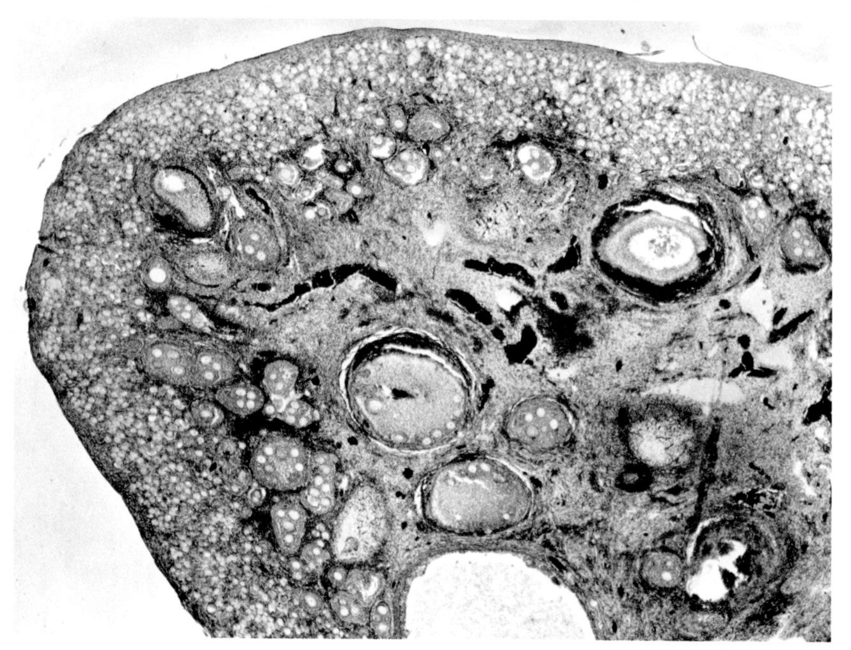

Fig. 7.3 Portion of the ovary of a juvenile rhesus monkey (*Macaca mullata*), weighing about 2.5 kg. The numerous polyovular follicles appear to have been derived from medullary cords. *M. mullata*, 3. × 16.

"result from the temporary incorporation within the follicular epithelium of retarded cells whose potentialities as future egg cells are fully determined or are produced by the subsequent oocytic differentiation of apparent follicle cells whose potentialities were not finally restricted at the time they became specifically oriented about the ovum."

In a combined descriptive and experimental study of the ovary of the striped skunk (*Mephitis mephitis*), Leach and Conaway (1963) discussed the origin and fate of polyovular follicles (Fig. 7.4). They stated that such follicles arose in the primary sex cords by the incorporation of two or more oocytes in a common follicular envelope. Such follicles were found in both the medulla and cortex, and degeneration was their usual fate. However, injections of exogenous gonadotropins, resulting in ovulation, involved some polyovular follicles. Leach and Conaway based their conclusion on the fact that the number of tubal ova and the number of corpora lutea did not correspond, that is, 15 tubal ova were counted while only 13 corpora lutea were identified. Mainland (1928)

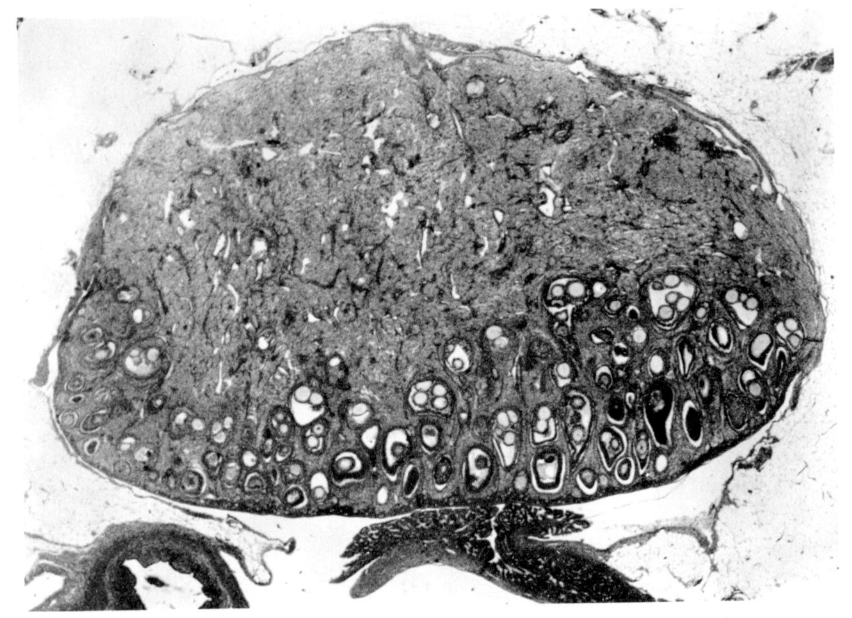

Fig. 7.4 Ovary of a juvenile striped skunk (*Mephitis mephitis*), showing numerous polyovular follicles. These are apparently in the cortex. *M. mephitis.* × 27. (Courtesy of Clinton H. Conaway.)

arrived at a similar conclusion from a study of polyovular follicles in the ovary of the European ferret. Such observations as these no doubt led some investigators to consider polyovular follicles as an adaptation to increase fecundity. Bodemer and Warnick (1961*b*) found that treatment of immature hamsters with gonadotropic hormones stimulated antral development in both polyovular and uniovular follicles. This suggests a similar or identical degree of competence in the response of polyovular and uniovular follicles to gonadotropins. A degree of this competence is clearly demonstrated in the ovaries of untreated juvenile rhesus monkeys (see Fig. 7.3).

Although the existence of polyovular follicles has been known for many years, present information on the subject leaves much to be desired. Why is the phenomenon so common in some mammals and rather uncommon in others? Is it species variation or lack of information, partly the result of not observing juvenile ovaries at the critical period in their development?

The relatively rare and sporadic occurrence of follicles of any size with two to four or five eggs is entirely understandable as being fortuitous, in view of the way oogonia and oocytes tend to be grouped at the time follicle formation starts. However, the regular occurrence of many such follicles, or of truly polyovular ones such as those of the juvenile rhesus monkey, which may contain literally dozens of ova, is not to be explained so simply. Two facts stand out: regular occurrence is always in juvenile animals, and such follicles are located primarily in the medulla. We have shown that medullary cords, presumably remnants of the primary sex cords, persist in many mammals at least into the juvenile period, and that they often contain ova. Parts of these commonly break up to form primary follicles, some of which, at least in one species of chipmunk (*Tamias striatus*), may even develop into functional ovulatory follicles. If these cords are relatively slender, that is, made up of only two or a few cell layers, and ova are scarce in them, then most of the medullary follicles formed from them will be uniovular. However, if these cords are thick, that is, consist of many cell layers, and if they contain numerous potential oocytes relatively crowded together, as seems to be the case in the rhesus monkey (see Fig. 7.3), then it becomes almost certain that segmental breakup of these cords to form follicles will result in most of them being polyovular.

We believe, therefore, that the rarely occurring large follicles with two to a few eggs in mature animals are essentially accidental, and do account for most of the cases where embryos have been reported to outnumber corpora lutea, although unrecognized or fused corpora lutea may explain some of these reports. We also believe that the large numbers of polyovular follicles reported in some species are formed from medullary cords, and are entirely normal. Since medullary follicles function as the first ovulatory follicles in at least one species, there is reason to believe that medullary follicles respond to gonadotropins and that treatment with such agents at the time these follicles are present can result in their ripening and ovulation, as has been actually demonstrated by Leach and Conaway (1963).

There seems to be little functional reason for polyovular follicles. To our knowledge, this occurs in no mammalian species as a way of increasing the number of eggs ovulated. Those that ovulate large numbers of eggs, such as the tenrecs (Tenrecidae), South African long-eared elephant shrew (*Elephantulus myurus*), and plains viscacha (*Lagostomus maximus*), do so as the result of great numbers of uniovular follicles. This situation would seem to require simpler anatomical and physiological mechanisms and to be more desirable, for it provides thecal and luteal glands in direct proportion to the number of eggs ovulated. Perhaps the numerous polyovular medullary follicles of such forms as the striped skunk and rhesus monkey are simply an innocuous inherited pattern and have no functional significance.

Homology of follicles with the coelom

The vesicular ovarian follicle of mammals and its method of ovulation are unique among vertebrates. A morphologist may well wonder what its antecedents were.

The nonvesicular ripe follicle of *Setifer* and a few other genera is obviously similar to that of reptiles, except in the proportionate size of the egg. The transition from this condition to a vesicular, liquid-filled follicle would seem to be a relatively simple evolutionary problem both morphologically and physiologically. A liquid-filled follicle is possibly in part an adaptation to assure freeing of the minute ovum with its cumulus into range of the ciliary action of the funnel of the oviduct. It is tempting to think of the follicles as isolated coelomic spaces. Three indications of this possibility are: the surface epithelium of the ovary is directly continuous with the peritoneal mesothelium; follicles develop more or less directly from the surface epithelium; and follicular liquor apparently bears a relation to the follicular wall comparable with that of coelomic fluid to the peritoneum. Figure 7.5 illustrates this concept. For a more detailed discussion of this matter, see Mossman (1938).

The origin of oogonia

Among the first definite histological observations of the mammalian ovary were those made in the nineteenth century by Schrön (1863), Pflüger (1863), Waldeyer-Hartz (1870), and Beneden (1880). The problem of the origin of ova and whether or not the original supply could be replenished was certainly considered by most of these, and an explanation attempted by some of them. Schrön (1863) surmised that additional ova (oocytes) were added during estrus in the rabbit and cat and during menstruation in humans. Pflüger (1863) reported periodic development of ova (oocytes) from epithelial tubules in the cat. He evidently also believed that additional ova were produced by division of existing ones. Ova were reported in the germinal epithelium by Waldeyer-Hartz (1870), Beneden (1880), and others. However, at that

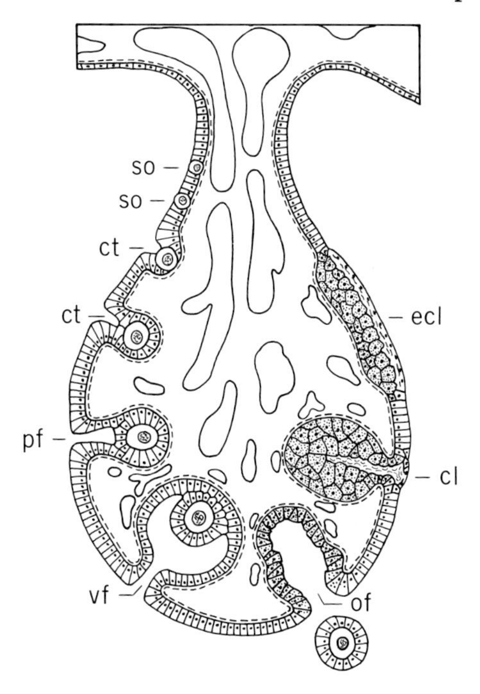

Fig. 7.5 Diagram of a mammalian ovary to illustrate the theoretical homology of the follicular epithelium and luteal tissue to the coelomic lining, and of the follicular cavity to the coelom. The only condition indicated on this diagram that does not occur normally in some mammal is the connection of the cavity of a vesicular follicle by an open tube to the peritoneal cavity. *cl*, corpus luteum. Its cavity is filled with fibroblasts which are taking part in healing the rupture point; *ct*, cortical tubes; *ecl*, everted corpus luteum with a fiibroblastic layer covering it. This fibrous tissue would ordinarily be enclosed within the corpus except at the region of the rupture point; *of*, ovulated follicle; *pf*, primary follicle at end of a cortical tube; *so*, surface ova; *vf*, vesicular follicle indicated as having retained continuity with a cortical tube.

time no methodical attack was made on the problem of the origin and fate of the germ cells.

Then in 1892 came Weismann's theory of the segregation and continuity of the germ plasm. His theory was based on conclusions he reached from a study of the formation of germ cells in several species of hydroid hydrozoans (1883). Because an eye ailment prohibited further microscopy, Weismann's final formulation of the theory was the result of a fair amount of speculation. Both his observations and speculations have been seriously questioned by Berrill and Liu (1948). The theory of the continuity of the germ plasm stimulated a flurry of activity to test its credibility, and it has had a marked effect on biological thought for decades.

Winiwarter (1901) conducted the first serious investigation of the development of the mammalian ovary wherein a large, closely graded, chronological series of developmental stages was collected and studied. His beautifully lithographed plates and text illustrated the developmental changes that occurred in the rabbit ovary from the time of sex differentiation to four weeks after birth. Winiwarter became intrigued with the unique chromatinic patterns of the oocyte nucleus during the meiotic prophase stages (mainly from birth to about 15 days). Earlier workers must have seen these structures, but he was the first to describe, figure, and name them (deutobroque, leptotene, synaptene, pachytene, and so forth). He concluded that these nuclear patterns were strictly characteristic of germ cells, and that the only proof of a cell being a germ cell was whether or not its nucleus had undergone these characteristic phases.

From 1901 to the present time, then, many investigators have made determined attempts to ascertain the origin and history of the ovarian germ cells. New and sophisticated tools and techniques have been used, as they became available, but there is not now, as there never has been, an explanation that is acceptable to all.

Two basic, but divergent, explanations, with minor variations within each, have been advanced: 1) the germ cells are segregated early in ontogeny, perhaps in early cleavage, and remain "sealed off" from the somatic cell line as primordial germ cells. Eventually, they migrate to the primitive gonad where they multiply further and differentiate into sex cells; 2) the germ cells arise by differentiation from somatic cells of the gonad.

The place of origin and manner of migration of the primordial germ cells of mammals have now been fairly well established. Ożdżeński (1967) has apparently been able to recognize them in the endoderm of the region of the primitive allantoic diverticulum in presomite mouse embryos. Vanneman (1917) saw large cells which she believed to be primordial germ cells in the endoderm of the "two-bud" blastocyst of the nine-banded armadillo, the earliest stage to which they have been traced in mammals. Since these cells are so large and distinct in the armadillo, it would be an ideal form on which to use the alkaline phosphatase and other modern techniques for identifying and tracing their migration and activities.

Most of the studies on the migration of primordial germ cells in mammals have been on mouse and human embryos (Everett, 1943, 1945; Witschi, 1948, 1963; McKay et al., 1953; Chiquoine, 1954; Mintz, 1957*a*, 1957*b*; Mintz and Russell, 1957; Pinkerton et al., 1961; Odor and Blandau, 1969*a*, 1969*b*). Blandau, White, and Rumery (1963) showed by cinephotomicrography of cell cultures that the primordial germ cells of mice and hamsters are motile. This has further substantiated the histological evidence that these cells migrate by amoeboid motility. Peters (1970) presented data on the time and duration of oogenesis and migration of germ cells in the laboratory rabbit, hamster, mouse, rat, and guinea pig, and in the European ferret, rhesus monkey, cat, cow, and man.

By comparing the gonadal development of a normal strain of mice with that of a mutant strain which exhibited spontaneous sterility (in progeny that are homozygous for certain genic loci), and by using staining techniques that visualize the cytoplasmic alkaline phosphatase present in the primordial germ cells, Mintz (1960, 1961) produced evidence that the sterility was caused by failure of germ cells to reach the gonad and to proliferate there. She stated that the germ cell line was normally established early in embryonic life — by the time of neural fold and somite formation — and that all subsequent reproductive cells were derived directly from this line. This theory was further tested by irradiating pregnant females of the normal strain between the 8th and 12th day of gestation (400 R delivered in 5 doses). This treatment resulted in the complete absence of germ cells in the gonad, as revealed at autopsy on the 14th day. The effect, if any, on other gonadal elements was not mentioned.

Jost and Prepin (1966) applied histochemical methods, similar to those used by the investigators studying the mouse and human, to 25–34-day bovine embryos, and obtained very similar results. Jost and Prepin, however, did see some alkaline phosphatase-positive cells within the fetal blood vessels, and raised the question of whether some bovine primordial germ cells migrate by the vascular route, as they have been shown to do in birds (Simon, 1960).

These anatomical and experimental procedures have left little reason to doubt that in mammals primordial germ cells arise from endoderm in the caudal portion of the very early yolk sac and migrate within the hindgut mesentery to the dorsal abdominal lining and thence laterally to the gonadal ridges, where they undergo further multiplication. There is also good experimental evidence that in the mouse early destruction of these primordial germ cells results in sterile gonads, and that all definitive germ cells of this species are therefore derived from the primordial ones. There is much evidence, however, that the latter may not be true of some other mammals, where neo-oogenesis, even in adult life, may occur.

Descriptive studies of postnatal and adult ovaries have yielded divergent conclusions with respect to the importance of the primordial germ cells. Hargitt (1930*a*, 1930*b*, 1930*c*), working with the rat, Sneider (1940) with the rat and cat, Duke (1941) with the rabbit, Bookhout (1945) with the guinea pig, and Barton (1945) with the dog all indicated varying degrees of oogenetic activity in the postnatal ovary. Sneider even cited evidence that prepuberal oogenesis is rhythmical. In addition, other investigators, incidental to other major research objectives, have reported evidence of neo-oogenesis: Harrison (1948*b*), working with the goat, R. K. Enders (1952) with the mink, Wimsatt and Trapido (1952) with the common vampire bat (*Desmodus rotundus*), and Wagenen and Simpson (1965) with the rhesus monkey and human. But Oehler (1951), after a study of human fetal ovaries and those of infants 4, 11, and 16 months of age, concluded that there was no evidence for postnatal oogenesis. With respect to ovaries older than 16 months, this conclusion is obviously unwarranted. A. C. Enders (1960) stated that, in the nine-banded armadillo, neoformation of oocytes, as revealed by the presence of meiotic activity, ceased at the end of the juvenile stage.

Harrison and Matthews (1951) examined a series of ovaries, representing some 98 mammalian genera, to determine the presence of invaginations of the germinal epithelium and to determine the relationship of these to postnatal oogenesis. Invaginations were reported in 26 genera, but no relationship to oogenesis was recognized. Since the ovarian picture, including the activity of the germinal epithelium, may vary with the stages of the reproductive cycle, it is worth pointing out that 38 of the genera were represented by a single specimen. We have seen such invaginations in adult ovaries of several genera, and in some, also included by Harrison and Matthews, there is rather good evidence to suggest a relationship to oogenesis — *Procyon* (raccoon), *Canis* (dog), *Tupaia* (tree shrew), and *Nycticebus* (slow loris). In others — *Tamias* (chipmunk), *Marmota* (marmot), *Erethizon* (porcupine), *Spermophilus* (ground squirrel), *Felis* (cat), and *Paradoxurus* (palm civet) — we have seen in the germinal epithelium probable transitional stages between epithelial cells and oocytes.

Using the rhesus monkey to test the theory of postnatal oogenesis, Vermande-Van Eck (1956) made a quantitative analysis of the number of normal and atretic oocytes (in both immature and mature females), and used X-irradiation as a means of determining the rate of follicular atresia. She concluded that at any time during the cycle 4.5% of the follicles of all sizes were in the process of acute degeneration. The X-rayed ovaries indicated that all follicles, except the primordial ones, had completely degenerated within 14 days after irradiation. Basing her conclusions on 1) the quantitative analysis of the total oocyte population of both immature and mature females, 2) the estimate that 4.5% of the follicular population was undergoing degeneration at all times, and 3) the rate of the degenerative process as determined by irradiation, Vermande-Van Eck theorized that the follicular supply of the rhesus monkey female would be exhausted within a two-year period unless there was postnatal oogenesis. She gave no histological evidence for such a process.

The existence of a fetal type of oogenesis in adult *Galago* ovaries was described by Gérard (1919–1920, 1932). Not only was the germinal epithelium active in proliferating cell cords into the cortex, but germ cells, including some in the meiotic prophase I, were in the cords and the germinal epithelium. Such invaginations often terminated in large nests of epithelial cells and germ cells. In 1932, he reiterated his earlier observations, and described similar activity in two other species of the same genus. Rao (1927) reported comparable oogenetic activity in the ovary of the adult slender loris (*Loris tardigradus*). Gérard and Herlant (1953) showed the same type of oogenesis in the ovary of the adult potto (*Perodicticus*). A study of the ovaries of the adult slow loris (*Nycticebus*) revealed oogonia and/or oocytes not only in tubular invaginations and in cortical egg nests but in the germinal epithelium as well (Duke, 1964, 1966*b*, 1967). *Loris*, *Nycticebus*, *Perodicticus*, and *Galago* are all prosimian primates of the family Lorisidae. Petter-Rousseaux and Bourlière (1965) studied the ovaries of 6 of the 10 genera of the Lemuriformes (omitting tupaiids), but found the fetal type of oogenesis in only one, *Daubentonia.*

Ioannou (1968) verified the presence of nests of oogonia in the ovaries of adult *Perodicticus*, *Loris*, and two species of *Galago.* In addition, he injected tritiated thymidine into two specimens of *Galago demidovii.* Some of the mitotically active oogonia and some oocytes in early stages of the meiotic prophase incorporated ^{3}H-thymidine. Ioannou concluded that active oogenesis was taking place, the source being preexisting oogonia that were just then undergoing division and growth.

Butler and Juma (1970) repeated this experiment on three *Galago senegalensis* known to be 2–2½ years old. In two animals, they found numerous heavily labeled cells in the cortical germinal cords of the ovaries removed four hours after injection. All had nuclei characteristic of oogonia. Cells with meiotic prophase nuclei were unlabeled. In the third animal, one ovary removed at four hours was the same, but the other, removed 10 days later, had no heavily labeled oogonia, but many lightly labeled ones. However, it did contain many heavily labeled oocytes with "zygotene/pachytene" nuclei. Butler and Juma concluded that some oogonia had continued to divide after labeling, while others had quickly entered the prophase of meiosis, resulting in heavily labeled oocytes. They decided that this was a strong indication of new formation of oocytes in the adult *Galago.*

In a histological study of fetal and postnatal ovaries of *Loris tardigradus* and of six postnatal *Nycticebus coucang*, Kumar (1968) made quantitive estimates of the number of germ cells of the ovaries of *Loris.* He found considerable individual variation in the total number of germ cells and in the number of cells undergoing mitosis and in the prophase of meiosis, and he interpreted this to mean "that oogenesis may occur continuously during postnatal life with spurts of intense activity at pro-oestrus, oestrus and early pregnancy" (p. 175). Referring to the use of tritiated thymidine, he said: "What is not known is whether these labelled germ cells would be incorporated into follicular envelopes if allowed to continue their development for a sufficiently long period" (p. 175). He was unable to ascertain the source of the germ cells formed during successive cycles, but felt that it was unlikely to be the germinal epithelium (surface epithelium). Although Ioannou (1968) mentioned the presence of an occasional oogonium in the germinal epithelium, he too felt there was "no reason to believe that such cells are derived from transformed epithelial cells" (p. 142), and that they were more likely the daughter cells of pre-existing oogonia.

Removal of the germinal epithelium in rats by tannic acid applications (Moore and Wang, 1947; Mandl and Zuckerman, 1951*b*) had no appreciable effect on the number or classes of follicles even after more than a year following treatment. Mandl and Zuckerman (1951*a*) also found that phenol injected into the ovarian bursa of rats left the epithelium mainly intact, yet there was a marked reduction in the number of follicles when compared with those of the control animals. Both groups of investigators concluded that if any oogenesis occurred during the course of these experiments it did so without benefit of the germinal epithelium. Yet Everett (1943), using ovarian transplants, found that if the germinal epithelium of the transplant was destroyed by fusion to the host tissue no new oocytes were formed, but that if the ovarian bursa was included in the transplant, insuring an intact germinal epithelium, then proliferation of oocytes continued from the epithelium. However, Everett reasoned that these oocytes arose from primordial germ cells, set aside early in ontogeny, which were so closely related to the somatic epithelial cells that they could not be distinguished morphologically.

Rudkin and Griech (1962) injected tritiated thymidine into pregnant mice midway through gestation. They found the label in the fetal oocytes undergoing the prophase of meiosis (two to four days after injection). They also found the tritium label in oocyte nuclei of litters six weeks after birth following similar treatment of pregnant females. They concluded that the label was picked up by the cells undergoing protein synthesis (preparation for chromosomal division), and that, since the oocytes do not divide after entering the meiotic prophase, the cells identified by the label (six weeks after birth) were actually labeled at midgestation. Kennelly and Foote (1966) repeated the same type of experiment on the rabbit. In this animal, the prophase I stages of meiosis do not appear in the ovary until a few days after birth. Thus, newborn females could be injected directly with the tritium label, and the status of the label determined at various ages, from 4 to 40 weeks after injection, by means of unilateral ovariectomies. It was calculated that 91% of the oocytes were still labeled at 40 weeks, well after the attainment of sexual maturity. Another series of doses was injected at four weeks of age and the females superovulated at 20 weeks of age, at which time it was determined that 82.5% of the recovered ova were tagged.

The authors said that their results "conclusively support the view that most, if not all definitive ova are formed at birth and *de novo* oocytogenesis does not occur in the post-pubertal rabbit" (p. 573). Crone and Peters (1968) reported the unusual incorporation of tritiated thymidine in the oocytes of mice injected shortly after birth. This they thought to be due to abnormal DNA synthesis in these cells.

Gaillard (1950) described the results of tissue culture experiments, using human fetal ovaries of various ages (14–36 weeks). The central core of the transplants invariably degenerated, but if the germinal epithelium remained attached to parts of the transplant surface, this epithelium proliferated cell cords which contained "a great number of newly formed genital cells and all stages between morphologically indifferent epithelial cells and differentiated genital cells . . ." (p. 28). He also reported that in some of the transplants from the older fetuses, a second source of epithelial proliferation was from surviving follicular cells.

On the basis of electron micrographs, Rhodin (1963) stated, without citing any experimental evidence, that the germinal epithelium of the mouse gives rise to oocytes throughout adult life. His electron micrographs of the germinal epithelium reveal a much clearer picture of the diversity of cellular size and shape (transitional stages?) in this layer than do light microscope photomicrographs.

From 1949 to 1962, Mandl and Zuckerman and associates published a series of experimental and statistical studies on the rat which strongly indicated the absence of oocyte formation after the immediately neonatal period. For an excellent marshaling of their work and other evidence against oocyte formation in the adult mammal, see Franchi, Mandl, and Zuckerman (1962) and Beaumont and Mandl (1962).

Neither descriptive nor experimental approaches to the problem of neo-oogenesis have as yet yielded sufficiently clear-cut evidence to generate a unanimity of opinion. Even Winiwarter (1942) saw evidence of neoformation of ova, but seemed to think it was an anomalous condition. There are still many questions that need answering: In an organ made up of tissues as plastic as those of the ovary, is the oogenetic capacity ever lost? If so, when and why? Is the maintenance or loss of this capacity the same in all mammalian genera? Are the germinal epithelium and stroma involved in oogenesis, or are all definitive oocytes direct descendants of primordial germ cells? What are the potentialities of the cells of the germinal epithelium? What is the effect of movement into the stroma on this potentiality? Can oogenesis occur after the prophase stages of meiosis? Can the cytological end results of meiosis be achieved without the conventional prophase stages? Is it not really the telephase of meiosis, when the first polar body is formed and the chromosomal number reduced, that is the real proof of a cell being a definitive female gamete? What evidence is there that any particular oocyte, in a primordial or primary follicle, persists in a semidormant state for periods up to fifty years and is then capable of completing successful maturation and ovulation? Is the potentiality of small ovarian follicles the same in short-lived and long-lived species? Of the hundreds of mammalian genera whose ovarian histology is unknown, how many, like the lorises, will show the fetal type of oogenesis in the adult? In adult ovaries, new primary follicles always appear first in the outer cortex. How have oocytes that were formed perinatally or even during the juvenile period been able to maintain this position when all other structures such as atretic follicles, corpora lutea, and corpora albicantia are gradually buried more deeply and moved toward the corticomedullary border by continuing hyperplasia of the outer cortex (see pp. 272–73)?

The need is apparent for histological studies of numerous mammalian genera plus a vigorous program of experimental research to clarify the many unsolved problems of oogenesis. It is also manifest that both morphological and experimental studies of this problem must consider the possibility that neo-oogenesis may occur only during a very short phase of each reproductive cycle. Bullough (1942) cautioned about this, for he found cycles of mitotic activity in mouse germinal epithelium which were correlated with phases of the estrous cycle. There is also the probability that, if oogenetic waves do occur in adult life, the actual number of cells involved at any one time may be relatively few, hence easily overlooked.

For further discussion of these problems, see Gropp and Ohno (1966), Edwards (1970), and Tarkowski (1970).

Ovarian regeneration

It is somewhat surprising that so few studies have been made of ovarian regeneration. Most were undertaken during the third decade of the current century: Davenport, 1925 (mouse); Parkes, Fielding, and Brambell, 1927 (mouse); Haterius, 1928 (mouse); Heys, 1929 (rat); and Pencharz, 1929 (mouse and rat).

Davenport (1925) reported 64% regeneration in mice following unilateral or bilateral ovariectomy; this was interpreted as regeneration from nonovarian tissue, not as hypertrophy of unremoved fragments. No doubt this report stimulated other similar studies. Parkes, Fielding, and Brambell (1927) and Haterius (1928) noted some regeneration in their studies of the mouse, but the percentage was much lower than that of Davenport: Parkes, 6%; Haterius, 5%. Heys (1929) and Haterius each concluded that most cases of regeneration occurred from unremoved ovarian fragments and not *de novo* from coelomic epithelium or other nonovarian tissue. Parkes, Fielding, and Brambell were of the opinion that new ovarian tissue derived from a nonovarian source was demonstrable in 8 of 121 cases. Young (1944) reported that four out of five ovariectomized chimpanzees later became cyclic. He concluded that functional ovaries could regenerate from "exceedingly small fragments of ovarian tissue."

A study by Butcher (1932) of adult rats, in which the blood supply to each ovary was ligated, involved regeneration of a different sort. Follicular degeneration occurred within 10 days of ligation. Revascularization was established within 10–20 days, and by the 30th day the ovaries appeared normal. Butcher attributed much of the reappearance of normal follicles to the activity of the germinal epithelium. As an ancillary part of this study, Butcher also transplanted small ovarian fragments to the perinephric fat of a series of rats. Again, within 9 days the follicles of the transplants had disappeared, yet the transplants increased in size. Butcher again attributed this to proliferation of germ cells from the germinal epithelium, followed by growth of the follicles thus developed.

Pansky and Mossman (1953) tested the regenerative capacity of different segments of the rabbit ovary by selective segmental removals, and found that the caudal portion of the ovary regenerated much better than the cephalic part. No explanation for this phenomenon was obtained from the study.

It is regrettable that no careful histological studies were made by Davenport and Haterius nor in the transplant experiments by Butcher. It is important that we know the types of cells and tissues elaborated in regeneration (and in hypertrophy), as well as their source. Pansky and Mossman mentioned the presence of follicles in all stages of development and atresia in the ovarian tissue of their experimental animals, but here again, careful histological study of a chronological series would have been helpful.

With improved methods and media for tissue culture, this technique should be used more extensively. Gaillard (1950) stressed the importance of an intact germinal epithelium for the successful regeneration and growth of transplants of small pieces of human fetal ovaries. The use of colchicine in regeneration experiments might be revealing. Investigators should also attempt working with animals having a more elaborate and active germinal epithelium, such as the raccoon, dog, and loris, both in regeneration and tissue culture experiments.

There seems to be no doubt that appreciable growth of the remaining portion of a partially extirpated ovary can occur, as Pansky and Mossman showed. In that sense, ovarian regeneration is certainly a fact. The unanswered and more important questions are: 1) Can new oogonia differentiate from the remaining portion of a partially extirpated ovary? 2) Can typical ovarian tissue — germ cells, epithelium, or gland cells — differentiate from originally nongonadal tissue of the gonadal area, such as the mesovarium or proper ovarian ligament?

Natural superovulation

In mammals, the average number of eggs ovulated exceeds the number of embryos implanted, and the average number implanted exceeds the number born. In other words, the factor of prenatal death begins to act at fertilization and continues until parturition is completed. Many species compensate for this loss in obvious ways, such as ovulation of a few more eggs than can be efficiently nourished in the uterus and quick recurrence of estrus after an early interruption of pregnancy. However, a few mammals seem to have far overcompensated, to the point where it is perhaps unreasonable to read into the situation any adaptive factor. Conceivably, the 5–8 eggs ovulated and fertilized by the pronghorn antelope (*Antilocapra americana*) (O'Gara, 1969) are insurance that at least one or two will successfully implant. But what of the 50–60 ovulated by the South African long-eared elephant shrew (Horst and Gillman, 1941) when only two can be accommodated, or of the 300–800 of the plains viscacha (*Lagostomus maximus*) (Weir, 1971) when only 7–8 are to be fertilized and only 2 will implant? It seems unlikely that natural selection would have proceeded to such extravagance merely to assure a litter of two. The more reasonable interpretation would seem to be that, if this degree of superovulation is adaptive at all, it is related to something other than litter size. It does provide large numbers of thecal glands and of corpora lutea; if large amounts of these glandular tissues are for some reason necessary in this species, this is one way to get them. In a similar way, atresia of late secondary and early vesicular follicles in many species assures a large supply of thecal type interstitial gland tissue.

The endocrinology of these natural superovulators should be investigated. Either their gonadotropins are extremely potent or at high levels, or their ovaries are unusually sensitive to them.

Morphologically, the ovaries of the pronghorn and of the elephant shrew differ little from those of closely allied species with a more typical ovulation number. However, the ovary of the plains viscacha is unique (Figs. 7.6–7.10). It is complexly lobed and convoluted, an obvious necessity if room is to be provided for hundreds of ripe follicles or corpora lutea without a biologically uneconomical three-dimensional expansion of the ovary. Similar demands have been met by this sort of lobulation in many other organs — compound glands, kidneys, some placentae, and the brains of higher mammals. Lobulation and convolu-

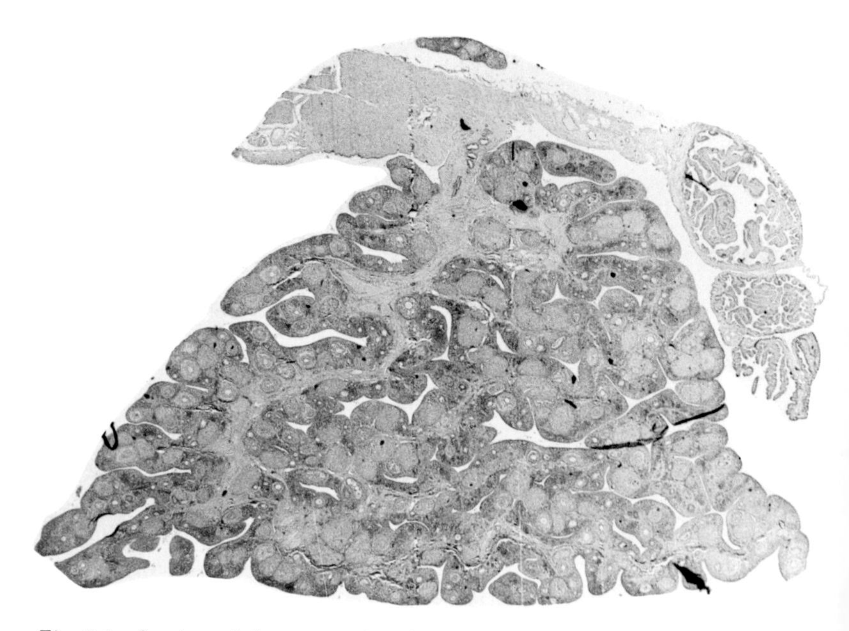

Fig. 7.6 Section of the ovary of a plains viscacha (*Lagostomus maximus*) in the middle trimester of gestation, showing its complex convolutions and lobulation. *L. maximus*. × 8. (Courtesy of Barbara Weir.)

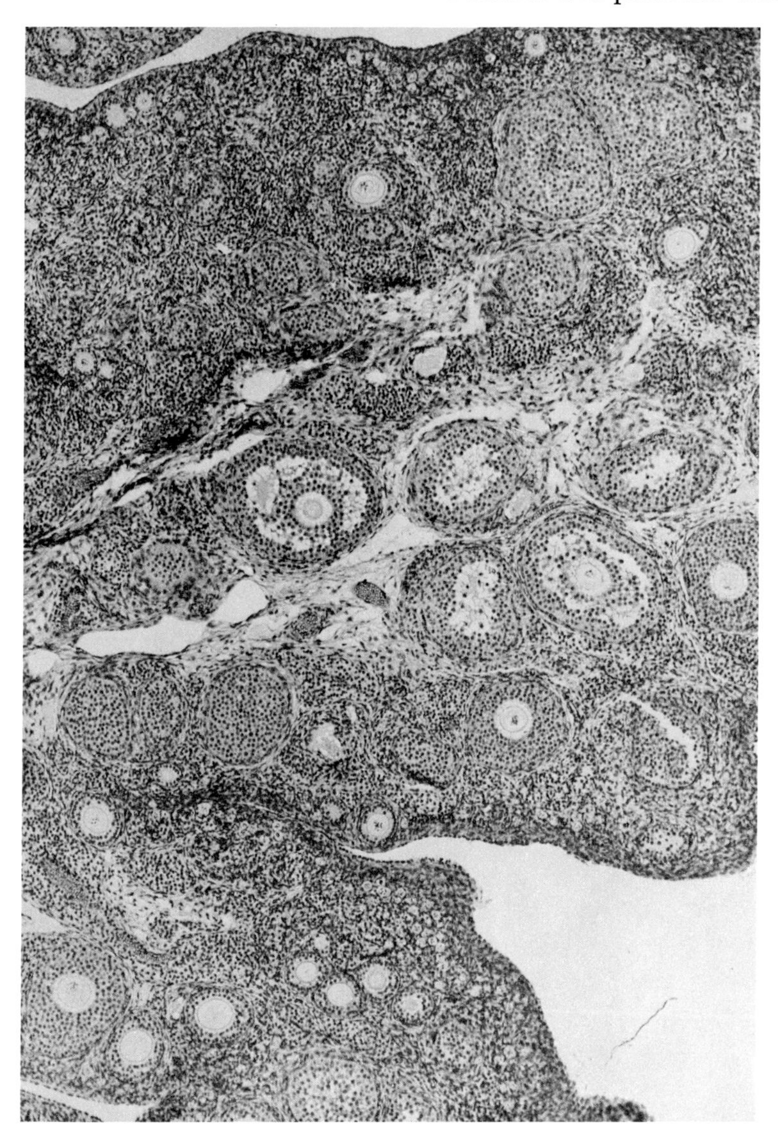

Fig. 7.7 Lobules of a proestrous ovary of the plains viscacha (*Lagostomus maximus*), showing the small size of the mature follicles and their antra. Scattered interstitial gland tissue is fairly abundant in the cortex, but no corpora atretica are obvious. *L. maximus*, 53. × 80. (Courtesy of Barbara Weir.)

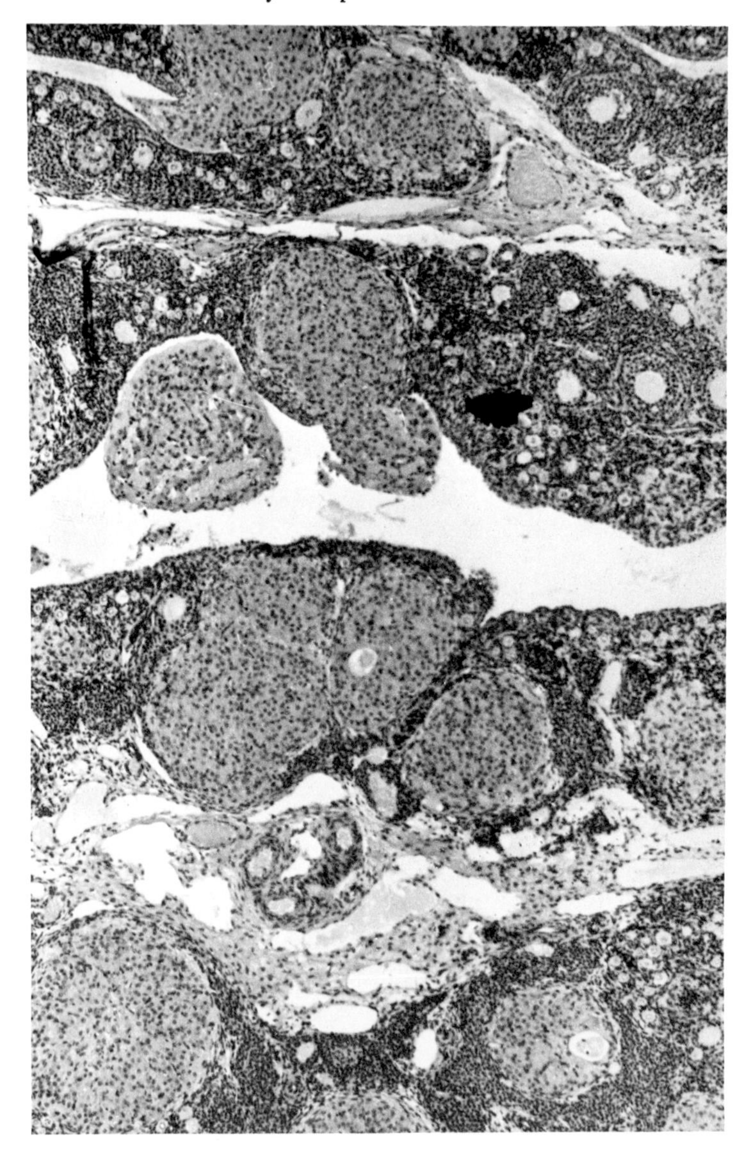

Fig. 7.8 Section of adjacent lobules of the postovulatory ovary of a plains viscacha (*Lagostomus maximus*), showing young primary and accessory corpora lutea, the latter with the retained ova. Corpora atretica with thecal type interstitial gland tissue and without luteal cells are not seen. *L. maximus*, 44. × 64. (Courtesy of Barbara Weir.)

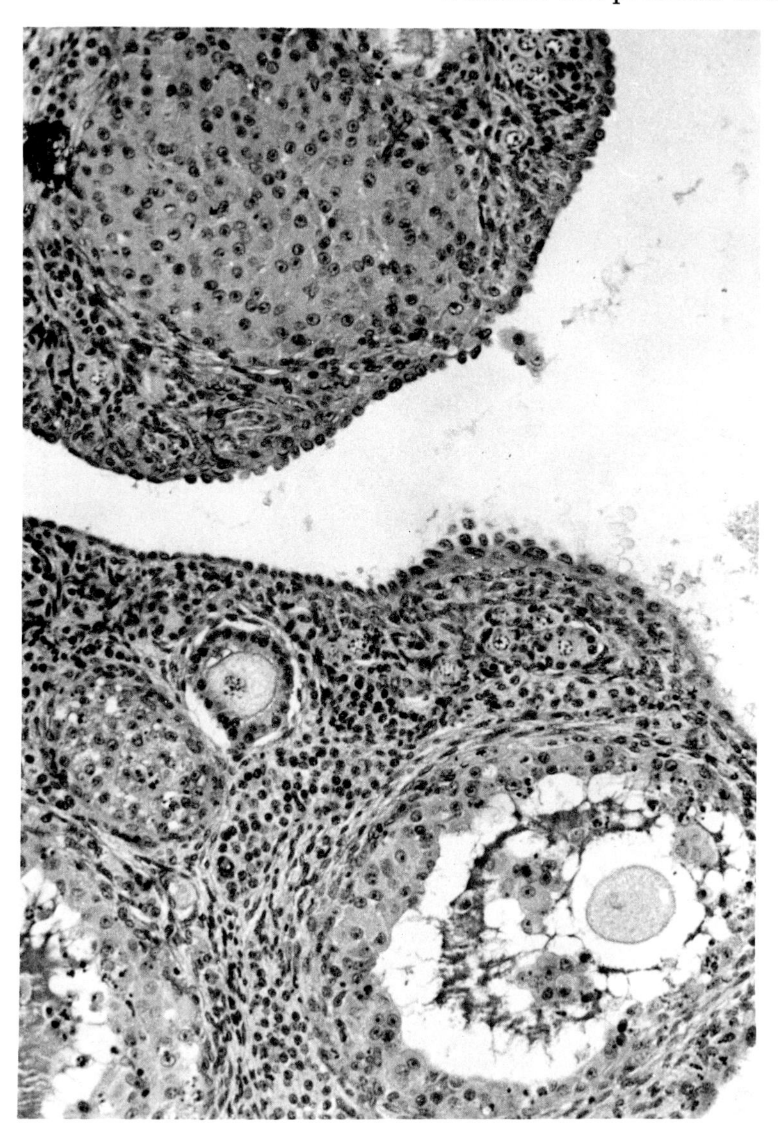

Fig. 7.9 A 12-hour postcoital ovary of a plains viscacha (*Lagostomus maximus*), showing obvious lutealization of the epithelium of two atretic follicles. Numerous small, relatively undifferentiated interstitial gland cells occur in the cortex, but seem to bear no relation to atretic follicles. *L. maximus*, 4. × 200. (Courtesy of Barbara Weir.)

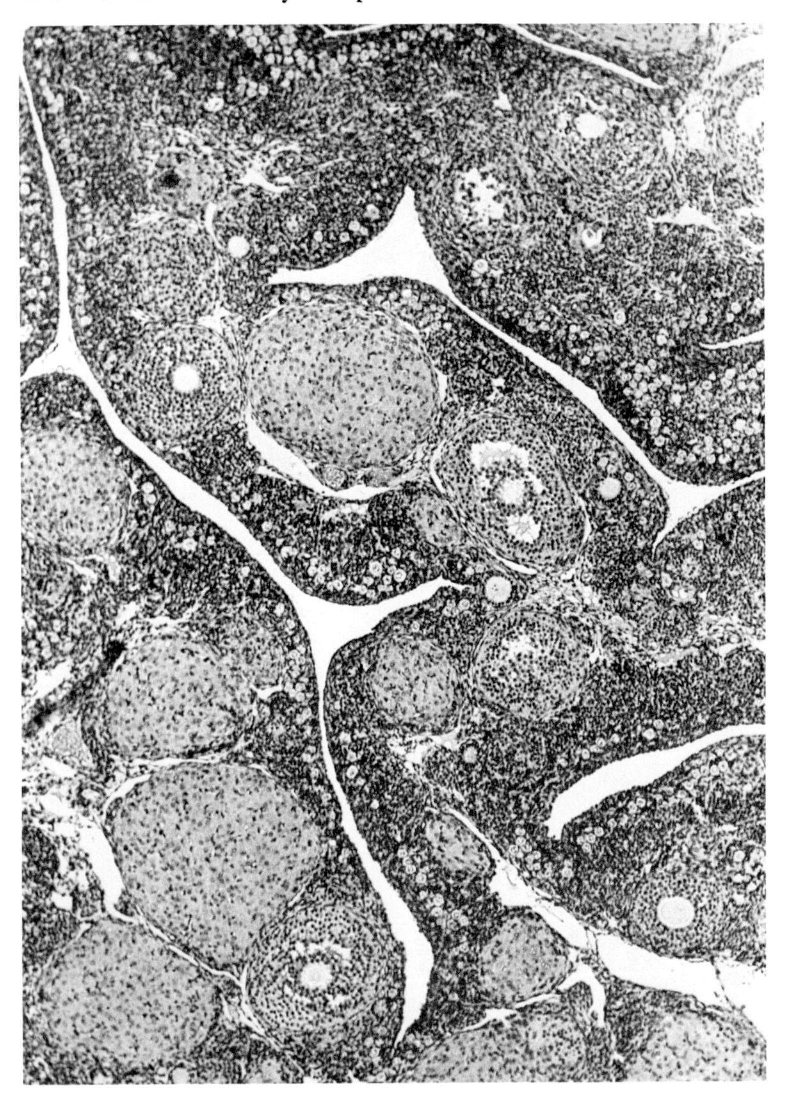

Fig. 7.10 Lobules and fissures of the ovary of a plains viscacha (*Lagostomus maximus*) at the 133d day of gestation, showing numerous corpora lutea, probably both primary and accessory. These animals are believed to ovulate a few follicles occasionally during pregnancy, which may account for the two nearly full-sized vesicular follicles, although large nonovulatory follicles are common during pregnancy in many mammals. *L. maximus*, 42. × 64. (Courtesy of Barbara Weir.)

tion are not only economical of space, but also of nerves, blood, and lymph vessels, which would need to be far longer to supply an equal area spread out as a broad smooth double-surfaced organ. It is also almost imperative that if large numbers of follicles and corpora lutea are to be accommodated, they must be relatively small. Actually, these structures of the plains viscacha are the smallest known among mammals: about 200 μ for the diameter of a mature follicle, and about 400 μ for the diameter of a corpus luteum that is fully developed. Compare the viscacha with the porcupine (*Erethizon dorsatum*), a hystricomorph of about the same size, whose ripe follicles and corpora lutea have a diameter of about 6 mm. As can be seen in Figure 7.7, the antra of the viscacha's follicles are unusually small. Its oocyte diameters, 70 μ, are also nearly comparable with those of most other mammals, which are usually between 100 and 200 μ. Follicular and luteal cell size is also about normal, in spite of the small size of the follicles and corpora.

Natural superovulation, particularly as it occurs in the plains viscacha, poses many intriguing problems, an understanding of which might well lead to practical applications. Why does so much cortex develop in the first place? Is there some inducing factor that stimulates not just oogonial multiplication and follicular formation, but hyperplasia of the whole cortical stroma as well? Is there possibly some obligatory relationship during development between number of germ cells and number of indifferent stromal cells? Why do the follicles of this species mature and ovulate at a size and cell number (Fig. 7.7) comparable with the early secondary follicles of most mammals? How do ovulated eggs escape from the numerous and complex interlobular fissures? Is there some form of ovarian motility that squeezes them out of these recesses, or is there enough ovarian fluid released to flush them out? Can the low percentage of fertilized eggs be accounted for by a high percentage of defective sperm or by defective eggs, or is there some unfavorable factor in the tubal environment? What are the special properties of that limited portion of the endometrium in each uterine horn which seems to be the only spot fitted to allow egg implantation?

The latter situation is not unique to the plains viscacha, but is very obvious in the elephant shrew, some African antelope (Child and Mossman, 1965), and probably is true to an appreciable extent in most mammals with large uterine horns, which normally accommodate only one embryo. Certainly the constancy with which blastocysts of a given mammalian group are limited in their attachment to one side and often to a particular proximodistal level of the uterine cornu strongly indicates a specific suitability of these areas not possessed by the rest of the uterine mucosa (Mossman, 1971).

Another angle to the problem of the number of eggs ovulated compared with number of embryos successfully gestated involves the anatomic capacity of the uterus and, more importantly, the physiological capability of the mother to maintain more than a certain number of fetuses to birth. This is manifest in the fact that litter size is largest in the prime period of reproductive life, and also in the higher incidence of prenatal death and premature delivery in cases of multiple conceptuses in such species as cattle, horses, and primates, in which single embryos are the rule. Regardless of the degree of superovulation, each female has a mechanism that limits the number of young she can successfully gestate.

Superovulation could therefore be defined as ovulation of a number of ova which is beyond the carrying capacity of the individual female in which it occurs. However, this is not the ordinary concept of superovulation, for, on the average, all species tend to ovulate more eggs than are normally carried to term. Superovulation is therefore commonly defined as the ovulation at one ovulation period of many more eggs than can be successfully gestated.

Intraovarian parthenogenesis

Certain tumors of the ovaries, particularly teratomas, may possibly arise by parthenogenetic development of ova; many instances have also been reported of apparent cleavage stages of embryos in atretic follicles of mammals (Loeb, 1905, 1912, 1930, 1932; Athias, 1929; Harman and Kirgis, 1938). The subject has been reviewed by Austin and Walton, 1960. Mossman has seen equally cleaved ova, even morulalike stages, in a few ovaries. There is no proof that these are anything other than oocytes that have undergone "fragmentation" during their degeneration. Only rarely would such a degenerative process accidentally result in equal-sized nucleated fragments. However, since masses of unequal fragments, some with nuclei, some without, are fairly common in atretic follicles, there is the possibility of purely accidental normal-appearing "cleavage stages," but this does not account for the teratomas. Some form of "embryonic" development and differentiation is necessary to explain these, especially the more circumscribed or cystic ones where direct metaplasia of pluripotential ovarian stromal cells is not likely. In such tumors, there is often no evidence of spreading differentiation through contiguous ovarian tissue comparable with that of cartilage or bone development in blastemal tissue of a limb bud. Instead, the differentiating tissue is confined within a cyst, as if it had been derived from a single small mass or a single cell and had grown as an embryo grows, by nutrition derived through the surrounding tissues, but not by direct differentiation from them.

Primary ovarian pregnancy has been known to occur in mammals, presumably because an egg has adhered to the stigma of the ruptured follicle, was fertilized there, and failed to enter the oviduct. These may become advanced normal embryos, so there is no reason why the ovary cannot be considered capable of supporting an embryo. Confinement within an atretic follicle could be the reason why presumed parthenogenetic development ends so quickly or in such abnormal growths. Experimental artificial rupture of numerous atretic follicles in ovaries might occasionally result in an advanced parthenogenetic embryo. To our knowledge, this has never been tried. Such an effort might be thwarted by adhesions, but techniques should be possible which would minimize this.

Induction and metaplasia in the ovary

Induction of one type of tissue or cell by another during ontogeny has been demonstrated experimentally many times. Hisaw (1947) suggested that inductor mechanisms might be responsible for the initial organization of primordial ovarian follicles. Ohno and Smith (1964) correlated contact of the oocyte with follicular cells and the ability of first meiotic prophase oocytes to enter the dictyotene phase. They also thought that the degeneration of early oocytes was the result of failure to contact follicular cells. The sequential development of follicular epithelium around ova, of theca interna around this epithelium, and of theca externa around the interna suggests that an inductor mechanism is involved. Perhaps the ripening of an egg, the rapid increase in follicular fluid during proestrus and estrus, and thecal gland differentiation are all part of this system. Even the differentiation of thecal type interstitial gland around atretic follicles could depend partly on induction.

El-Fouly et al. (1970) removed ova from large follicles of rabbits. The granulosa of these follicles then differentiated into true luteal cells. Normally occurring accessory corpora lutea often retain the ova for at least a few days, as in the porcupine. This is a clear indication that in some species lutealization is not inhibited by the oocyte. Almost certainly these processes are influenced by gonadotropic hormone levels and may have an obligatory relation to this factor.

Metaplasia as a normal nonpathological process is common in the ovary. Theca interna cells of early vesicular follicles appear to be well on their way to becoming thecal gland cells, yet when such follicles begin to undergo atresia their thecal cells change course and become interstitial gland cells. Eventually these dedifferentiate and lose their identity among the undifferentiated stromal cells. While there is no experimental proof that these former interstitial gland cells are ever again pluripotential, their gradual disappearance without indication of cytolysis and the gradual reappearance from such indifferent stroma of new interstitial gland cells, new thecal gland cells, and even of paraluteal and luteal cells are strong circumstantial evidence to justify this interpretation. Ovaries certainly contain pluripotential cells which, given the right tissue environment, may be induced to differentiate not only into "climax" types, such as thecal and luteal gland cells which degenerate after serving their function, but also into "temporary" functional cells, such as thecal type interstitial gland cells, which after serving their function dedifferentiate and await another induction period to again become either interstitial cells, or one of the climax types. This ability to dedifferentiate and redifferentiate qualifies many ovarian stromal cells as potentially metaplastic.

Ovaries are unusually subject to teratoma formation, especially dermoid cysts, which are usually bilateral. Mossman has seen these in two pairs of ovaries of the white-tailed deer (*Odocoileus virginianus*), while Duke has noted membranous bone plaques associated with atretic follicles in the mouse deer (*Tragulus javanicus*), as well as masses of epithelium and cartilage in the ovary of a gibbon (*Hylobates lar*). Whether these were derived by metaplasia or by parthenogenetic development of germ cells is unknown, but the parthenogenetic derivation seriously proposed for these teratomas is certainly reasonable for the more cystic type, especially in view of the rarity of such tumors elsewhere in adults.

Grob (1971) has devised a monolayer method of cultivation of ovarian cells which would seem to be adaptable for the study of inductor mechanisms between ovarian tissues.

Endocrine tumors of the ovary

We make no pretense of being qualified to contribute much to this subject, but we do know that ovarian tumors, especially the so-called functional or endocrine secreting types, have been notoriously difficult to classify and that there has been much confusion in their identification. We have one suggestion concerning this problem which may have some significance.

In view of the probability of the action of inductor mechanisms in the ovary and the certainty that much of the ovary is composed of "embryonic" or pluripotential cells, why should abnormal inductor mechanisms always be expected to produce identifiable normal cell types, such as granulosa cells or luteal cells? Is it not more reasonable that only occasionally would typical normal types result, that more commonly almost any intermediate, and sometimes even unknown, radically different types might appear? These could be intermediate or unique both morphologically and physiologically. Even if these assumptions are true, they will not solve ovarian tumor classification problems, but they should make it possible to devise some form of graded classification into which most functional tumors could be fitted without pretending an exactness which experience with such material has proved and will continue to prove unrealistic. It does not make sense to classify a tumor cell as luteal when it has characteristics intermediate between a luteal and an interstitial cell and when the patient exhibits signs of excess estrogens rather than progestins. With the present imperfections in our ability to classify cells by their cytological characteristics, it is probable that functional manifestations in the patient should be weighted more heavily than cytological characteristics, provided the tumor can be pinpointed as the source of the disturbance. However, hilus cell tumors always seem to be masculinizing, but some are clearly derived from adneural cells and others from gonadal adrenal cells; in other words, cells of different origin and type may have the same physiological effect. For a good discussion of hilus cells, see Sternberg, Segaloff, and Gaskill (1953) and Green and Maqueo (1966).

For information on ovarian tumors, consult Selye (1946), Novak and Woodruff (1967), and Teilum (1971).

Permanent ovarian asymmetry

Permanent asymmetry in size and function of the ovaries is typical of most birds, but is unusual in other vertebrate groups. It occurs in certain fish (myxinoids and some elasmobranches and teleosts), a few reptiles, and in the duck-billed platypus (*Ornithorhynchus anatinus*), but not in the spiny anteater (*Tachyglossus*). It is rare in marsupials and eutherians, although minor differences in average weights of the two ovaries have been demonstrated in a few species. For excellent reviews of this subject, consult the chapters by Franchi and by Matthews in Zuckerman (1962).

One of the most remarkable cases of ovarian asymmetry in eutherian mammals is that of the mountain viscacha (*Lagidium peruanum*) of South America (O. P. Pearson, 1949). Its left ovary and uterine horn function only if the functional right ovary is removed.

Matthews (1937) showed that the left ovary of the horseshoe bat (*Rhinolophus*) rarely has large follicles and that these never ovulate. Ramaswamy (1961) described similar phenomena in the Indian false vampire bat (*Megaderma*). In both of these, the uterine horn on the side of the nonfunctional ovary is appreciably smaller than the one on the side of the functional ovary. Both are monotocous species, but there are so many monotocous mammals without asymmetry of the ovaries that this correlation can be ruled out as a likely causal factor. The fact that the uterine cornu on the side of the nonfunctional ovary is congenitally smaller than the one on the functional side suggests the hypothesis that there may be an ipsilateral utero-ovarian interdependence controlling the degree of function of both uterus and ovary. This idea may be worth testing experimentally, but consideration should also be given to the fact that a congenitally smaller uterine horn is not always associated with lesser ovarian function on that side. This is well illustrated by at least four species of African antelope (*Aepyceros melampus, Kobus kob, K. lechee, Sylvicapra grimmia*) that have congenitally smaller left uterine horns but equal function of each ovary, although implantation always occurs in the right horn regardless of which ovary ovulates (Buechner, 1961; A. S. Mossman and H. W. Mossman, 1962; Robinette and Child, 1964; Child and Mossman, 1965). One of the long-winged bats (*Miniopterus dasythrix*) has equal-sized ovaries, yet ovulation is always from the left and gestation always in the right cornu (Matthews, 1941). O. P. Pearson's (1949) experiments, in which removal of the functional ovary in the mountain viscacha resulted in development of function of the inactive one, strongly suggest that there is a congenital difference in sensitivity to gonad-stimulating hormones, and that removal of the target organ with the lowest threshold allows circulating gonadotropins to increase until they can activate the less sensitive ovary.

Whatever the mechanism controlling asymmetry in any species, the value of such a condition to the species is difficult to understand. Perhaps it has been of little or no selective value in evolution, but has been perpetuated because it happened to be linked with some other genetic factors in the parent stock which were valuable. One functional ovary and one functional uterine horn would certainly seem to be adequate for a monotocous species.

Smooth muscle associated with the ovary

Since the midnineteenth century, it has been known that certain "ligaments" of the internal genitals of female mammals contain smooth muscle. Hansen (1957) has reviewed the literature, and has described lucidly the distribution of this musculature in man. Besides the delicate intervascular and perivascular sheath musculature associated with most uterine and ovarian blood vessels, heavier strands of so-called subperitoneal musculature make up much of the substance of the suspensory and proper ligaments of the ovary. (Similar musculature also occupies much of the uterine ligaments.) That of the proper ligament of the ovary continues into the mesovarium and tends to ensheath the vessels and other structures of the ovarian hilus. According to Hansen, it also gives off a band, the transverse musculature of the mesovarium ("musculus transversus mesovarii"), which extends along the base of the mesovarium. The function of all this smooth muscle, other than to maintain tonus in the ligaments, is conjectural. The muscle of the suspensory ligament could draw the ovary, oviduct, and uterus cephalad, and that of the proper ligament of the ovary could approximate the ovary and tubo-uterine junction area. A small strand, "musculus attrahens tubae," in the "ligamentum infundibulo-ovaricum," which attaches one area of the edge of the infundibulum to the ovary in man and many other mammals, could pull the infundibulum closer to the ovary and perhaps could thus be part of the mechanism for favorably positioning the infundibulum for reception of an ovulated egg. But these are only guesses, and there is obvious need for study of this strong and definitely patterned system of smooth muscle associated with these crucial reproductive organs.

The existence of contractile cells in the follicular wall and within the cortical stroma has long been controversial. Clearly identifiable trabecular extensions of smooth muscle from the hilus into the adjacent medulla are common in many species, both large and small. These strands have no obvious specific attachment to particular vascular or other ovarian elements, and we believe they serve simply as insertions of the extra-ovarian musculature into the ovary as a whole.

We have not seen typical smooth muscle cells in the cortical stroma or in the theca externa of follicles. However, electron microscopy has clearly demonstrated cells with typical smooth muscle myofilaments in the theca externa of large follicles in the laboratory rabbit, rat, rhesus monkey, and the sheep (Espey et al., 1965; O'Shea, 1970, 1971; Osvaldo-Decima, 1970). Also Lipner and Maxwell (1960) have shown that strips of ovarian stroma respond by contraction to stimuli suitable for activation of smooth muscle. Fink and Schofield (1971), studying innervation of the ovary of the domestic cat, saw unequivocal smooth muscle only in the theca externa. We conclude, therefore, that contractile cells do exist in the cortical stroma and theca externa of some mammals, but that they are not always highly enough differentiated to be identifiable microscopically as typical smooth muscle cells.

It seems almost certain that the contractile cells of the cortex and theca externa play a functional role in ovulation, but just how and to what extent they are necessary is not clear. It must also be kept in mind that they have been demonstrated in only a few species, and that they may not occur in all mammals.

For an excellent discussion of the whole problem of the mechanism of ovulation, see Rondell (1970).

The apparent migration of cortical elements toward the medulla

Normally, the older an atretic follicle, corpus luteum, or corpus albicans is, the deeper is its position in the ovary. This rule applies to normal follicles as well, if one considers the position of their centers. Since these structures have no intrinsic migratory capacity, what accounts for their shifting position? The only explanation seems to be continuous hyperplasia of the cortex, especially of its more peripheral portions. But if this is true, why does not the cortex of the ovary reach a proportionately much greater size than it does? Barring fluctuations associated with reproductive cycles, ovaries do gradually increase in size throughout the active reproductive life of a mammal. However, from the size at the first estrus to that at the end of the active reproductive years, only about a doubling or at most a tripling of linear dimensions occurs.

A number of elements that contribute significantly to temporary increase in ovarian size more or less completely disappear in many mammals during the nonbreeding season. These include the follicular fluid of ovulatory follicles, follicular fluid and epithelium of atretic follicles, and luteal cells. There is also some cyclic decrease in volume by involution of interstitial gland cells, but this is not permanent, for most of them are destined to redifferentiate into some form of active glandular cell. In fact, it appears that in many mammals the number of interstitial gland cells gradually increase over the years. In some species, such as the rabbits and hares (Leporidae), these would account for much of the increase in size with age, but in those that have little interstitial gland tissue there is another factor, an increase in number of stromal cells of the cortex. Hyperplasia of these "embryonic" cells is slow and inconspicuous, but it certainly does take place, most noticeably in that zone of the cortex just beneath the surface epithelium, or tunica albuginea if one is present. The significance of this hyperplasia is not generally recognized. It should be studied and measured. It is analogous to the growth of the so-called secondary cortex of the fetal ovary, and is probably actually a continuation of that process, a phenomenon that does not occur in the testis and that even in the embryo is an important distinguishing feature between male and female gonads.

In the adult ovary, much as in the fetal ovary, peripheral growth thus tends continually to bury such elements as the follicles and corpora lutea. To what extent the buried older stromal cells themselves degenerate is not apparent. Many, especially the outer ones, differentiate into follicular epithelium and luteal cells, both of which eventually disappear. The proportional relation of this degeneration to the hyperplasia is unknown, but it would seem to be the major check on an undue increase in ovarian size.

For a discussion of cortical growth in relation to neo-oogenesis, see p. 257.

Eight

Retrospect and prospect

To attempt to record knowledge is to uncover ignorance. As we consider what we have said and what we have had to leave unsaid in this book, we are amazed at how little is definitely known about ovarian morphology. In the literature, our own as well as that of others, many features are inadequately described and many completely neglected.

Figures 8.1 and 8.2 represent data sheets suggested for recording data on the morphology of a mammalian ovary. They include the items that would give a reasonably complete anatomical characterization of the ovary of a species. We have used similar data sheets for the ovaries we have studied, but rarely has our material been adequate for determining all of the items, nor do we know of any species for which we could fully complete such data sheets from published accounts. The abbreviated nature of our synoptic tables, which are in the next section, and the numerous blank spaces in them are thus understandable. We present these sheets in the hope that future investigators undertaking morphological studies will make use of them or a similar systematic scheme for recording data. Workers concerned with the biology of reproduction need this type of information.

The most obvious deficiency in our knowledge of ovarian morphology is the lack of information on numerous taxonomic groups. Of the 121 families of living mammals (Simpson, 1945), there is a significant amount of information on the ovarian morphology of 29. These are so distributed that the ovarian morphology of at least half of the 18 orders of living mammals is very poorly known. Some of these are comprised of only one to three families (Dermoptera, Edentata, Pholidota, Tubuli-

MORPHOLOGY OF ADULT MAMMALIAN OVARY

OVARIAN BURSA
Presence: Type: Orifice:

OVARY
Location: Shape: Lobes: Fissures:
Surface epithelium: (nature of)
Cortical cords, tubes, and crypts:
Evidence of oogenesis:
Tunica albuginea: (fibrous or cellular, thickness)
Medullary cords or tubules: (presence or absence, nature)
Rete: (size, location, morphology, degree of epithelial specialization)
Epoophoron: (diameter of tubules)
Mature follicles
Number at estrus: Litter size: Diameter: (within basement membrane)
Antrum diameter: (cumulus to opposite wall)
Trabeculae: (present or absent) Thecal gland: (thickness)
Thecal gland cells: (size, degree of differentiation)
Interstitial gland types (presence or absence, location, amount)
Thecal:
Stromal:
Medullary cord:
Gonadal adrenal
at epoophoron:
in medulla:
in cortex:
accessory adrenal bodies:
Adneural:
Luteal glands
Primary: (number, diameter) Accessory: (number, diameter)
Secondary: (number, diameter)
Number of basic cell types:
Persistence of luteal cells:
Corpora albicantia of luteal origin: (fibrosity, pigmentation)
Atretic follicles
Relative abundance:
Corpora atretica: (abundance, degree of development)
Corpora albicantia of follicular origin: (presence or absence)

Fig. 8.1 Sample data sheet.

DISTINCTIVE CHARACTERISTICS OF THE MAMMALIAN OVARY
AT THE VARIOUS PHASES OF LIFE AND OF THE REPRODUCTIVE CYCLE

Late fetus and newborn:

Infant:

Juvenile:

Proestrus
- in adolescent:
- in mature:

Estrus
- in adolescent:
- in mature:

Preimplantation pregnancy
- tubal period:
- uterine period:

Implantation period:

Early placentation:

Midplacentation:

Late placentation and partus:

Lactation
- no postpartum estrus:
- postpartum estrus and no pregnancy:
- postpartum estrus and pregnancy:

Anestrus or diestrus:

Senility:

Fig. 8.2 Sample data sheet.

dentata, Sirenia, Perissodactyla); others include many families of which too few have been investigated to give much assurance that they are reasonably representative of the whole order (Chiroptera, Primates, Cetacea).

A significant amount of information on the development of the ovary from the embryo through the juvenile period to puberty is available for only seven species (man, rhesus monkey, rabbit, rat, mouse, guinea pig, and cat). Embryonic development has been described in considerable detail in several other species, but seldom has postnatal development been considered; in fact, no completely adequate account covers this period for any mammalian species.

Another poorly understood general subject concerns the origin and fate of the various types of interstitial gland cells. This problem has been studied almost solely by only the older histological methods, which often fail to give the clear-cut evidence that it is possible to obtain experimentally. Careful observations on cell origin, their degree of differentiation or involution and degeneration in relation to the reproductive period, if coupled with the proper experimental techniques, could clear up many points concerning these important elements of the ovary. Until much more accurate information is available, the anatomical classification of the types of interstitial gland cells must remain tentative. And until endocrinologists can be provided with reasonably precise information on the location and identification of these cells, there is little hope that their specific functions can be clearly demonstrated.

Anatomical interpretations in regard to the involvement of primordial germ cells and so-called germinal epithelium in the origin of oogonia have been recurring subjects of controversy for about a century. Modern techniques, such as labeling, must be applied to more species which show histological evidence of neo-oogenesis in the adult. Perhaps some species need to provide so few new oogonia at each cycle that finding and identifying the critical nuclear stages may be practically impossible. In other words, statements that all germ cells are in the prophase of the first meiotic division after the early infantile period in a given species may be based on failure to find the few that are not, simply because they are indistinguishable from the multitude of stromal cells among which they lie. Absence of oogenesis in the adult should not be based on decreasing numbers of follicles as animals age. A few hundred new oocytes could be formed during each reproductive cycle and not result in a detectable fluctuation in the steady decrease in oocytes present, if either the percentage or the absolute rate of atresia were steadily increasing.

Examination of embryonic gonads gives little indication that the primary sex cords originate from the surface epithelium. In fact, even at the time of formation of the ovarian cortex, it is difficult to demonstrate that the secondary cords or any other tissue of the cortex result from a proliferation of surface epithelium. How important, then, is the surface epithelium? Is it a germinal epithelium or only a special type of mesothelium with no more germ-cell-forming capacity than the gonadal mesenchyme? The ovaries of some juvenile and even of some adult mammals, such as those of raccoons (Procyonidae), seals (Phocidae), prosimians, and the porcupine (*Erethizon dorsatum*), give evidence that the surface epithelium is germinal, but what of the many species in which oogenic-like processes have never been seen in the epithelium? Could it be that cells separate and mingle with the underlying stroma before they can be identified as oogonia, or could it be that new oogonia simply differentiate from the stromal cells themselves? At the moment, these questions may not be of much practical importance, but they are of great theoretical interest.

Figure 8.3 is an attempt to summarize in diagrammatical form our concepts of the developmental origin and time of occurrence of the major elements of the mammalian ovary. We make no claim that it is correct in every detail, but we believe it presents the most probable hypotheses based on the available evidence, much of which is from our own observations. The questions it raises must be tested by further investigation.

Ovarian characters and taxonomy

Another problem concerns the value of the ovary, in relation to other anatomical features, as a criterion for taxonomic purposes. Our present knowledge indicates that ovarian characters are of slight value because of both convergent and parallel evolution. This points up the fact that we have little understanding of the relation between ovarian morphology and the reproductive physiology of a species. Some correlations are, of course, quite obvious: at any single reproductive period, species usually ripen slightly more follicles and develop slightly more corpora lutea than the number of young born; mammals with long anestrous periods have relatively little follicular activity during these periods; and interstitial gland tissue is usually highly differentiated only from proestrus through about the first third of gestation.

While these and several other similar correlations are common, there are many exceptions to all of them. Why? We have no idea of the basic biological reason why some species ovulate many more eggs than they implant, or why ovulation of a single egg is followed by identical twinning in certain species of armadillos (Dasypodidae) and not in others. No correlations have ever been made between reproductive pattern and such things as the type or amount of interstitial gland tissue, or the formation of numerous accessory corpora lutea. What are the correlations between ovarian morphology and length of gestation, length of lactation, absence or length of hibernation, longevity, and so forth? Is the presence of apparently the same two distinct types of luteal gland cells throughout the corpora lutea of perissodactyls, artiodactyls, and cetaceans an indication of phylogenetic relationship or merely of convergent evolution because of some similarity in reproductive pattern? When such questions are answered, we may be able to apply them as correction factors to arrive at an estimate of the basic morphological nature of the ovaries of a given family or order, whose members exhibit several of these variables. Only then could we intelligently use ovarian characters as criteria of taxonomic relationship.

Quantification in ovarian morphology

A major problem in descriptive morphology is quantification, especially when dealing with scattered cells or intergrading cell types, irregular shapes, and elements that are difficult to isolate sufficiently for accurate weighing or measurement. Such conditions are common with ovarian material, and our synoptic tables are a victim of these difficulties. It would be desirable to have numerical measurements of many items which we have had to designate in subjective terms. However, in descriptive anatomy, qualitative data have priority over quantitative data: one must know what is present in order to know what to measure. Since we were first chiefly concerned with learning what was present, we fixed our material as quickly as possible, and handled it carefully to avoid distortion of normal relationships and damage to delicate structures. This precluded weighings and most other measurements of gross material. The new scanning instrumentation techniques can be applied to microscopic sections, and should be accurate and rapid enough to make practical much data-gathering which has not been possible previously.

Divergences between the testis and ovary

Consideration of the morphogenesis of the testis and ovary of eutherian mammals clearly shows that they are homologous organs, although in the adults they differ widely in both structure and function. It is interesting to speculate on the reasons for this divergence. It is also practical to do so, for it brings to attention certain qualities of the ovary that should always be considered in interpreting either its function or morphology.

In lower vertebrates (Anamniota), the chief differences between the ovary and testis are due to differences in size and number of germ cells produced, for the endocrine tissues are structurally similar in both. In general, the cyclic changes in the two gonads are also closely parallel.

In reptiles and birds (Sauropsida), the female reproductive pattern is usually more complex than that of the male. These divergences in the female — internal fertilization, viviparity in some reptiles, and brooding in birds — are more or less correlated with ovarian hormone mechanisms that have no counterparts in the males. Thecal and luteal gland

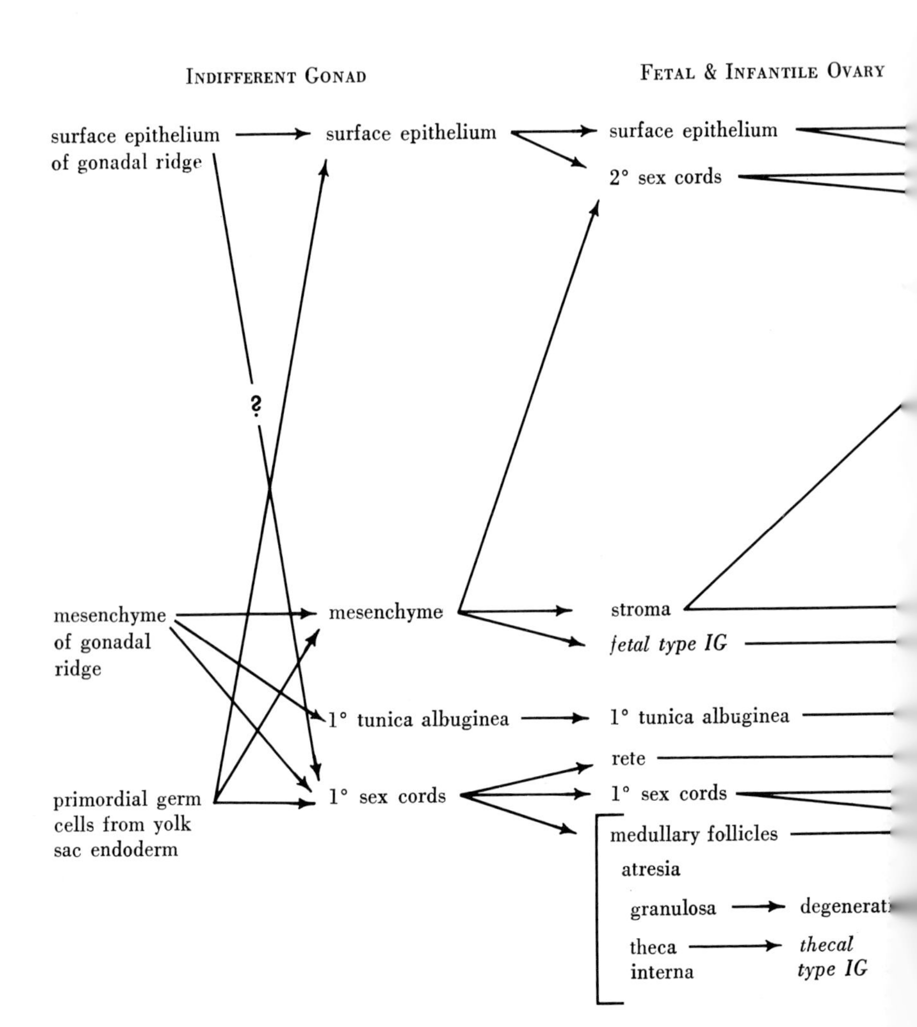

Fig. 8.3 Schematic summary of the origin of components of the mammalian ovary. Parentheses indicate that the component may be absent. Glandular tissues are italicized. *IG*, interstitial gland. Follicular thecae and, in most

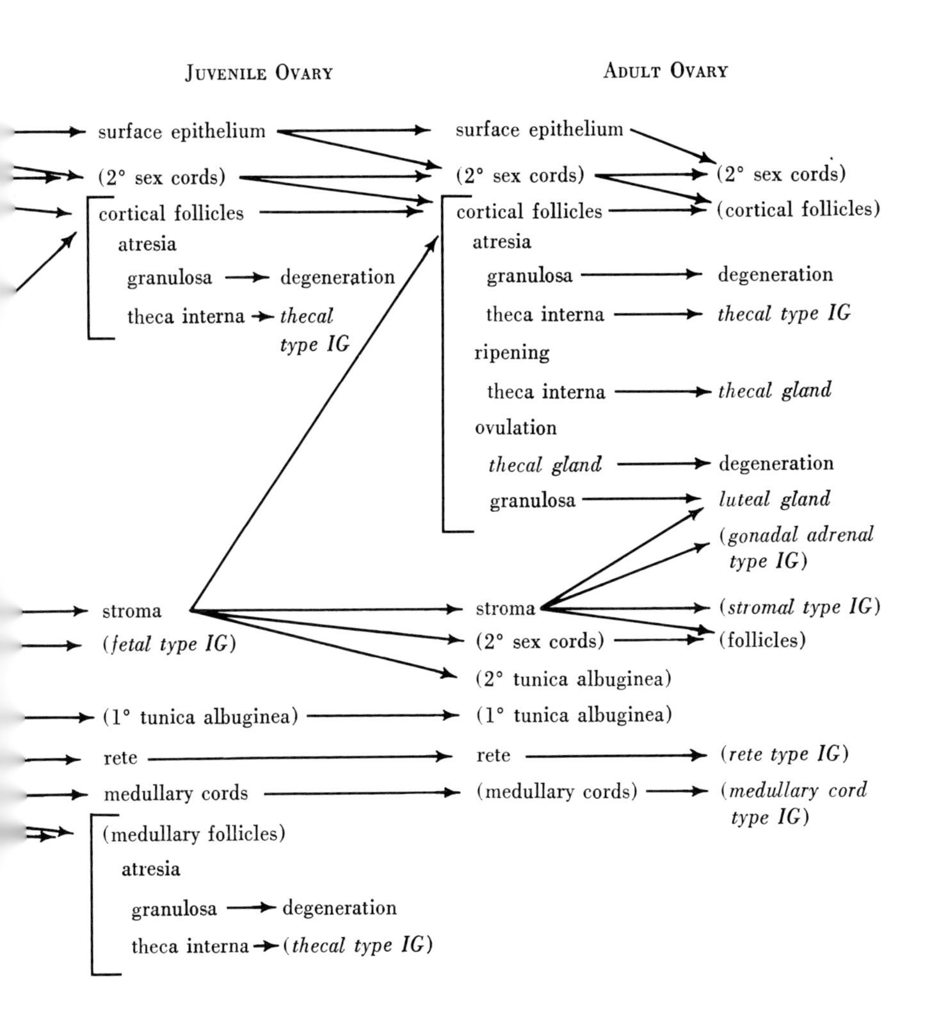

cases, the granulosa arise from stroma. We believe that the oogonia of the cortex and possibly some of those of the medulla may be derived from pluripotential stromal and epithelial cells rather than from primordial germ cells.

tissues are often well developed in these groups. Also, since many more follicles are stimulated to undergo partial development than are brought to ovulation, there is much follicular atresia. These are all common features of the ovaries of Amniota, but are rarely met with in Anamniota.

As would be anticipated, the widest basic differences between testis and ovary occur in eutherian mammals, and are correlated primarily with viviparity and lactation, and secondarily with such variables as cycle length and frequency, litter size, delayed implantation, postpartum estrus, and so forth. To be able to adjust and function in harmony with these cyclic phenomena, the adult eutherian ovary has to be a much more labile organ than the testis. Nevertheless, certain processes are common to both, but usually in varying degree.

At the end of rut, widespread and sudden degeneration of germ cells occurs in the testis. Atresia of germ cells also takes place in the ovary; however, because ovum degeneration often occurs in follicles whose neighbors are maturing normal ova, a selective mechanism must be involved which is either absent or minor in the testis. Alternate hypertrophy and involution of interstitial gland tissue also takes place in both testis and ovary, but there is more than one variety of this tissue in the ovary, and its integration with the reproductive cycle is more complex and is not the same in all species. But the most striking difference is the ability of the ovary to develop new structures which eventually degenerate, thus making way for repetition of the process. Actually, this process begins in the fetal period and continues throughout life. The growth and atresia of follicles is conspicuous in most mammals long before birth, especially in the medulla. At first, atresia overtakes only early follicular stages, but later many follicles develop to the vesicular stage with well-developed thecae before atresia starts. Whether atresia of these older follicles occurs prenatally or postnatally, their thecae internae commonly are converted into thecal type interstitial gland tissue. This means that large numbers of undifferentiated thecal cells are converted into secretory cells. During adult life, some follicles mature and their thecae internae become active thecal gland tissue. When these ovulate, the thecal gland degenerates, and another new but temporary organ, the corpus luteum, develops. Thus, in the mammalian ovary there is need, almost throughout life, for an ample source of pluripotential cells. In the adult, the chief supply of these comes from the cortical stroma, although the surface and follicular epithelium are also relatively pluripotential.

While the difference between the cortex and medulla is often indistinct, it is certain that the cortex forms in the late embryonic and early fetal periods as a new portion of the gonad not represented in the male, and that it retains its embryonic nature throughout life to a considerably greater degree than does the medulla. This concept of the ovarian cortex as an embryonic tissue is of great importance in attempting to understand and interpret both the normal and abnormal morphology and physiology of the ovary. Pathologists must take into account both "germ cell tumors" derived directly from oogonia and oocytes, and "gonadal stromal tumors" derived from pluripotential stromal and epithelial cells (Teilum, 1971).

Evaluation

We hope we have demonstrated that there are major deficiencies in our knowledge of ovarian morphology, and that the incompleteness and lack of depth of this knowledge are severe handicaps to investigations of the biology of reproduction in mammals. We also hope that the data presented are convincing evidence of the multitude of interesting and biologically important things that remain to be discovered by anatomical methods.

Morphological and functional studies complement each other. This is true of the whole range from ultrastructure and molecular biology through gross anatomy and physiology to the structure of environment and the behavior of its populations. Just as it is necessary to conserve environments and populations for future study, so it is essential to conserve morphological materials. New functional concepts require new morphological studies, and new morphological concepts require new functional studies. Each is limited by the other.

Through lack of understanding of these complementary relationships in the past, large amounts of valuable and often irreplaceable anatomical material have been discarded. Microscopic slides of organs of rare and even of almost extinct mammals have been consigned to the ash cans. It is impractical, in fact usually impossible, to record in writing or by illustrative techniques all of the useful information that can be obtained from an anatomical specimen. Furthermore, new concepts in biology are continually arising and these require restudy of the morphology involved from entirely new points of view. Reference collections of anatomical material must be built up, cataloged, properly housed, and preserved in accessible form for future investigators (Mossman, 1969). They are as necessary as libraries. Instead of providing already recorded observations, they will be potent sources of new information.

REFERENCE MATTER

The synoptic tables

This section summarizes morphological data on the ovaries of various taxonomic groups of mammals. The characteristics selected for record represent a compromise which is far from ideal. Information from our own material and from that which we examined in various other collections was often incomplete for a variety of reasons, including insufficient numbers of specimens, poor distribution throughout the reproductive season, absence of the mesovarium from the sections, and lack of gross specimens for examination of the bursa and of ovarian size, shape, and location. Also, when one depends on the literature for much information, one must use that which the authors have provided. In most cases, much of the information that we consider important simply was not mentioned, although occasionally some of it could be gleaned from the illustrations or by "reading between the lines." The outlines provided in Chapter 8, Figs. 8.1 and 8.2, are examples of the type and amount of information that would be more ideal. It is hoped that in the future those working on the ovarian morphology of a species will attempt to record such basic information. Many of the items suggested in these outlines are not in our tables, either because such complete information was available for only a few of the species listed in the tables, or because some of the items were too numerous and detailed for inclusion in a book that treats ovarian morphology at the relatively intermediate level necessary here.

For the investigator of female reproductive biology, we realize that it would have been helpful to have included in our synoptic tables much other closely correlated data, such as breeding seasons, length of estrous and pregnancy cycles, age at sexual maturity, whether ovulation is spontaneous or induced, and so forth. We have omitted this type of data because that which is known is readily available in Asdell's *Patterns of Mammalian Reproduction.* In fact, we have leaned heavily on this excellent compendium for some of the data in our synoptic tables, especially that regarding number of ovulatory follicles, litter size, diameter of ripe follicles, and persistence of corpora lutea. Rather than list Asdell or the individual authors cited by him as authorities under dozens of these species, we take this opportunity to acknowledge our indebtedness to him and to them.

It was impracticable to designate in these tables the exact source of each item of information. Actually, in many cases an individual item may represent a compromise between statements of the different authorities cited, or between them and our own observations. Hence, we assume full responsibility for the statements, and hope that no one will cite any of the information as coming directly from any of the sources given until they have consulted the original work.

The collections of material cited in the tables are located as follows: Conaway, Department of Zoology, University of Missouri, Columbia, Missouri; Gropp, Institute of Pathology, University of Bonn, Bonn/Rhein, Germany; Hamilton, Department of Anatomy, Charing Cross Hospital Medical School, London, England; Hartman, Department of Anatomy, University of Wisconsin Medical School, Madison, Wisconsin; Hill, Hubrecht Laboratory, Utrecht, Holland; Horst, Department of Zoology, University of Witwatersrand, Johannesburg, South Africa; Hubrecht, Hubrecht Laboratory, Utrecht, Holland; Keith, Department of Wildlife Ecology, University of Wisconsin, Madison, Wisconsin; Orsini, Department of Anatomy, University of Wisconsin Medical School, Madison, Wisconsin; Sinha, Veterans' Administration Hospital, Minneapolis, Minnesota; Wright, Department of Zoology, University of Montana, Missoula, Montana.

One of the major problems in presenting data of this sort is the absence of quantitative criteria and the need to use subjective terms such as "medium," "much," and so forth. There are no quantitative measurements for most of these items and usually, considering the nature of the material, they would be impossible or at least difficult to make, and in many cases of little real value. Although it should be obvious, it is probably best to point out that the subjective term used for any quantitative item is related first to the species and second to the relative size or number of that particular item throughout the mammalian group. For instance, the thecal glands of the mouse and cow are both listed as "thin" because of their relation to the follicular diameter in each, in spite of the fact that in terms of absolute thickness that of the cow is much greater. However, the number of accessory corpora lutea in *Erethizon* is indicated as "very many," because the number in other species is rarely as great.

From some viewpoints, certain items such as epoophoron, rete, shape, lobes, fissures, and tunica albuginea may not have enough significance to warrant their inclusion in these tables. However, they have been included for two reasons: 1) they tend to characterize the ovaries of certain taxa; 2) although the epoophoron and rete are usually thought to be vestigial structures which are often absent, we know that they are of constant occurrence and in many species are relatively well developed and specialized structures which may eventually be shown to play at least a minor functional role in reproductive biology.

The number of individual specimens examined, as given under each species, is not always a strictly proportional indication of the reliability of the data. The distribution of the specimens at various ages and reproductive phases is more important, but this information must be sought in the original papers. In the case of our own material, the number of items that we have recorded for any species is some measure of our estimate of the adequacy of our material for that species. However, especially in the cases of rare species, we have recorded items based on inadequate material in order to give some indication of the probable nature of the ovary of the species.

We hope that these tables will be useful as a source of information on various taxa, and that the blanks and the absence of any information on numerous taxonomic groups will give a perspective on the incompleteness of our knowledge and will stimulate investigation of significant groups about which information is inadequate.

We urge consultation of the supplementary notations concerning each group in the section that follows the synoptic tables, because these often contain qualifications of the tabulated data (usually indicated in the tables by a "?" mark). They also contain additional data and literature references, as well as references to figures or to specific parts of the text. The abbreviation "t. i." after the estimate of the size of the thecal gland indicates that no preovulatory follicle was seen, but that the estimate is based on the theca interna (undifferentiated thecal gland) of earlier stages of vesicular follicles. We know from experience with many species that this is legitimate.

Abbreviations in the tables:

irreg. = irregular
ovul. = ovulation
rud. = rudimentary
t. i. = theca interna
— = no reliable information
0 = none present
> = more than
? = for number of individuals, sample adequate but exact number unknown; otherwise, item qualified in supplementary notes.

Sometimes the number of individual animals examined is indicated by the surname initial of the investigator named in the source: e.g., under *Eptesicus*, the abbreviation "W—41, C—45, M—12" means that Wimsatt examined 41, Christian 45, Mossman 12, a total of 98 specimens.

Mammalian classification in the synoptic tables and supplementary notes

Major groups are based on Simpson (1945). Generic names follow Walker (1964), except for some which have been updated on advice from the U.S. National Museum. Page numbers in boldface type refer to the synoptic tables, those in roman type to the supplementary notes.

Class MAMMALIA
  Subclass PROTOTHERIA
    Order MONOTREMATA
      Family Tachyglossidae (spiny anteaters) **298**, 348
        Tachyglossus, **298**, 348 (1 other genus)
      Family Ornithorhynchidae (duck-billed platypuses) **298**, 348
        Ornithorhynchus, **298**, 348 (no other genera)
  Subclass THERIA
    Infraclass METATHERIA
      Order MARSUPIALIA
        Superfamily Didelphoidea
          Family Didelphidae (opossums) **299**, 349
            Didelphis, **299**, 349 (11 other genera)
        Superfamily Dasyuroidea
          Family Dasyuridae (pouched "mice," native "cats," etc.) 350
            Dasyurus, 350 (18 other genera)
        Superfamily Phalangeroidea
          Family Phalangeridae (phalangers, koalas, etc.) 350
            Trichosurus, 350 (16 other genera)
          Family Macropodidae (kangaroos, wallabies) 350
            (17 genera)
      (This order includes 15 other genera in 5 additional families)
    Infraclass EUTHERIA
      Order INSECTIVORA
        Superfamily Tenrecoidea
          Family Tenrecidae (tenrecs and Madagascar "hedgehogs") **299**, 350
            Tenrec, 350; *Setifer*, **299**, 350; *Hemicentetes*, 350 (7 other genera)
        Superfamily Chrysochloroidea
          Family Chrysochloridae (golden moles) **299**, 351
            Amblysomus, 351; *Calcochloris*, 351; *Eremitalpa*, **299**, 351 (4 other genera)
        Superfamily Erinaceoidea
          Family Erinaceidae (moon rats, gymnures, hedgehogs) **300**, 351
            Echinosorex, **300**, 351; *Hylomys*, **300**, 351; *Erinaceus*, **301**, 351 (5 other genera)
        Superfamily Macroscelidoidea

Family Macroscelididae (elephant shrews) **301**, 352
Elephantulus, **301**, 352; *Petrodromus*, **301**, 352 (3 other genera)
Superfamily Soricoidea
Family Soricidae (shrews) **302**, 353
Sorex, **302**, 353; *Neomys*, **303**, 353; *Blarima*, **303**, 353; *Crocidura*, **303**, 353; *Suncus*, **304**, 353 (17 other genera)
Family Talpidae (moles, shrew-moles, desmans) **304**, 353
Galemys, **304**, 353; *Talpa*, **305**, 354; *Neurotrichus*, **305**, 354; *Scapanus*, **306**, 353; *Scalopus*, **306**, 354; *Condylura*, **307**, 353 (6 other genera)
(This order includes 3 other genera in 2 additional families)
Order DERMOPTERA
Family Cynocephalidae (gliding lemurs) **307**, 355
Cynocephalus, **307**, 355 (no other genera)
Order CHIROPTERA
Suborder MEGACHIROPTERA
Family Pteropodidae (Old World fruit bats, flying "foxes") **307**, 355
Cynopterus, 355; *Ptenochirus*, 355; *Rousettus*, 355; *Pteropus*, **307**, 355; *Epomophorus*, 355; *Eonycteris*, 355; *Macroglossus*, 355; *Harpyionycteris*, 355 (31 other genera)
Suborder MICROCHIROPTERA
Superfamily Rhinolophoidea
Family Rhinolophidae (horseshoe bats)
Rhinolophus, 357 (1 other genus)
Superfamily Phyllostomatoidea
Family Phyllostomatidae (American leaf-nosed bats) **308**, 355
Carollia, **308**, 355; *Uroderma*, **308**, 356 (49 other genera)
Family Desmodontidae (vampire bats) **309**, 355
Desmodus, **309**, 356 (2 other genera)
Superfamily Vespertilionoidea
Family Vespertilionidae (brown, hoary, long-eared bats, etc.) **309**, 356
Myotis, **309**, 356; *Pipistrellus*, 357; *Eptesicus*, **309**, 356; *Lasiurus*, 356; *Plecotus*, **310**, 356 (33 other genera)
Family Molossidae (free-tailed, mastiff bats, etc.) **310**, 356
Eumops, 357; *Tadarida*, **310**, 356 (10 other genera)
(This order includes 34 other genera in 11 additional families)
Order PRIMATES
Suborder PROSIMII
Infraorder LEMURIFORMES
Superfamily Tupaioidea
Family Tupaiidae (tree shrews) **311**, 358
Tupaia, **311**, 358; *Dendrogale*, 358; *Urogale*, 358 (2 other genera)
Superfamily Lemuroidea
Family Lemuridae (lemurs)

Hapalemur, 358; *Lemur*, 359; *Lepilemur*, 358; *Cheirogaleus*, 359; *Microcebus*, 359 (1 other genus)
Family Indridae (avahis, indris, etc.) 358
Avahi, 359 (2 other genera)
Infraorder LORISIFORMES
Family Lorisidae (lorises, pottos, bush babies, etc.) **311**, 359
Loris, **311**, 359; *Nycticebus*, **311**, 359; *Galago*, **312**, 359 (3 other genera)
Infraorder TARSIIFORMES
Family Tarsiidae (tarsiers) **312**, 359
Tarsius, **312**, 359 (no other genera)
Suborder ANTHROPOIDEA
Superfamily Ceboidea
Family Cebidae (New World monkeys) **313**, 360
Cacajao, 360; *Alouatta*, **313**, 360; *Saimiri*, 360; *Ateles*, **313**, 360 (8 other genera)
Family Callithricidae (marmosets, tamarins) **313**, 360
Saguinus, **313**, 360 (3 other genera)
Superfamily Cercopithecoidea
Family Cercopithecidae (Old World monkeys and baboons) **313**, 360
Macaca, **313**, 360; *Papio*, 361; *Erythrocebus*, 360 (15 other genera)
Superfamily Hominoidea
Family Pongidae (apes) **314**, 361
Hylobates, **314**, 361; *Pongo*, 361; *Pan*, 361; *Gorilla*, 361 (1 other genus)
Family Hominidae (man) **314**, 361
Homo, **314** (no other genera)
(This order includes 1 other genus in 1 other family)
Order EDENTATA
Suborder XENARTHRA
Infraorder PILOSA
Superfamily Myrmecophagoidea
Family Myrmecophagidae (anteaters) **315**, 362
Tamandua, **315**, 362 (2 other genera)
Superfamily Bradypodoidea
Family Bradypodidae (sloths) **315**, 362
Choloepus, **315** (1 other genus)
Infraorder CINGULATA
Superfamily Dasypodoidea
Family Dasypodidae (armadillos) **315**, 362
Dasypus, **315**, 362 (8 other genera)
Order PHOLIDOTA
Family Manidae (pangolins) **316**, 363
Manis, **316**, 363 (no other genera)
Order LAGOMORPHA
Family Ochotonidae (pikas) **316**, 364

Ochotona, **316**, 364 (no other genera)
Family Leporidae (hares and rabbits) **317**, 364
Lepus, **317**, 364; *Sylvilagus*, **317**, 364; *Oryctolagus*, **317**, 364 (6 other genera)
Order RODENTIA
Suborder SCIUROMORPHA
Superfamily Aplodontoidea
Family Aplodontidae (sewellels) **318**, 365
Aplodontia, **318**, 365 (no other genera)
Superfamily Sciuroidea
Family Sciuridae (squirrels, marmots, etc.) **318**, 365
Sciurus, **318**, 366; *Tamiasciurus*, **319**, 365; *Funambulus*, **319**, 365; *Marmota*, 366; *Cynomys*, 366; *Spermophilus*, **319**, 366; *Tamias*, **320**, 366; *Glaucomys*, 366 (44 other genera)
Superfamily Geomyoidea
Family Geomyidae (pocket gophers) **320**, 366
Geomys, **320**, 366 (7 other genera)
Family Heteromyidae (pocket mice, kangaroo rats, etc.) **321**, 366
Perognathus, 366; *Microdipodops*, 366; *Dipodomys*, **321**, 366 (2 other genera)
Superfamily Castoroidea
Family Castoridae (beavers) **321**, 366
Castor, **321**, 366 (no other genera)
Superfamily Anomaluroidea
Family Pedetidae (springhares) **321**, 366
Pedetes, **321**, 366 (no other genera)
Suborder MYOMORPHA
Superfamily Muroidea
Family Cricetidae (mice, rats, voles, etc.) **322**, 367
Reithrodontomys, 367; *Peromyscus*, 367; *Neotoma*, **322**; *Mesocricetus*, **322**, 367; *Clethrionomys*, 367; *Ondatra*, 367; *Microtus*, 367 (94 other genera)
Family Muridae (Old World rats and mice) **323**, 367
Rattus, **323**, 367; *Mus*, **323**, 367 (98 other genera)
Superfamily Gliroidea
Family Gliridae (dormice) 368
Eliomys, 368 (6 other genera)
Superfamily Dipodoidea
Family Zapodidae (birch mice, jumping mice) **323**, 368
Zapus, **323**, 368; *Napaeozapus*, **323**, 368 (2 other genera)
Suborder HYSTRICOMORPHA
Superfamily Hystricoidea
Family Hystricidae (Old World porcupines) **324**, 369
Thecurus, **324**, 369; *Hystrix*, **324**, 369; *Atherurus*, **325**, 369 (1 other genus)
Superfamily Erethizontoidea

Family Erethizontidae (New World porcupines) — **325**, 369
Erethizon, **325**, 369 (3 other genera)
Superfamily Cavioidea
Family Caviidae (guinea pigs, cavies) — **326**, 369
Cavia, **326**, 369; *Galea*, **326**, 370 (2 other genera)
Family Hydrochoeridae (capybaras) — 370
Hydrochoerus, 370 (no other genera)
Family Dasyproctidae (pacas, agoutis) — **327**, 370
Dasyprocta, **327**, 370; *Myoprocta*, **327**, 370 (1 other genus)
Superfamily Chinchilloidea
Family Chinchillidae (viscachas, chinchillas) — **327**, 370
Lagostomus, **327**, 370; *Lagidium*, **328**, 371; *Chinchilla*, **329**, 371 (no other genera)
Superfamily Octodontoidea
Family Capromyidae (hutias, nutrias) — **329**, 371
Myocastor, **329**, 371 (3 other genera)
Family Octodontidae (degus) — **329**, 371
Octodon, **329**, 371; *Octodontomys*, **330**, 371 (3 other genera)
Family Ctenomyidae (tucu-tucos) — **330**, 371
Ctenomys, **330**, 371 (no other genera)
Family Echimyidae (spiny rats) — **331**, 371
Proechimys, **331**, 371 (14 other genera)
Superfamily Bathyergoidea
Family Bathyergidae (African mole rats) — **331**, 371
Cryptomys, **331**, 371; *Bathyergus*, **331** (3 other genera)
(This order includes 29 other genera in 11 additional families)
Order CETACEA
Suborder ODONTOCETI
Superfamily Physeteroidea
Family Ziphiidae (beaked whales) — 373
Mesoplodon, 373 (4 other genera)
Superfamily Delphinoidea
Family Monodontidae (white whales and narwhales) — **332**, 372
Delphinapterus, **332**, 372 (1 other genus)
Family Delphinidae (dolphins, porpoises, etc.) — **332**, 372
Delphinus, 373; *Pseudorca*, 373; *Globicephala*, **332**, 372 (16 other genera)
Suborder MYSTICETI
Family Balaenopteridae (finback, humpback, and blue whales) — 373
Balaenoptera, 373; *Megaptera*, 373; *Sibbaldus*, 373 (no other genera)
Family Balaenidae (right and bow-head whales) — 373
Balaena, 373 (2 other genera)
(This order includes 7 other genera in 3 additional families)

Order CARNIVORA

Suborder FISSIPEDIA

Superfamily Canoidea

Family Canidae (dogs, wolves, foxes, etc.) **333**, 373

Canis, **333**, 373; *Vulpes*, **333**, 374; *Urocyon*, **333**, 374 (11 other genera)

Family Ursidae (bears) **334**, 374

Ursus, **334**, 374 (6 other genera)

Family Procyonidae (raccoons, pandas, etc.) **334**, 374

Bassariscus, 374; *Procyon*, **334**, 374; *Nasua*, 374; *Potos*, 374; *Bassaricyon*, 374; *Ailurus*, 374 (3 other genera)

Family Mustelidae (weasels, skunks, badgers, etc.) **335**, 375

Mustela, **335**, 375; *Taxidea*, **336**, 375; *Mephitis*, 376; *Spilogale*, 376; *Enhydra*, **336**, 376 (21 other genera)

Superfamily Feloidea

Family Viverridae (genets, civets, mongooses, etc.) **337**, 376

Genetta, **337**, 376; *Viverra*, 377; *Prionodon*, 377; *Nandinia*, 377; *Arctogalidia*, **337**, 377; *Paradoxurus*, **337**, 377; *Paguma*, **338**, 377; *Suricata*, 377; *Herpestes*, **338**, 376; *Helogale*, 377; *Mungos*, 377; *Ichneumia*, 377; *Paracynictis*, 377 (24 other genera)

Family Hyaenidae (aardwolves, hyenas) **339**, 377

Proteles, **339**, 377; *Crocuta*, **339**, 377 (1 other genus)

Family Felidae (cats) **339**, 377

Felis, **339**, 377; *Lynx*, 377 (4 other genera)

Suborder PINNIPEDIA

Family Otariidae (eared seals) **340**, 378

Arctocephalus, **340**, 378; *Callorhinus*, **340**, 378 (4 other genera)

Family Phocidae (earless or true seals) 378

Phoca, 378; *Halichoerus*, 378; *Leptonychotes*, 378; *Mirounga*, 378 (9 other genera)

(This order includes 1 other genus in 1 additional family)

Order TUBULIDENTATA

Family Orycteropodidae (aardvarks) **341**, 379

Orycteropus, **341**, 379 (no other genera)

Order PROBOSCIDEA

Suborder ELEPHANTOIDEA

Family Elephantidae (elephants) **341**, 379

Loxodonta, **341**, 379; *Elephas*, **341**, 379 (no other genera)

Order HYRACOIDEA

Family Procaviidae (hyraxes) **342**, 380

Dendrohyrax, **342**; *Heterohyrax*, **342**; *Procavia*, **343** (no other genera)

Order SIRENIA

Suborder TRICHECHIFORMES
Family Dugongidae (dugongs) — 381
Dugong, 381 (no other genera)
Family Trichechidae (manatees) — 381
Trichechus, 381 (no other genera)
Order PERISSODACTYLA
Suborder HIPPOMORPHA
Superfamily Equoidea
Family Equidae (horses, zebras, etc.) — **343**, 381
Equus, **343**, 381 (no other genera)
Suborder CERATOMORPHA
Superfamily Tapiroidea
Family Tapiridae (tapirs) — 382
Tapirus, 382 (no other genera)
Superfamily Rhinocerotoidea
Family Rhinocerotidae (rhinoceroses) — 383
Rhinoceros, 383 (3 other genera)
Order ARTIODACTYLA
Suborder SUIFORMES
Infraorder SUINA
Superfamily Suoidea
Family Suidae (pigs) — **344**, 383
Sus, **344**, 383; *Phacochoerus*, 383 (3 other genera)
Family Tayassuidae (peccaries) — 383
Tayassu, 383 (no other genera)
Infraorder ANCODONTA
Superfamily Anthracotherioidea
Family Hippopotamidae (hippopotamuses) — 384
Hippopotamus, 384 (1 other genus)
Suborder TYLOPODA
Family Camelidae (camels, llamas, etc.) — 384
Lama, 384; *Camelus*, 384 (1 other genus)
Suborder RUMINANTIA
Infraorder TRAGULINA
Superfamily Traguloidea
Family Tragulidae (chevrotains) — **344**, 384
Tragulus, **344**, 384 (1 other genus)
Infraorder PECORA
Superfamily Cervoidea
Family Cervidae (deer, elk, moose, etc.) — **345**, 385
Cervus, 385; *Odocoileus*, **345**, 385; *Alces*, 385; *Capreolus*, 385 (13 other genera)
Superfamily Giraffoidea
Family Giraffidae (giraffes, okapis) — 385
Giraffa, 385 (1 other genus)
Superfamily Bovoidea
Family Antilocapridae (pronghorns) — **345**, 386

Antilocapra, **345**, 386 (no other genera)

Family Bovidae (antelope, buffalo, cattle, goats, sheep, etc.) **345**, 386

Tragelaphus, 386; *Taurotragus*, 386; *Bos*, **345**, 386; *Syncerus*, 386; *Bison*, **346**, 386; *Sylvicapra*, 386; *Kobus*, **346**, 386; *Redunca*, 386; *Hippotragus*, 386; *Connochaetes*, 386; *Aepyceros*, 386; *Antidorcas*, 386; *Capra*, **347**; Ovis, **347**, 387 (35 other genera)

Order	MONOTREMATA	MONOTREMATA
Family	Tachyglossidae	Ornithorhynchidae
Genus	*Tachyglossus*	*Ornithorhynchus*
Species	sp.	*anatinus*
No. of individuals	?	?
Bursa ovarica	—	—
Orifice	—	—
Epoophoron	—	—
Ovary		
Location	lumbar	lumbar
Shape	irregular	irregular
Lobes	many	many
Fissures	many	many
Tunica albuginea	—	—
Rete	—	—
Ripe follicles		
No. & (litter size)	1 (1)	2 (2)
Diameter (mm)	4.0	4.5
Antrum	0	0
Thecal gland	thick	thick
Interstitial gland type		
Thecal	—	—
Stromal	—	—
Medullary cord	—	—
Gonadal adrenal		
in cortex	—	—
in medulla	—	—
at epoophoron	—	—
accessory adrenal	—	—
Luteal glands		
No. of cell types	1	1
Accessory corpora	0	0
Persistence	into incubation	into incubation
Sources	J. P. Hill & Gatenby, 1926; Garde, 1930; Broek, 1931; Flynn & Hill, 1939; Harrison, 1948*a*	J. P. Hill & Gatenby, 1926; Flynn & Hill, 1939; Harrison, 1948*a*

MARSUPIALIA	INSECTIVORA	INSECTIVORA
DIDELPHIDAE	TENRECIDAE	CHRYSOCHLORIDAE
Didelphis	*Setifer*	*Eremitalpa*
marsupialis	*setosus*	*granti*
(M) — > 20	20	> 150
?	complete	complete
?	porelike	porelike
—	medium	medium
lumbar	lumbar	lumbar
globose	reniform	ellipsoidal
rarely lobed	0	0
0	0	0
thin	thin	0
—	medium	medium
22 (9)	4 (4)	— (—)
about 1.7	0.5	—
medium	0	—
thin; interrupted	thin	thick (t. i.)
little	medium	little
—	—	—
—	0	0
—	—	—
—	—	—
—	—	little
—	—	—
1	1	1
0	—	0
into lactation	—	—
Hartman, 1923; Martínez-Esteve, 1942; Hartman Coll. (Mossman)	Goetz, 1937; Strauss, 1938, 1939	Horst Coll. (Mossman)

Order	INSECTIVORA	INSECTIVORA
Family	ERINACEIDAE	ERINACEIDAE
Genus	*Echinosorex*	*Hylomys*
Species	*gymnurus*	*suillus*
No. of individuals	15	3
BURSA OVARICA	complete	—
Orifice	—	—
EPOOPHORON	medium	—
OVARY		
Location	—	—
Shape	reniform	moruloid
Lobes	several	several
Fissures	several	several
Tunica albuginea	0 to very thin	0 to very thin
Rete	medium	medium
Ripe follicles		
No. & (litter size)	1 (1)	3 (?)
Diameter (mm)	0.6	0.5
Antrum	0?	0?
Thecal gland	thick	thick
Interstitial gland type		
Thecal	—	—
Stromal	—	—
Medullary cord	—	—
Gonadal adrenal		
in cortex	—	—
in medulla	—	—
at epoophoron	—	—
accessory adrenal	—	—
Luteal glands		
No. of cell types	1	1
Accessory corpora	0	0
Persistence	into lactation	into lactation
SOURCES	Duke	Duke

INSECTIVORA	INSECTIVORA	INSECTIVORA
ERINACEIDAE	MACROSCELIDIDAE	MACROSCELIDIDAE
Erinaceus	*Elephantulus*	*Petrodromus*
europaeus	*myurus*	*tetradactylus*
?	> 1,000	4
complete	complete	complete
porelike	porelike	porelike
medium	large	large
—	lumbar	lumbar
globose	irregular	angular
many	several	few
many	several	few
0 to very thin	0	0
—	medium?	medium
5 (5)	120 (2)	2 (1)
1.3	0.4	—
medium; trabeculae	small	—
thick	thin	—
—	little	much
—	0	much
—	0	?
—	—	0
—	—	much
—	?	much
—	—	—
1	1	1
0	0	—
into lactation	into lactation	—
Deanesly, 1934	Horst, 1942, 1945, 1954; Horst & Gillman, 1940, 1941, 1942, 1945, 1946; Horst Coll. (Mossman)	Horst Coll. (Mossman)

Order	INSECTIVORA	INSECTIVORA
Family	Soricidae	Soricidae
Genus	*Sorex*	*Sorex*
Species	*araneus*	*vagrans*
No. of individuals	487	24
Bursa ovarica	complete	complete
Orifice	0	—
Epoophoron	—	large
Ovary		
Location	lumbar	lumbar
Shape	moruloid	reniform
Lobes	0	0
Fissures	0	0 to few
Tunica albuginea	—	0
Rete	—	medium
Ripe follicles		
No. & (litter size)	7 (7)	— (5)
Diameter (mm)	0.4	0.4
Antrum	small; trabeculae	small
Thecal gland	thick	thick
Interstitial gland type		
Thecal	little	little
Stromal	—	0
Medullary cord	—	"testis cords"
Gonadal adrenal		
in cortex	—	0
in medulla	—	0
at epoophoron	—	0
accessory adrenal	—	—
Luteal glands		
No. of cell types	1	1
Accessory corpora	—	0
Persistence	term	into 2d gestation
Sources	Brambell, 1935	Wilcox & Mossman, 1945; Mossman

INSECTIVORA	INSECTIVORA	INSECTIVORA
SORICIDAE	SORICIDAE	SORICIDAE
Neomys	*Blarina*	*Crocidura*
fodiens	*brevicauda*	sp.
98	> 150	6
complete	complete	complete
—	porelike	—
—	—	—
—	lumbar	—
globose	reniform	ellipsoidal
0	0	0
0	0	0
0 to very thin	0	0
—	—	—
— (7)	6 (5)	— (—)
0.4	0.6	0.4
small	small	small
thick	thick	thin
much?	little	—
very much?	0	—
0	0	—
—	—	—
—	—	—
—	—	—
—	—	—
1	1	1
—	—	0
term?	term	into 2d gestation
Price, 1953; Mossman	O. P. Pearson, 1944; Mossman	Hill Coll. (Mossman)

Order	INSECTIVORA	INSECTIVORA
Family	SORICIDAE	TALPIDAE
Genus	*Suncus*	*Galemys*
Species	*murinus*	*pyrenaicus*
No. of individuals	23	?
BURSA OVARICA	—	complete
Orifice	—	porelike
EPOOPHORON	—	—
OVARY		
Location	—	—
Shape	globose	irregular ellipsoidal
Lobes	0	0
Fissures	0	—
Tunica albuginea	—	—
Rete	—	—
Ripe follicles		
No. & (litter size)	— (—)	— (4)
Diameter (mm)	0.4	—
Antrum	very narrow	large
Thecal gland	thin (t. i.)	—
Interstitial gland type		
Thecal	—	medium
Stromal	—	—
Medullary cord	—	"testis cords"
Gonadal adrenal		
in cortex	—	—
in medulla	—	medium
at epoophoron	—	—
accessory adrenal	—	—
Luteal glands		
No. of cell types	1	—
Accessory corpora	few	—
Persistence	to 1 week postpartum	—
SOURCES	Dryden, 1969	Peyre, 1953

INSECTIVORA	INSECTIVORA	INSECTIVORA
TALPIDAE	TALPIDAE	TALPIDAE
Talpa	*Talpa*	*Neurotrichus*
europaea	*leucurus*	*gibbsi*
(M) — 15	1 (early pregnancy)	4
complete	complete	complete
medium	—	0
medium	—	small
—	—	lumbar
globose	ellipsoidal	globose
0	0	0 to few
0	0	0
0 to very thin	0	0
medium	medium	medium
— (4)	3 (3)	4 (4)
0.7	—	—
small; trabeculae	—	small
thick (t. i.)	thick (t. i.)	thin (t. i.)
little	little	little
0	0	0
little?	"testis cords"	little?
0	0	0
very much	much	medium
very much	much	medium
0	—	0
2?	1	1
0	—	0
term	—	—
Popoff, 1911; Altmann, 1927; Harrison-Matthews, 1935; Deanesly, 1966; Hubrecht Coll. (Mossman); Gropp Coll. (Mossman)	Duke, 1966*a*	Mossman

Order	INSECTIVORA	INSECTIVORA
Family	TALPIDAE	TALPIDAE
Genus	*Scapanus*	*Scalopus*
Species	*latimanus*	*aquaticus*
No. of individuals	1	35
BURSA OVARICA	complete	complete
Orifice	porelike	porelike
EPOOPHORON	—	small
OVARY		
Location	—	lumbar
Shape	irregular reniform	irregular reniform
Lobes	0	0
Fissures	0	0
Tunica albuginea	thin	thin
Rete	small	small
Ripe follicles		
No. & (litter size)	— (3)	4 (4)
Diameter (mm)	—	—
Antrum	—	small; trabeculae
Thecal gland	thin (t. i.)	thin
Interstitial gland type		
Thecal	—	medium
Stromal	—	0
Medullary cord	0	0
Gonadal adrenal		
in cortex	0	0
in medulla	0?	little
at epoophoron	0?	0
accessory adrenal	0?	0
Luteal glands		
No. of cell types	—	1
Accessory corpora	—	0
Persistence	—	term
SOURCES	Mossman	Conaway Coll. (Mossman); Mossman

INSECTIVORA	DERMOPTERA	CHIROPTERA
TALPIDAE	CYNOCEPHALIDAE	PTEROPODIDAE
Condylura cristata	*Cynocephalus variegatus*	*Pteropus vampyrus*
16	40	17
complete	complete	complete
porelike	medium	—
small	—	medium
lumbar	lumbopelvic	—
irregular reniform	ellipsoidal	globose
0	0	0
0	0	0
0	thin; fibrous	thin
medium	large	medium
5 (5)	1 (1)	1 (1)
—	2.0	—
small	large	large
thin	thick	thin
little	little	little
0	little	little
0	—	—
0	—	—
very much	—	—
very much	—	—
0	—	—
1	1	1
0	0	0
term	mid-gestation	mid-gestation
Mossman	Duke	Duke

Order	CHIROPTERA	CHIROPTERA
Family	PHYLLOSTOMATIDAE	PHYLLOSTOMATIDAE
Genus	*Carollia*	*Uroderma*
Species	sp.	*bilobatum*
No. of individuals	7	13
BURSA OVARICA	—	—
Orifice	—	—
EPOOPHORON	medium	large
OVARY		
Location	—	—
Shape	ellipsoidal	ellipsoidal
Lobes	0	0
Fissures	0	0
Tunica albuginea	0	0
Rete	medium	medium?
Ripe follicles		
No. & (litter size)	— (1)	— (1)
Diameter (mm)	—	—
Antrum	—	—
Thecal gland	—	—
Interstitial gland type		
Thecal	little	little
Stromal	0	0
Medullary cord	0	0
Gonadal adrenal		
in cortex	little	0
in medulla	little	0
at epoophoron	0	—
accessory adrenal	0	0
Luteal glands		
No. of cell types	—	1
Accessory corpora	—	—
Persistence	—	through gestation
SOURCES	Mossman	Mossman

CHIROPTERA	CHIROPTERA	CHIROPTERA
DESMODONTIDAE	VESPERTILIONIDAE	VESPERTILIONIDAE
Desmodus	*Myotis*	*Eptesicus*
rotundus	*lucifugus* & *grisescens*	*fuscus*
38	G & J — 175, W — 310	W — 41, C — 45, M — 12
—	complete	—
—	—	—
—	medium	—
—	—	—
ellipsoidal	ellipsoidal	ellipsoidal
0	0	0
0	0	0
0	0	0
—	medium	large
— (1)	1 (1)	4 (3)
—	0.4	—
—	small	medium
—	thin	thin
medium	much	much
—	much	much
—	0	0
—	little	0
—	medium	0
—	much	0
—	0	0
1	1	1
—	0	0
into early lactation	through gestation	into lactation
Wimsatt & Trapido, 1952	Guthrie & Jeffers, 1938; Wimsatt, 1944; Mossman	Wimsatt, 1944; Christian, 1956; Mossman

Order	CHIROPTERA	CHIROPTERA
Family	Vespertilionidae	Molossidae
Genus	*Plecotus*	*Tadarida*
Species	*rafinesquii*	*brasiliensis*
No. of individuals	288	26
Bursa ovarica	complete	complete
Orifice	—	porelike
Epoophoron	medium	small
Ovary		
Location	—	—
Shape	ellipsoidal	ellipsoidal
Lobes	0	0
Fissures	0	0
Tunica albuginea	0	0
Rete	medium	large
Ripe follicles		
No. & (litter size)	1 (1)	1 (1)
Diameter (mm)	0.4	—
Antrum	small; trabeculae	—
Thecal gland	thin	thin (t. i.)
Interstitial gland type		
Thecal	much	little
Stromal	much	little?
Medullary cord	0	0
Gonadal adrenal		
in cortex	little	0
in medulla	medium	medium
at epoophoron	much	medium
accessory adrenal	0	0
Luteal glands		
No. of cell types	1	1
Accessory corpora	0	—
Persistence	through gestation	late gestation
Sources	O. P. Pearson, Koford & Pearson, 1952; Mossman	Sherman, 1937; Hartman Coll. (Mossman)

PRIMATES	PRIMATES	PRIMATES
TUPAIIDAE	LORISIDAE	LORISIDAE
Tupaia	*Loris*	*Nycticebus*
javanica	*tardigradus*	*coucang*
> 20	?	> 20
complete	—	partial
large	—	large
large	—	—
lumbar	—	—
ellipsoidal	ellipsoidal	ellipsoidal
few	0	0
few	—	few
0	—	thin
large	—	large
5 (2)	1–2 (1–2)	— (1)
1.0	—	0.9
large	—	large
thick	thick (t. i.)	thick
little	little	—
—	—	—
"testis cords"	—	—
little	—	—
little	—	—
little	—	—
—	—	—
1	1	1
few	0	0
term	through gestation	early gestation?
Duke & Luckett, 1965; Luckett, personal communications	Rao, 1927; Ramaswami & Kumar, 1965	Duke, 1967

Order	PRIMATES	PRIMATES
Family	LORISIDAE	TARSIIDAE
Genus	*Galago*	*Tarsius*
Species	*senegalensis* & *demidovii*	sp.
No. of individuals	> 17	36
BURSA OVARICA	complete	0
Orifice	—	(no bursa)
EPOOPHORON	very large	—
OVARY		
Location	—	—
Shape	ellipsoidal	globose
Lobes	few	0
Fissures	few	0
Tunica albuginea	thin	thin
Rete	very large	large
Ripe follicles		
No. & (litter size)	1 (1)	1 (1)
Diameter (mm)	0.7	1.2
Antrum	—	large
Thecal gland	—	thick
Interstitial gland type		
Thecal	little	little
Stromal	little	medium
Medullary cord	—	0
Gonadal adrenal		
in cortex	—	—
in medulla	—	—
at epoophoron	—	little
accessory adrenal	—	—
Luteal glands		
No. of cell types	1	1
Accessory corpora	—	0
Persistence	term	term
SOURCES	Gérard, 1932; Butler, 1967; Horst Coll. (Mossman)	Hubrecht Coll. (Duke)

PRIMATES	PRIMATES	PRIMATES
CEBIDAE	CALLITHRICIDAE	CERCOPITHECIDAE
Alouatta *Ateles*	*Saguinus*	*Macaca*
palliata *geoffroyi*	*geoffroyi*	*mulatta*
D—> 50 W—> 40, M—1	W—> 19, M—1	K—> 27, M—> 25
0 (rud. tubal memb.)	0?	partial
(no bursa)	(no bursa)	very large
—	small	medium
—	—	pelvic
ellipsoidal	amygdaloidal	amygdaloidal
0	0	0
0	0	0
thin	thin	thin
medium	large	medium
1 (1)	2 (2)	1 (1)
6.0	—	6.0
large	large	large
thick (t. i.)	thick (t. i.)	thick
very much	very much	medium
medium	—	0
—	—	0
—	—	—
—	—	little
little	—	medium
—	—	0
1	1	1
—	0	0
—	—	into lactation
Wislocki, 1930; Dempsey, 1939; Mossman	Wislocki, 1939; Mossman	Corner, 1945; Sturgis, 1949; Koering, 1969; Mossman

Order	PRIMATES	PRIMATES
Family	Pongidae	Hominidae
Genus	*Hylobates*	*Homo*
Species	*lar*	*sapiens*
No. of individuals	16	M—> 20
Bursa ovarica	0	0
Orifice	(no bursa)	(no bursa)
Epoophoron	—	medium
Ovary		
Location	pelvic	pelvic
Shape	ellipsoidal	amygdaloidal
Lobes	0	0
Fissures	0	0
Tunica albuginea	thick	thick
Rete	medium	small
Ripe follicles		
No. & (litter size)	1 (1)	1 (1)
Diameter (mm)	6.0	10.0–15.0
Antrum	large	large
Thecal gland	thick	thick
Interstitial gland type		
Thecal	much	medium
Stromal	little	0
Medullary cord	—	0
Gonadal adrenal		
in cortex	—	0
in medulla	—	0
at epoophoron	—	little
accessory adrenal	—	rare
Luteal glands		
No. of cell types	1	1
Accessory corpora	0	0 to few
Persistence	—	into lactation
Sources	Saglik, 1938; Dempsey, 1940; Duke	Boyd & Hamilton, 1955; Watzka, 1957; Forleo, 1961; Hibbard, 1961; Mossman, Koering & Ferry, 1964; Mossman

EDENTATA	EDENTATA	EDENTATA
MYRMECOPHAGIDAE	BRADYPODIDAE	DASYPODIDAE
Tamandua	*Choloepus*	*Dasypus*
tetradactyla	*hoffmanni*	*novemcinctus*
1	2	> 200
0	partial	partial
(no bursa)	slit	large
—	very large	large
—	—	pelvic
fusiform	reniform	ellipsoidal; ovul. pit
0	0	0
0	0	0
thick	thick	thick
—	medium	large
— (1)	— (1)	1 (4 monovular)
—	—	—
—	—	large
—	thick (t. i.)	thick
little	little	much
much	0	0
0	0	0
—	0	medium
—	0	medium
—	0	medium
—	0	common
—	1	1
—	—	rare
—	—	into lactation
Mossman	Mossman	A. C. Enders & Buchanan, 1959; Mossman

Order	PHOLIDOTA		LAGOMORPHA
Family	Manidae		Ochotonidae
Genus	*Manis*	*Manis*	*Ochotona*
Species	*javanica*	*temminckii*	*princeps*
No. of individuals	D—6	M—3	> 100
Bursa ovarica	0		partial
Orifice	(no bursa)		large
Epoophoron	very large		medium
Ovary			
Location	pelvic		lumbar
Shape	irregular ellipsoidal		reniform
Lobes	0		few
Fissures	0		few
Tunica albuginea	thin?		thin
Rete	very large		medium
Ripe follicles			
No. & (litter size)	—	(1)	3 (3)
Diameter (mm)	> 2.5		0.8
Antrum	large		medium; trabeculae
Thecal gland	thick (t. i.)		thin; interrupted
Interstitial gland type			
Thecal	much		little
Stromal	much?		0
Medullary cord	much?		0
Gonadal adrenal			
in cortex	much?		0
in medulla	much		little to 0
at epoophoron	much		medium
accessory adrenal	common?		0
Luteal glands			
No. of cell types	1		1
Accessory corpora	0		0
Persistence	into 2d gestation?		into lactation
Sources	Mossman & Duke		Duke, 1952; Mossman

LAGOMORPHA	LAGOMORPHA	LAGOMORPHA
LEPORIDAE	LEPORIDAE	LEPORIDAE
Lepus	*Sylvilagus*	*Oryctolagus*
americanus	*floridanus*	*cuniculus*
> 20	8	M—> 10
partial	partial	partial
very large	very large	very large
medium	medium	medium
lumbar	lumbar	lumbar
fusiform	fusiform	fusiform
0	0	0
0	0	0
thin	thin	thin
small	small	small
— (3)	— (5)	5 (5)
1.5	—	1.8
medium; trabeculae thick	medium; trabeculae thick	medium; trabeculae thick
much	much	much
very much	much	very much
0	0	0
0	0	0
0	0	0
—	—	—
—	—	common
1	1	1
0	0	0
into 2d gestation	—	into 2d gestation
Keith Coll. (Mossman); Mossman	Mossman	Hammond & Marshall, 1925; Brambell, 1944; Duke, 1947; Mossman

Order	RODENTIA	RODENTIA
Family	APLODONTIDAE	SCIURIDAE
Genus	*Aplodontia*	*Sciurus*
Species	*rufa*	*carolinensis* & *niger*
No. of individuals	P — 289, M — 4	M—> 100
BURSA OVARICA	—	complete
Orifice	—	large
EPOOPHORON	medium	large
OVARY		
Location	lumbar	lumbar
Shape	ellipsoidal	globose
Lobes	0	0
Fissures	0	0
Tunica albuginea	thin	thin
Rete	?	medium
Ripe follicles		
No. & (litter size)	3 (2)	4 (3)
Diameter (mm)	3.0	1.1
Antrum	large	large
Thecal gland	thin	thin
Interstitial gland type		
Thecal	medium	much
Stromal	much	little
Medullary cord	0	0
Gonadal adrenal		
in cortex	0	0
in medulla	0	0
at epoophoron	—	medium
accessory adrenal	—	0
Luteal glands		
No. of cell types	1	1
Accessory corpora	0	0
Persistence	into gestation	into 2d gestation
SOURCES	Pfeiffer, 1958; Mossman	Deanesly & Parkes, 1933; Mossman

RODENTIA	RODENTIA	RODENTIA
SCIURIDAE	SCIURIDAE	SCIURIDAE
Tamiasciurus hudsonicus	*Funambulus pennanti*	*Spermophilus tridecemlineatus*
> 100	300	M—> 150
complete	—	complete
medium	—	small
large	—	medium
lumbar	—	lumbar
globose	globose	globose
0	0	0
0	0	0 to few
thin	thin	thin
medium	—	medium
4 (4)	3 (3)	8 (8)
1.2	—	0.7
large	medium	medium
thin	thin	thick
much	much	much
little	little	little
0	—	0
0	—	0
0	—	much
0	—	medium
0	—	0
1	1	1
0	—	0
into 2d lactation	into lactation	into lactation
Mossman	Seth & Prasad, 1967, 1969	Foster, 1934; Mossman

Order	RODENTIA	RODENTIA
Family	SCIURIDAE	GEOMYIDAE
Genus	*Tamias*	*Geomys*
Species	*striatus*	*bursarius*
No. of individuals	> 25	> 25
BURSA OVARICA	complete	0
Orifice	medium	(no bursa)
EPOOPHORON	medium	medium
OVARY		
Location	lumbar	lumbar
Shape	globose	globose
Lobes	0	0
Fissures	0	0
Tunica albuginea	thin	thick
Rete	medium	large
Ripe follicles		
No. & (litter size)	4 (4)	3 (3)
Diameter (mm)	1.0	1.0
Antrum	medium	medium
Thecal gland	thin	very thick
Interstitial gland type		
Thecal	much	much
Stromal	0	much
Medullary cord	0	0
Gonadal adrenal		
in cortex	—	0
in medulla	—	0
at epoophoron	medium	little
accessory adrenal	0	0
Luteal glands		
No. of cell types	1	1
Accessory corpora	0	0
Persistence	into lactation	through gestation
SOURCES	Mossman, 1966; Mossman	Mossman, 1937; Mossman

RODENTIA	RODENTIA	RODENTIA
HETEROMYIDAE	CASTORIDAE	PEDETIDAE
Dipodomys	*Castor*	*Pedetes*
ordii	*canadensis*	*capensis*
D — 42, M — 3	5	9
0	partial	complete
(no bursa)	long slit	medium
—	medium	small
lumbar	lumbar	lumbar
globose	amygdaloidal	globose
0	0	0
0	0	0 to few
thin	thick; fibrous	thick; fibrous
—	—	small
— (—)	— (4)	— (1)
—	6.5	—
medium	large	large
very thick (t. i.)	thick (t. i.)	thick
much	medium	much
little	medium	much
—	0	0
—	0	0
—	0	0
—	little	medium
—	—	—
1	2	2
0	0	0
term	—	—
Duke, 1940; Mossman	Provost, 1962; Mossman	Horst Coll. (Mossman); Mossman

Order	RODENTIA	RODENTIA
Family	CRICETIDAE	CRICETIDAE
Genus	*Neotoma*	*Mesocricetus*
Species	sp.	*auratus*
No. of individuals	4	(M)—> 20
BURSA OVARICA	complete	complete
Orifice	—	0
EPOOPHORON	medium	medium
OVARY		
Location	lumbar	lumbar
Shape	globose	globose
Lobes	few	several
Fissures	few	few
Tunica albuginea	thin	thin
Rete	small	small
Ripe follicles		
No. & (litter size)	— (4)	— (6)
Diameter (mm)	—	—
Antrum	medium	medium
Thecal gland	thick (t. i.)	thin
Interstitial gland type		
Thecal	medium	medium
Stromal	much	much
Medullary cord	0	0
Gonadal adrenal		
in cortex	—	0
in medulla	—	0
at epoophoron	—	—
accessory adrenal	—	—
Luteal glands		
No. of cell types	1	1
Accessory corpora	—	0
Persistence	—	into lactation
SOURCES	Mossman	Orsini Coll. (Mossman)

RODENTIA	RODENTIA	RODENTIA	
MURIDAE	MURIDAE	ZAPODIDAE	
Rattus	*Mus*	*Zapus*	*Napaeozapus*
norvegicus	*musculus*	sp.	*insignis*
M—> 10	M—> 10	2	3
complete	complete	complete	
porelike	porelike	—	
—	—	medium	
lumbar	lumbar	lumbar	
globose	globose	globose	
several	several	0	
0	0	0	
thin	thin	0	
small	small	large	
10 (9)	8 (6)	—	(5)
0.9	0.7	—	
medium	medium	medium	
thin	thin	thick (t. i.)	
much	much	much	
little	little	little	
0	0	0	
—	—	—	
—	—	—	
—	—	medium	
—	—	—	
1	1	1	
0 to few	0	—	
into lactation	into lactation	—	
Long & Evans, 1922; Alden, 1942; Mossman	Snell, 1941; Rugh, 1968; Mossman	Mossman	

Order	RODENTIA		RODENTIA	
Family	HYSTRICIDAE		HYSTRICIDAE	
Genus	*Thecurus*		*Hystrix*	
Species	*pumilis*		*brachyurum*	
No. of individuals	1		3	
BURSA OVARICA	—		—	
Orifice	—		—	
EPOOPHORON	—		—	
OVARY				
Location	—		—	
Shape	—		—	
Lobes	0		—	
Fissures	0		—	
Tunica albuginea	thick		thick	
Rete	—		—	
Ripe follicles				
No. & (litter size)	—	(—)	—	(—)
Diameter (mm)	—		—	
Antrum	large		large	
Thecal gland	thick		thick	
Interstitial gland type				
Thecal	much		medium	
Stromal	much		medium	
Medullary cord	—		—	
Gonadal adrenal				
in cortex	—		—	
in medulla	—		—	
at epoophoron	—		—	
accessory adrenal	—		—	
Luteal glands				
No. of cell types	1		1	
Accessory corpora	—		—	
Persistence	—		—	
SOURCES	Duke		Duke	

RODENTIA	RODENTIA	RODENTIA
HYSTRICIDAE	HYSTRICIDAE	ERETHIZONTIDAE
Hystrix	*Atherurus*	*Erethizon*
cristata	*macrourus*	*dorsatum*
3	3	305
partial	—	partial
large	—	large
—	—	medium
lumbar	—	lumbar
amygdaloidal	—	amygdaloidal
0	—	0
0	—	0
thick	thin	thin
—	—	large
— (1–4)	— (—)	1 (1)
—	—	7.0
very large	—	large
thick	thick	thin
much	medium	medium
much	medium	much
0	—	0
—	—	0
—	—	0
—	—	medium
—	—	0
1	1	1
several	—	very many
—	—	> 1 yr.; into 2d gestation
Mossman	Duke	Mossman & Judas, 1949; Mossman

Order	RODENTIA	RODENTIA
Family	CAVIIDAE	CAVIIDAE
Genus	*Cavia*	*Galea*
Species	*porcellus*	*musteloides*
No. of individuals	R — 62, M — 19	150
BURSA OVARICA	partial	partial
Orifice	large	large
EPOOPHORON	—	—
OVARY		
Location	lumbar	lumbar
Shape	amygdaloidal	reniform
Lobes	0	—
Fissures	0	—
Tunica albuginea	thin	thick
Rete	very large	—
Ripe follicles		
No. & (litter size)	3 (3)	6 (3)
Diameter (mm)	0.8	0.8
Antrum	large	large
Thecal gland	thick	thin
Interstitial gland type		
Thecal	much	medium
Stromal	medium	medium
Medullary cord	0	—
Gonadal adrenal		
in cortex	0	—
in medulla	—	—
at epoophoron	medium	—
accessory adrenal	—	—
Luteal glands		
No. of cell types	1	1
Accessory corpora	rare	rare
Persistence	early lactation	—
SOURCES	Rowlands, 1956; Mossman	Rowlands & Weir, unpublished

RODENTIA	RODENTIA	RODENTIA
DASYPROCTIDAE	DASYPROCTIDAE	CHINCHILLIDAE
Dasyprocta	*Myoprocta*	*Lagostomus*
aguti	*pratti*	*maximus*
W — 15, M — 3	21	67
partial	partial	partial
long slit	—	large
—	—	—
lumbar	lumbar	lumbar
amygdaloidal	ellipsoidal	irregular
0	0	very many
0	0	very many
thin	thin	0
medium	medium	medium
3 (2)	2 (2)	300–800 (2)
> 0.8	0.7	0.3
medium	medium	small; trabeculae
thin	thick	thin
much	medium	medium
much	—	little
"testis cords"	"testis cords"	—
—	—	—
—	—	—
medium	—	—
—	—	—
1	1	1
very many	very many	very many
—	—	into lactation
Weir, 1971*a*; Mossman	Weir, 1967, 1970; Rowlands, Tam & Kleiman, 1970	Weir, 1971*b*

Order	RODENTIA	RODENTIA
Family	CHINCHILLIDAE	CHINCHILLIDAE
Genus	*Lagidium*	*Lagidium*
Species	*peruanum*	*boxi*
No. of individuals	P — 473, M — 1	1
BURSA OVARICA	—	—
Orifice	—	—
EPOOPHORON	—	—
OVARY		
Location	lumbar	lumbar
Shape	amygdaloidal	fusiform
Lobes	0	0
Fissures	0	few
Tunica albuginea	thin	thin
Rete	large	medium
Ripe follicles		
No. & (litter size)	1 (1)	1 (1)
Diameter (mm)	1.4	—
Antrum	medium	medium
Thecal gland	thick	thick
Interstitial gland type		
Thecal	medium	medium
Stromal	very much	much
Medullary cord	0	—
Gonadal adrenal		
in cortex	—	—
in medulla	—	—
at epoophoron	medium	—
accessory adrenal	—	—
Luteal glands		
No. of cell types	—	—
Accessory corpora	few	few
Persistence	into lactation	—
SOURCES	O. P. Pearson, 1949; Mossman	Weir, 1971c and unpublished

RODENTIA	RODENTIA	RODENTIA
CHINCHILLIDAE	CAPROMYIDAE	OCTODONTIDAE
Chinchilla	*Myocastor*	*Octodon*
laniger	*coypus*	*degus*
H & T—87, W—73, M—5	H—12, R & H—47, M—1	55
partial	partial	partial
large	large	—
—	—	—
lumbar	lumbar	lumbar
fusiform	amygdaloidal	ellipsoidal
0	0	few?
0	0	few?
thin	thin	thin
large	small	small
4 (3)	9 (5)	— (4)
1.2	1.8	0.6
medium	medium	medium
thick	thick	thin
much	much	medium
much	medium	much
—	—	—
—	—	—
—	—	—
—	—	—
—	—	—
1	1	1
very many	—	rare
into lactation	into lactation	—
Hillemann & Tibbitts, 1956; Weir, 1967 & personal communication; Mossman	Hillemann, Gaynor & Stanley, 1958; Rowlands & Heap, 1966; Mossman	Weir, unpublished

Order	RODENTIA	RODENTIA
Family	OCTODONTIDAE	CTENOMYIDAE
Genus	*Octodontomys*	*Ctenomys*
Species	*gliroides*	*torquatus*
No. of individuals	7	120
BURSA OVARICA	partial	partial
Orifice	—	—
EPOOPHORON	—	—
OVARY		
Location	lumbar	lumbar
Shape	ellipsoidal	ellipsoidal
Lobes	0	few?
Fissures	0	few?
Tunica albuginea	thin	thin
Rete	small	medium
Ripe follicles		
No. & (litter size)	— (2)	— (4)
Diameter (mm)	1.0	0.8
Antrum	medium	medium
Thecal gland	thin	thin
Interstitial gland type		
Thecal	0	little
Stromal	0	little
Medullary cord	—	—
Gonadal adrenal		
in cortex	—	—
in medulla	—	—
at epoophoron	—	—
accessory adrenal	—	—
Luteal glands		
No. of cell types	1	1
Accessory corpora	0	rare
Persistence	—	—
SOURCES	Weir, unpublished	Weir, unpublished

RODENTIA	RODENTIA	RODENTIA
ECHIMYIDAE	BATHYERGIDAE	BATHYERGIDAE
Proechimys	*Cryptomys*	*Bathyergus*
sp.	*natalensis*	*suillus*
4	15	6
partial	partial	partial
—	very large	very large
—	medium	medium
lumbar	lumbar	lumbar
amygdaloidal	amygdaloidal	amygdaloidal
0	0	0
0	0	0
thin	0	thin
small	—	medium
— (4)	— (5)	— (—)
—	—	—
medium	—	—
thick	thick (t. i.)	—
medium	very much	much
medium	0	0
—	0	0
—	0	0
—	0	0
—	little	little
—	0	0
1	1	1
many	0	0
—	—	—
Weir, unpublished	Horst Coll. (Mossman)	Horst Coll. (Mossman)

Order	CETACEA	CETACEA
Family	MONODONTIDAE	DELPHINIDAE
Genus	*Delphinapterus*	*Globicephala*
Species	*leucas*	*melaena*
No. of individuals	18	5
BURSA OVARICA	partial	partial
Orifice	very large	very large
EPOOPHORON	—	—
OVARY		
Location	lumbar	—
Shape	irregular reniform	irregular reniform
Lobes	few	few
Fissures	few	few
Tunica albuginea	thick	thick
Rete	medium	—
Ripe follicles		
No. & (litter size)	1 (1)	1–3 (1)
Diameter (mm)	> 11.0	30.0
Antrum	very large	very large
Thecal gland	thick (t. i.)	thick (t. i.)
Interstitial gland type		
Thecal	medium	medium
Stromal	?	?
Medullary cord	0	0
Gonadal adrenal		
in cortex	—	—
in medulla	—	—
at epoophoron	—	—
accessory adrenal	—	—
Luteal glands		
No. of cell types	2	2
Accessory corpora	rare	0
Persistence	late lactation?	—
SOURCES	Daudt, 1898; James Brooks, personal communications & specimens; Mossman	Harrison, 1949; Sergeant, 1962; Mossman

CARNIVORA	CARNIVORA	CARNIVORA
CANIDAE	CANIDAE	CANIDAE
Canis	*Vulpes*	*Urocyon*
familiaris	*fulva*	sp.
M — 11	M & D — 4	15
complete	complete	complete
medium	medium	medium
—	—	—
lumbar 3	lumbar	lumbar
irregular ellipsoidal	ellipsoidal	ellipsoidal
0	0	0
0	0	0
thick	thick	thick
large	medium	medium
6 (6)	6 (5)	— (5)
6.0	7.0	—
very large	—	—
thick	—	thick (t. i.)
medium	much	medium
medium	much	medium
little?	—	—
—	—	—
—	—	—
—	—	—
—	—	—
1	1	1
0	0	0
into lactation	into lactation	into lactation
Jonckheere, 1930; Evans & Cole, 1931; A. C. Anderson & Simpson, 1971; Mossman	Rowlands & Parkes, 1935; O. P. Pearson & Enders, 1943; Duke & Mossman	Duke

Order	CARNIVORA		CARNIVORA	
Family	URSIDAE		PROCYONIDAE	
Genus	*Ursus*		*Procyon*	
Species	*americanus*		*lotor*	
No. of individuals	M—> 18		3	
BURSA OVARICA	complete		complete	
Orifice	small		small	
EPOOPHORON	medium		—	
OVARY				
Location	—		lumbar	
Shape	globose		globose	
Lobes	0 to few		0	
Fissures	few		0	
Tunica albuginea	thick		thick	
Rete	medium		medium	
Ripe follicles				
No. & (litter size)	3	(2)	—	(4)
Diameter (mm)	10.0		—	
Antrum	very large		—	
Thecal gland	—		—	
Interstitial gland type				
Thecal	much		little	
Stromal	much		little	
Medullary cord	medium?		much	
Gonadal adrenal				
in cortex	—		—	
in medulla	—		—	
at epoophoron	—		—	
accessory adrenal	—		—	
Luteal glands				
No. of cell types	1		—	
Accessory corpora	0		—	
Persistence	into lactation		—	
SOURCES	Wimsatt, 1963; Mossman		Mossman	

CARNIVORA	CARNIVORA	CARNIVORA
MUSTELIDAE	MUSTELIDAE	MUSTELIDAE
Mustela	*Mustela*	*Mustela*
erminea & *frenata*	*vison*	*putorius*
(M) — 50	M — 22	(M) — 16
complete	complete	complete
small	small	small
—	medium	—
lumbar	lumbar	lumbar
oblate	oblate	irregular oblate
0	0	0
0	0	few
thin	thin	thin
large	medium	medium
— (5)	10 (5)	— (9)
0.8	1.0	1.4
medium	medium	medium
thin	thick	thick
much	medium	medium
medium	?	?
much	very much	much
—	—	—
—	—	—
—	—	—
—	—	—
1	1	1
0	few	few
through gestation	—	through gestation
Watzka, 1940, 1949; Wright, 1942, 1963; Wright Coll. (Mossman)	Hansson, 1947; R. K. Enders, 1952; Mossman	Hammond & Marshall, 1930; Hamilton & Gould, 1939; Hamilton Coll. (Mossman)

Order	CARNIVORA	CARNIVORA
Family	MUSTELIDAE	MUSTELIDAE
Genus	*Taxidea*	*Enhydra*
Species	*taxus*	*lutris*
No. of individuals	42	136
BURSA OVARICA	complete	complete
Orifice	small	0
EPOOPHORON	—	—
OVARY		
Location	lumbar	lumbar
Shape	oblate	irregular oblate
Lobes	0	several
Fissures	0	several
Tunica albuginea	thin	thin
Rete	medium	—
Ripe follicles		
No. & (litter size)	3 (3)	1 (1)
Diameter (mm)	—	8.0
Antrum	—	large
Thecal gland	thin (t. i.)	thick
Interstitial gland type		
Thecal	much	much
Stromal	much	?
Medullary cord	very much	medium
Gonadal adrenal		
in cortex	—	—
in medulla	—	—
at epoophoron	—	—
accessory adrenal	—	—
Luteal glands		
No. of cell types	1	1
Accessory corpora	0	0
Persistence	—	through gestation
SOURCES	Wright, 1966; Wright Coll. (Mossman)	Sinha, Conaway & Kenyon, 1966; Sinha Coll. (Mossman)

CARNIVORA	CARNIVORA	CARNIVORA
VIVERRIDAE	VIVERRIDAE	VIVERRIDAE
Genetta	*Arctogalidia*	*Paradoxurus*
genetta	*trivirgata*	*hermaphroditus*
1	4	12
partial	—	—
very large	—	—
—	—	—
—	—	—
globose	irregular globose	irregular globose
0	few	few
0	few	few
thin	thick	thick
—	small	medium
— (2)	2–3 (—)	2–3 (—)
—	3.0	3.0
—	very thick	thick
—	medium	medium
much	medium	medium
medium	medium	medium
0	—	—
—	—	—
—	—	—
—	—	—
—	—	—
1	1	1
0	0	—
into 2d gestation	into 2d gestation	into 2d gestation
Mossman	Duke	Duke

Order	CARNIVORA	CARNIVORA
Family	VIVERRIDAE	VIVERRIDAE
Genus	*Paguma*	*Herpestes*
Species	*larvata*	*auropunctatus*
No. of individuals	2	P & B — 77, D — 3, M — 1
BURSA OVARICA	—	complete
Orifice	—	small
EPOOPHORON	—	—
OVARY		
Location	—	lumbar
Shape	irregular globose	irregular globose
Lobes	—	—
Fissures	—	—
Tunica albuginea	thin	thin
Rete	medium	medium
Ripe follicles		
No. & (litter size)	— (3)	3 (3)
Diameter (mm)	—	> 1.4
Antrum	large	large
Thecal gland	thick	thin
Interstitial gland type		
Thecal	much	little
Stromal	medium	little
Medullary cord	—	—
Gonadal adrenal		
in cortex	—	—
in medulla	—	—
at epoophoron	—	—
accessory adrenal	—	—
Luteal glands		
No. of cell types	—	—
Accessory corpora	—	—
Persistence	—	into lactation
SOURCES	Duke	O. P. Pearson & Baldwin, 1953; Duke & Mossman

CARNIVORA	CARNIVORA	CARNIVORA
HYAENIDAE	HYAENIDAE	FELIDAE
Proteles	*Crocuta*	*Felis*
cristatus	*crocuta*	*catus*
3	40	D — 20
partial	partial	partial
very large	very large	very large
—	—	—
—	—	lumbar
ellipsoidal	—	ellipsoidal
0	—	0
0	—	0
thin	thin	thick
very large	—	large
— (3)	— (—)	4 (4)
—	10.0?	—
large	—	large
thick (t. i.)	—	thick
much	much	much
much	much	medium
much	—	medium
—	—	—
—	—	—
—	—	—
—	—	—
1	1	1
few	—	0
into 2d gestation	beyond lactation	into 2d lactation
Mossman	Matthews, 1939	Kingsbury, 1914, 1939; Dawson & Friedgood, 1940; Dawson, 1941, 1946; Duke & Mossman

Order	CARNIVORA		CARNIVORA	
Family	OTARIIDAE		OTARIIDAE	
Genus	*Arctocephalus*		*Callorhinus*	
Species	*pusillus*		*ursinus*	
No. of individuals	1		?	
BURSA OVARICA	complete		complete	
Orifice	small		small	
EPOOPHORON	—		—	
OVARY				
Location	—		—	
Shape	globose		globose	
Lobes	0		0	
Fissures	0		0	
Tunica albuginea	thick		thick	
Rete	medium		small	
Ripe follicles				
No. & (litter size)	—	(1)	1	(1)
Diameter (mm)	—		—	
Antrum	large		large	
Thecal gland	thick		—	
Interstitial gland type				
Thecal	—		—	
Stromal	—		—	
Medullary cord	—		—	
Gonadal adrenal				
in cortex	—		—	
in medulla	—		—	
at epoophoron	—		—	
accessory adrenal	—		—	
Luteal glands				
No. of cell types	1		1	
Accessory corpora	0		0	
Persistence	into lactation		into lactation	
SOURCES	Horst Coll. (Mossman)		R. K. Enders, Pearson & Pearson, 1946; A. K. Pearson & Enders, 1951	

TUBULIDENTATA	PROBOSCIDEA	PROBOSCIDEA
Orycteropodidae	Elephantidae	Elephantidae
Orycteropus	*Loxodonta*	*Elephas*
afer	*africana*	*maximus*
H — 1, M — 4	> 80	2
complete	partial	partial
—	large	large
large	large	large
—	pelvic	pelvic
very irregular	globose	globose
many	0	0
many	few	few
thin	—	—
—	—	—
4? (1)	— (1)	— (1)
—	> 20.0	—
—	large	—
very thick (t. i.)	—	—
much	medium	—
medium	—	0
0	—	0
0	—	—
0	—	—
much	—	—
0	—	?
1	1	1
0?	many?	many?
—	into lactation	—
Horst, 1949; Mossman	Perry, 1953; Short & Buss, 1965; Short, 1966, 1969	Mossman

Order	HYRACOIDEA	HYRACOIDEA
Family	PROCAVIIDAE	PROCAVIIDAE
Genus	*Dendrohyrax*	*Heterohyrax*
Species	*arboreus*	*brucei*
No. of individuals	48	W & W — 4, M — 5
BURSA OVARICA	partial	partial
Orifice	—	very large
EPOOPHORON	—	medium
OVARY		
Location	—	—
Shape	reniform	irregular globose
Lobes	several	several
Fissures	—	many
Tunica albuginea	—	thin
Rete	—	large
Ripe follicles		
No. & (litter size)	2 (2)	— (2)
Diameter (mm)	1.7	1.7
Antrum	—	—
Thecal gland	thick	thick (t. i.)
Interstitial gland type		
Thecal	—	little
Stromal	—	0
Medullary cord	—	0
Gonadal adrenal		
in cortex	—	0
in medulla	very much	very much
at epoophoron	—	little
accessory adrenal	—	—
Luteal glands		
No. of cell types	1	1
Accessory corpora	0	0
Persistence	life?	—
SOURCES	P. N. O'Donoghue, 1963	Wislocki & Westhuysen, 1940; Mossman

HYRACOIDEA	PERISSODACTYLA	PERISSODACTYLA
PROCAVIIDAE	EQUIDAE	EQUIDAE
Procavia	*Equus*	*Equus*
capensis	*caballus*	*burchelli*
W & W — 14, M — 13	K—>100,M(fromWa)—48	4
partial	partial	partial
very large	very large	very large
—	—	—
—	lumbar 4	—
irregular globose	ellipsoidal; ovul. pit	ellipsoidal; ovul. pit
several	0	0
many	0	0
thin	thick	thick
large	—	—
— (2)	1 (1)	1 (1)
—	40.0	—
medium; trabeculae	very large	—
thick	thick	thick (t. i.)
medium	much	medium
0	—	—
0	—	—
0	—	—
very much	—	—
little	often much	—
—	common	—
1	2	2
0	?	0
—	into late gestation	term?
Wislocki & Westhuysen, 1940; Mossman	Kupfer, 1928; Hammond & Wodzicki, 1941; Ono et al., 1969; Warszawsky et al., 1971; Mossman	Mossman

Order	ARTIODACTYLA	ARTIODACTYLA
Family	Suidae	Tragulidae
Genus	*Sus*	*Tragulus*
Species	*scrofa*	*javanicus*
No. of individuals	M — 10	D — 10, M — 3
Bursa ovarica	partial	partial
Orifice	very large	very large
Epoophoron	—	large
Ovary		
Location	lumbar 6	—
Shape	irregular ellipsoidal	amygdaloidal
Lobes	0	0
Fissures	0	0
Tunica albuginea	thin	thin
Rete	—	large
Ripe follicles		
No. & (litter size)	10 (8)	1 (1)
Diameter (mm)	10.0	> 1.5
Antrum	large	large
Thecal gland	thin	thin (t. i.)
Interstitial gland type		
Thecal	little	little
Stromal	0	0
Medullary cord	0	0
Gonadal adrenal		
in cortex	—	0
in medulla	—	0
at epoophoron	—	—
accessory adrenal	—	—
Luteal glands		
No. of cell types	2	2
Accessory corpora	0	0
Persistence	into lactation	through gestation
Sources	Clark, 1898, 1899; Corner, 1915, 1919, 1921; Mossman	Duke & Mossman

ARTIODACTYLA	ARTIODACTYLA	ARTIODACTYLA
CERVIDAE	ANTILOCAPRIDAE	BOVIDAE
Odocoileus	*Antilocapra*	*Bos*
virginianus	*americana*	*taurus*
> 75	O'G — 137, D — 2, M — 1	M — 7
partial	—	partial
very large	—	very large
—	—	—
lumbar	—	lumbar 6
amygdaloidal	globose	amygdaloidal
0	0	0
0	0	0
thick	thick	thick
medium	medium	large
2 (2)	3–7 (2)	1 (1)
—	—	17.0
large	large	large
thin	thick	thin
medium	little	little
0	—	0
0	—	0
0	—	0
0	—	0
—	—	—
—	—	—
2	2	2
rare	0	0
into lactation	—	into lactation
Sinha, Seal, & Doe, 1971*b*; Mossman	O'Gara, 1969; Duke & Mossman	Hammond, 1923; Höfliger, 1948; Donaldson & Hansel, 1965; Mossman

Order	ARTIODACTYLA	ARTIODACTYLA
Family	BOVIDAE	BOVIDAE
Genus	*Bison*	*Kobus*
Species	*bison*	*kob*
No. of individuals	3	> 340
BURSA OVARICA	partial	—
Orifice	very large	—
EPOOPHORON	—	—
OVARY		
Location	—	—
Shape	amygdaloidal	amygdaloidal
Lobes	0	0
Fissures	0	0
Tunica albuginea	thick	thick
Rete	—	—
Ripe follicles		
No. & (litter size)	— (1)	1 (1)
Diameter (mm)	—	11.0
Antrum	large	large
Thecal gland	thin	thin
Interstitial gland type		
Thecal	little	little
Stromal	0	0
Medullary cord	0	0
Gonadal adrenal		
in cortex	0	0
in medulla	0	0
at epoophoron	—	—
accessory adrenal	—	—
Luteal glands		
No. of cell types	2	2
Accessory corpora	0	0
Persistence	—	into lactation
SOURCES	Mossman	Morrison, 1971

ARTIODACTYLA	ARTIODACTYLA
Bovidae	Bovidae
Capra	*Ovis*
hircus	*aries*
?	(M) — 8
partial	0 (rud. tubal memb.)
very large	(no bursa)
—	—
—	—
reniform	—
0	0
0	0
thick	thick
—	—
2 (2)	— (2)
10.0	9.5
large	large
thick	thin
—	little
—	0
—	0
—	0
—	0
—	—
—	—
2	2
0	0
into lactation	into lactation
Harrison, 1948*b*	Grant, 1934; Warbritton, 1934; Deane et al., 1966; Hamilton Coll. (Mossman)

Supplementary notes

Class MAMMALIA

Subclass PROTOTHERIA

Order MONOTREMATA

*Tachyglossidae and *Ornithorhynchidae. We have no firsthand knowledge of these egg-laying mammals. An ovarian bursa appears to be absent, but Broek (1931, Fig. 2) showed what appears to be a prominent ventral peritoneal fold along the infundibular portion of the oviduct. This could be a homolog, or at least an analog, of the tubal membrane of Eutheria. Both ovaries are functional in the spiny anteater (*Tachyglossus*), but only the left functions in the duck-billed platypus (*Ornithorhynchus anatinus*). Since there is no follicular antrum, the follicular diameter is greater than that of the ovum by only the thickness of the zona pellucida and follicular epithelium, plus the thin film of fluid between the zona and the epithelium. The theca interna cells hypertrophy to produce a typical rather thin thecal gland surrounding the ripe follicle. Follicular rupture at ovulation is followed by formation of luteal cells from the follicular epithelium. The thecal gland cells apparently persist throughout the life of the corpus luteum, which begins to degenerate before oviposition, but still contains luteal cells well into the incubation period. After ovulation there is no evidence of degeneration of the thecal gland cells and of their replacement by paraluteal cells, as occurs in Eutheria, but it must be borne in mind that the degeneration of thecal cells at this time is not easy to detect and that many investigators

*An asterisk preceding a family name indicates that that family is represented by one or more genera in the synoptic tables. The absence of an asterisk indicates that too little information was available to warrant a column in the tables.

would still deny that it takes place in Eutheria. Atresia of small follicles takes place without rupture, but larger follicles extrude the egg contents either into the peritoneal cavity or into the adjacent intra-ovarian lymph spaces which are so prominent in these ovaries. The corpora atretica are very fibrous, but they contain an inner, less fibrous zone of small cells derived from the theca interna. There is no evidence that these are thecal type interstitial gland cells, as in Eutheria. Harrison (1948*a*) should be consulted for a comprehensive discussion of the corpora lutea of lower vertebrates and for a comparison with those of mammals.

In summary, the ovaries of monotremes are more like those of reptiles and birds than of marsupial or eutherian mammals.

Subclass THERIA

Infraclass METATHERIA

Order MARSUPIALIA

Superfamily Didelphoidea

*Didelphidae. It is questionable whether one should consider that the opossum (*Didelphis marsupialis*) has an ovarian bursa. Each oviduct has a well-developed tubal membrane which continues along the antimesometrial margins of the two uteri as an even longer *uterine membrane* (*membrana uterina*). The two uterine portions are broadly continuous with each other at the midline and, since the uteri curve dorsally, they form a membranous sheet dorsal to the uteri. Thus, a single tubo-uterine pouch extends across the midline from one oviduct to the other. The ovaries are attached by their mesovaria to the dorsomedial surface of the broad ligaments far enough caudolateral to the tubo-uterine junction to be approximately at the free edge of the *tubo-uterine membrane* (*membrana tubo-uterina*), and therefore not actually in the pouch. During pregnancy, expansion of the uteri obliterates the lateral portions of the uterine membrane and leaves the ovaries in deep recesses between the uteri and the broad ligament. Certainly, the tubal membrane of the opossum is homologous to that of eutherian mammals, and the uterine membrane corresponds to the rudimentary one present in the nine-banded armadillo (*Dasypus novemcinctus*) (see Fig. 1.15). However, the tubo-uterine pouch is not directly related to the ovary, and it is not certain that it can serve the same purposes as the typical eutherian ovarian bursa.

Published information on ovarian morphology of the opossum is fragmentary, but is more complete than for any other marsupial. Although we have histological material from several dozen animals, its condition is such that we are uncertain of the nature of the interstitial gland tissue. Martínez-Esteve (1942) shows what appears to be a considerable amount of probable thecal type interstitial gland in a few ovaries, but he usually found none.

Superfamilies Dasyuroidea and Phalangeroidea

Dasyuridae, Phalangeridae, and Macropodidae. C. H. O'Donoghue (1916) saw interstitial tissue corresponding to that of Eutheria in some pouch young and a few older specimens in certain of the several genera of marsupials, including Didelphidae, that he examined. Pilton and Sharman (1962) wrote of corpora lutea atretica in the vulpine phalanger (*Trichosurus vulpecula*). These were presumably masses of thecal type interstitial gland tissue. O'Donoghue (1912) described large follicles in the Australian native "cat" (*Dasyurus quoll*) as having thecae in which the externa and interna could not be distinguished from one another. He also believed that this species often ovulated more than one viable egg from a single follicle. There is general agreement that the luteal cells of marsupials are derived solely from follicular epithelium and that the corpora lutea persist through pregnancy, but degenerate while the young are in the pouch, usually rather early in that period.

In summary, marsupial ovaries are similar to those of eutherians in most respects. However, they are reported to have little or no thecal and interstitial gland tissue. Thus, there is a question as to the source of estrogens.

Infraclass EUTHERIA

Order INSECTIVORA

Superfamily Tenrecoidea

*Tenrecidae. Of the 10 genera of tenrecs, the ovaries of only *Tenrec*, *Hemicentetes*, and *Setifer* have been studied. Their ovaries are not lobed or fissured, but do have rough surfaces because of numerous projecting follicles and corpora lutea. Bluntschli (1937) found a maximum of 10 fetuses in the streaked tenrec (*Hemicentetes semispinosus*), but had evidence that many more eggs were ovulated. In the tenrec (*Tenrec ecaudatus*), he found 40 blastocysts in one tract, and 32 fetuses in another. He gives 60 μ as the diameter of ripe ova of *Tenrec ecaudatus*, and Strauss (1938) gives 84 μ for those of the large Madagascar "hedgehog" (*Setifer setosus*), and 60 μ for *Hemicentetes semispinosus*; so both the eggs and follicles of this family are unusually small. However, the most unique feature of this family are its ripe follicles, none of which have antra (see Fig. 6.26). Spermatozoa penetrate these solid follicles and fertilize the egg before it is ovulated (see Figs. 6.27 and 6.28). Probably because of the lack of an antrum, the fertilized egg is shed by withdrawal of the surface epithelium and the very thin tunica albuginea toward the circumference of the follicle. This process exposes and tends to evert the many-layered granulosa, which in turn moves the egg to the surface, free even of its corona radiata cells and covered only by its zona pellucida (see Figs. 6.29 and 6.30). As the granulosa cells lutealize, they bulge outward and to all sides of the stigma, thus forming an everted corpus luteum. Later, the still well developed corpora round over and gradually become completely buried in the cortex. No data are available on the interstitial gland tissue, but published illustrations of ovarian sections indicate that it is plentiful in the cortex, and that it is probably of the thecal type. Landau (1938) showed that the ovarian bursae of *Setifer* and *Hemicentetes* usually have one small slitlike opening and several minute pores as well. He also described a "strong musculus retractor bursae" which "radiates into that part of the bursa which lies near the infundibulum." This muscle and the "retractor ovarii" which enters the hilus of the ovary "develop from the longitudinal muscular apparatus of the tubular portion of the uterine cornu and reach their termination without connection with the tube" (p. 263). Landau also wrote of a vigorous secretion of the canaliculi of the epoophoron during tubal passage of the ova, but which ceased after implantation. Feremutsch (1948) and Feremutsch and Strauss (1949) described in further detail the female genital tracts and reproductive cycles of tenrecs.

Superfamily Chrysochloroidea

*Chrysochloridae. In addition to *Eremitalpa granti*, the Horst Collection contains sectioned material of two other golden moles (*Calcochloris obtusirostris* and *Amblysomus hottentotus*). Three juvenile ovaries of the first and one adult ovary of the second were like those of *Eremitalpa granti*, except that the rete was large and extended into the center of the medulla and that there was much thecal type interstitial gland tissue. Stromal type interstitial gland tissue is probably present in *Calcochloris* and *Amblysomus*, but it was impossible to be sure of this. No ripe follicles were seen, but the fact that the corpora lutea were everted is good evidence that they have either no antrum or a very small one. Since the largest follicles had a thick theca interna, a thecal gland is surely present around ripe follicles.

Superfamily Erinaceoidea

*Erinaceidae. The surface epithelium of the moon rat (*Echinosorex gymnurus*) and the lesser gymnure (*Hylomys suillus*) is apparently inactive, as it showed no invaginations and little or no cell enlargement. The largest follicles were 500–700 μ in diameter and had no antrum, although there was some cell dissolution around the centrally located ovum and corona radiata. Since fully developed corpora lutea had a diameter of only about 1100 μ, it is likely that ripe follicles are not much more than 700 μ in diameter and that they have either no antrum or a very small one. The corpora lutea were seldom everted and, in fact, had a small cavity, both of which suggest that a small follicular antrum is present. Interstitial gland cells were not seen.

Sections from three specimens of African hedgehog (*Erinaceus frontalis*) in the Horst Collection were studied by Mossman. Their ovaries are very irregular in shape, much lobed and fissured, with some of the central lobes very thin and leaflike. The mesovarium is very short and thick, and the ovary and oviduct are closely united to the cephalic end of the uterus and its mesometrium. There is little interstitial gland tissue. It is of the thecal type, but poorly differentiated in the specimens available. Only the younger corpora lutea have paraluteal cells. The corpora lutea are like those described for the Eurasian hedgehog (*E. europaeus*) (Deanesly, 1934). They persist into the lactation period. The corpora albicantia are distinct, small, hyalinized masses. The relation of the corpus luteum to the nature of the endometrium in *E. europaeus* was studied by Personen (1942). Niklaus (1950) studied the size of nuclei of follicular epithelium during the sexual cycle.

Superfamily Macroscelidoidea

*Macroscelididae. The Horst Collection at the University of Witwatersrand includes excellently prepared and cataloged serial sections of ovaries, uteri, and placentae of well over a thousand specimens of the South African long-eared elephant shrew (*Elephantulus myurus*), together with quantities of well-preserved gross material of this and related species. Horst (1942, 1945, 1949, 1954) and Horst and Gillman (1940, 1941, 1942, 1945, 1946) published papers on the biology of reproduction, ovary, and placentation of this species. Of these, only the major papers pertaining to the ovary are cited here. Unfortunately, their illustrations are mainly drawings, so that morphological features in which they were not interested, such as interstitial gland cells, are usually neither described nor clearly illustrated. In view of this species' unusual reproductive biology, a modern study should be undertaken of the glandular tissues of its ovary, using photomicrographs and electron micrographs.

In several gross specimens examined by Mossman, the ovaries of *E. myurus* were unusually far cephalic, opposite the hilus or at least the caudal half of the kidneys. Thecal gland and thecal type interstitial gland tissue was relatively minimal and the cells were small. Of the few dozen series of slides examined, only one had gonadal adrenal cells associated with the epoophoron. This animal had early uterine blastocysts. The forest elephant shrew (*Petrodromus tetradactylus*) has much more interstitial gland tissue than *E. myurus*, but the cells are also small. *E. myurus* corpora lutea are at first everted, as is common when the ripe follicles are so small. Both species have more fibrous and more conspicuous corpora albicantia than is usual in such small mammals. The remarkable differences in number of eggs ovulated by *Petrodromus tetradactylus* and *Elephantulus myurus*, even among the six different species of *Elephantulus*, would seem to warrant comparative studies of gonadotropins and the ovulatory mechanisms of this group. In fact, Tripp (1971) has started a reinvestigation of reproduction in this group, and has shown a continuous gradient in number of eggs ovulated in a series of species of macroscelids rather than a sharp division into two groups, as Horst (1944) concluded. The rete of *E. myurus* extends the length of the core of the medulla.

Superfamily Soricoidea

*Soricidae. The Old World water shrew (*Neomys fodiens*) is unique in that during tubal transport of the embryos the thecal type interstitial tissue and the remaining stroma transform into cells identical in appearance with those of the corpora lutea. The borders of the original corpora lutea soon become indistinguishable, and the whole ovary, except its vascular core and a few small follicles, seems to be made up of luteal cells. These completely disappear during anestrus. This is one case where the use of the term "luteinization" to describe the hypertrophy of interstitial cells may be technically correct, but Price's (1953) study of the ovarian morphology is inadequate to settle this point unequivocally. Therefore, in the table we have tentatively classed this tissue as thecal and stromal interstitial gland. The unusual differentiation of medullary cords to form "testis cords" during pregnancy has been described in the vagrant shrew (*Sorex vagrans*) (Wilcox and Mossman, 1945) (see Figs. 6.16 and 6.17), but is unknown in other soricids. One specimen each of Trowbridge's shrew (*S. trowbridgii*) and the masked shrew (*S. cinereus*) indicates that their ovaries are essentially like those of *S. vagrans* and the short-tailed shrew (*Blarina brevicauda*).

We have a few photomicrographs of the Indian white-toothed shrew (*Crocidura* sp.) from the Hill Collection. These suggest that the ovarian morphology of the Crocidurinae closely resembles that of the Soricinae. However, their follicles have a very narrow crescentic antrum, and the early corpora lutea are all everted, but later become embedded in the cortex in the same manner as those of *Setifer*. Dryden (1969) indicated that the cumulus cells of the musk shrew (*Suncus murinus*) hypertrophy at estrus, as in other shrews (see Fig. 6.31), but that the very narrow cleftlike antrum between the cumulus and the outer epithelium does not appear until copulation has occurred. Dryden also showed partial true lutealization of follicles adjacent to primary corpora lutea, and occasional complete fusion of corpora.

Other papers on reproduction of soricids, but which contain little on ovarian morphology, are: Brambell and Hall (1936); Conaway (1952); and Tarkowski (1957).

*Talpidae. Although data on the Pyrenean desman (*Galemys pyrenaicus*) and the Amercan broad-footed mole (*Scapanus latimanus*) are inadequate, it seems quite probable that, except for the star-nosed mole (*Condylura cristata*) (see Figs. 2.7, 6.19, 6.32, 6.80, and 6.81), the ovaries of Old World and New World moles differ in that the latter do not have the relatively large amount of gonadal adrenal type interstitial gland tissue characteristic of the former. The shrew-mole (*Neurotrichus gibbsi*) is intermediate between the two, the interstitial gland tissue of its medulla being clearly continuous with, and the same as, that associated with its epoophoron, and being definitely of the gonadal adrenal type (see Figs. 6.82 and 6.83). Numerous medullary cords, or segements of them, are always scattered throughout the glandular tissue of the medulla. In the common European mole (*Talpa europaea*) and *Neurotrichus* during midgestation, the medullary cord epithelium hypertrophies and intergrades with the gonadal adrenal type cells to such an extent that it must also be considered glandular. The only other peculiarity reported in talpid ovaries is the presence of "testis cords" in *Galemys* and in the mole from southeast Asia (*Talpa leucurus*). *T. europaea* sections examined by Mossman appeared to have two distinct types of luteal cells, but more and better preserved material is necessary to confirm this. The corpora albicantia of the eastern mole (*Scalopus aquaticus*) and of *Condylura* are surprisingly fibrous and hence unusually conspicuous for such small mammals. Adneural gland cells were not found in *Scalopus*, although a careful search was made for them. The oviducts of all talpids examined have numerous tubular glands in their mucosae. Kohn (1921) discussed the development of the ovary of *Talpa europaea*, and published good photomicrographs. Godet (1949) described very completely the development of both the testis and ovary of *T. europaea*, but said little about the adult ovary.

Summary of the order INSECTIVORA

We have presented information on six of the eight recognized families of living insectivores. No data are available on the relatively unknown families Solenodontidae and Potamogalidae. A true picture of the relative amount of information at hand is provided by the fact that practically nothing is known about the ovaries of 46 out of the 67 recent genera recognized by Walker (1964) and Meester (1968). The five superfamilies of the order Insectivora are also very much more diverse and widely unrelated than are the superfamilies of most other orders. A much more thorough knowledge of each is therefore needed before many reliable generalizations can be made about the order as a whole. However, the following common insectivore characteristics do stand out from data presented in the tables: a complete ovarian bursa; little or no tunica albuginea; average or less than average development of the thecal gland; and probably a single type of luteal cell. Except for *Galemys*, the mature follicles have very small antra, or none at all, as in the tenrecs. Probably because of this, eversion of corpora lutea is common. "Testis cords" occur in some of the Soricoidea. The most outstanding peculiarities within the order are the intrafollicular fertilization in *Setifer* and the excess of eggs ovulated over embryos implanted in *Elephantulus*.

Order DERMOPTERA

*Cynocephalidae. The gliding lemurs (*Cynocephalus variegatus* and *C. volans*) are the only living representatives of this aberrant order. Very little is known about their reproduction. Ovaries measure about 3 × 2 × 1.5 mm, and are located well back in the lumbar region. The ovarian bursa is thin and has a slitlike opening. It often harbors parasitic roundworms. The germinal epithelium is low cuboidal, and inconspicuous invaginations into the cortex occur. The tunica albuginea is a definite fibrous layer 30–100 μ thick. Interstitial gland cells are scarce or possibly absent. Usually, one follicle matures during each cycle (one case of two and twin embryos). The antrum is large. The theca interna is well developed, although cellular hypertrophy is not marked, so possibly the follicles seen were not preovulatory. A few gonadal adrenal cells are occasionally seen. The corpus luteum (about 3 × 2 mm) consists of large vacuolated cells during early gestation, but these atrophy and are largely replaced by connective tissue near midgestation. The corpus albicans persists into the succeeding cycle. The fetal membranes resemble those of soricids, but there is nothing about the ovary to suggest clear affinity to this group.

Order CHIROPTERA

Suborder MEGACHIROPTERA

*Pteropodidae. Besides those of the flying "fox" (*Pteropus vampyrus*), a few ovaries of other fruit bats (*Cynopterus*, *Ptenochirus*, *Rousettus*, *Eonycteris*, *Macroglossus*, and *Harpionycteris*) have also been examined. There is little difference among them, except that in *Rousettus* a large corpus luteum was found persisting into a following pregnancy. Melanocytes were present in the ovary in varying numbers in *Pteropus vampyrus*. There is a marked tendency in all genera for the primordial and primary follicles to occur in a relatively localized mass in a limited area of the cortex. The attachment to the mesovarium is as broad as the whole ovary. The medulla is exceptionally small compared with the cortex. A description of the ovary of *Epomophorus anurus* was given by Herlant (1953). He found few interstitial gland cells. Large amounts of gonadal adrenal tissue occur in the ovarian medulla of *Cynopterus*.

Suborder MICROCHIROPTERA

Superfamily Phyllostomatoidea

*Phyllostomatidae and *Desmodontidae. Of the 54 genera in these families (Walker, 1964), we have only relatively scanty data on three. Thecal type interstitial gland tissue is relatively scarce in all three. The short-tailed leaf-nosed bat (*Carollia* sp.) has scattered single cells and groups of cells in both cortex and medulla which are tentatively presumed to be gonadal adrenal cells, because their cytoplasm is packed with uniform small vacuoles. However, no glandular cells of this type were seen in the more typical position adjacent to the epoophoron. The most unique ovarian character in this group is the definitely glandular nature of the rete epithelium in the tent-building bat (*Uroderma bilobatum*) (cf. pp. 207–8 and Fig. 6.92). Published information on the common vampire bat (*Desmodus rotundus*) is inadequate for determining many of the points listed in our synoptic data.

Superfamily Vespertilionoidea

*Vespertilionidae. The ripe follicles of all of the Microchiroptera so far studied have relatively small antra and hypertrophied cumulus cells (see Figs. 6.34 and 6.35) similar to those of shrews. Interstitial gland tissue of the thecal type is very plentiful and highly differentiated in the mouse-eared bats (*Myotis*) (see Figs. 6.58, 6.59, and 6.60), the big brown bat (*Eptesicus fuscus*), Rafinesque's big-eared bat (*Plecotus rafinesquii*), and in one pregnant specimen of the hoary bat (*Lasiurus cinereus*) in our collection. *Myotis* and *Lasiurus* also have typical gonadal adrenal cells associated with the epoophoron. Similar cells are scattered throughout the interstitial gland tissue of the medulla and to a lesser extent in the cortex, but it could not be determined whether these were actually gonadal adrenal cells or merely modifications of thecal or stromal type interstitial gland cells. Guthrie and Jeffers (1938) showed much interstitial gland tissue in newborn *Myotis*, both in the medulla and cortex. This is probably of the fetal type, but could be either of the thecal, stromal, or even gonadal adrenal type. Guthrie and Jeffers even suggested that these cells may possibly originate from the medullary cords, which are decreasing in prominence during this period. Interstitial gland cells of vespertilionids are most numerous and most highly differentiated during early pregnancy in our material, but O. P. Pearson, Koford, and Pearson (1952) found them best developed during the first postpartum month in *Plecotus*. In *Myotis*, as in other mammals with very small ovaries such as shrews, the surface epithelium is often indistinguishable in paraffin sections from the underlying stroma. Medullary cords are prominent in juveniles and occasionally in adults. In juveniles, they contain oocytes and sometimes extend well into the zone which would otherwise normally be considered cortex. The rete of *Eptesicus* and *Myotis* has a high columnar epithelium and narrow lumen (see Fig. 6.91). In *Eptesicus*, the rete is very large. Also in *Eptesicus*, the paraluteal cells intergrade with the surrounding interstitial cells.

*Molossidae. Of 31 Brazilian free-tailed bats (*Tadarida brasiliensis*) examined by Sherman (1937), 30 had a single corpus luteum in the right ovary, and 1 had two in the right ovary. Unilateral ovarian function is known in several other chiropterans (Matthews, 1937, 1941; Ramaswamy, 1961). (See Chapter 7, pp. 269–71, for discussion of this in bats and other mammals.) The rete of *Tadarida* and of the one probably juvenile mastiff bat (*Eumops* sp.) available to us has such high epithelium that the lumen is obscured or absent, much as in *Eptesicus*. No ripe follicles were seen in *Tadarida*, but the theca interna of immature vesicular follicles is thin; it is thus probable that the thecal gland is also minimal. Interstitial gland tissue is scarce in the cortex and medulla. Some is of the thecal type, but it is unclear whether any is of the stromal type. The epoophoron has only two or three efferent ductules, but they are of unusually wide caliber.

Summary of the order CHIROPTERA

There are 17 families, including 177 living genera, of Chiroptera (Walker, 1964). Of these, we know a significant amount about the ovaries of only 8 genera distributed in 5 families. There is a wide anatomical divergence between the fruit bats (Megachiroptera) and the others (Microchiroptera). There are also wide biological differences of various sorts among the families of Microchiroptera. It is obvious, therefore, that only the most tentative generalizations about the ovaries of either of the two suborders can be made from the very limited data now available. The often bizarre reproductive phenomena in this order, such as insemination followed by a long delay in fertilization, unilateral ovarian function, and the glandular nature of the rete in at least one genus, call for much more study. The interstitial gland tissue is in need of further investigation, first to clear up the problems of its classification, and then to demonstrate its functions. Athias (1920) described the ovarian interstitial cells of a number of bats, including 5 species of big brown bats (*Eptesicus*), 3 species of horseshoe bats (*Rhinolophus*), 2 species of mouse-eared bats (*Myotis*), and a big-eared bat (*Plecotus auritus*). He found interstitial gland well developed in all except *Rhinolophus*, and said that it is maximal during pregnancy and lactation. Guthrie and Jeffers (1938) confirmed this in *Myotis*. Stricht (1912) described in detail both the interstitial gland and the corpus luteum of some European pipistrelles (*Pipistrellus*). Guraya (1967) has made cytochemical studies of the bat ovary. Wimsatt (1944), Wimsatt and Trapido (1952), Wimsatt and Kallen (1957), and Wimsatt and Parks (1966) have reported on ovarian phenomena in relation to the reproductive cycle and to hibernation in various bats. The book *Biology of Bats*, edited by Wimsatt (1970), should also be consulted, as well as the paper by J. R. Baker and Bird (1936), which is a study of *Myotis myotis* in a uniform tropical climate, and which also has a good review of the various breeding patterns known in bats at that time.

Order PRIMATES

Suborder PROSIMII

It is only for the sake of conforming to a well-known and standard mammalian classification that we follow Simpson (1945) in grouping tupaiids (tree shrews), lemurids (lemurs), lorisids (lorises), and tarsiids (tarsiers) together under this heading. The fetal membranes of tupaiids show distinct affinities to the soricoids (shrews and moles) and carnivores (Luckett, 1969); those of lemurids and lorisids are of the same type as those of perissodactyls (horses); and only those of *Tarsius* bear any significant resemblance to the membranes of anthropoids. We have, therefore, presented the data on tupaiids, lemurids and lorisids, and *Tarsius* as three distinct unrelated groups.

Infraorder LEMURIFORMES

Superfamily Tupaioidea

*Tupaiidae. The ovary of the tree shrews (*Tupaia*) is small (about 2.5 × 1.5 × 1.5 mm). The germinal epithelium is cuboidal and invaginations of this layer in the form of cords, crypts, and folds are common. Spherical cells, intermediate in size between germinal epithelium cells and small oocytes, are seen in both the surface epithelium and the invaginations. Stratz (1898) saw these cells and called them "Eizellen." Corpora lutea of the nonpregnant cycle are smaller than those of pregnancy. The latter persist as well-vascularized bodies until parturition. A few accessory corpora lutea are fairly common. The presence of preovulatory follicles during late pregnancy and at parturition suggests that a postpartum estrus occurs. Interstitial gland cells are rare, and probably very transitory. Medullary cords frequently become modified into typical "testis cords" such as those of *Sorex vagrans*. The Mindanao tree shrew (*Urogale everetti*) is similar to *Tupaia*. In a personal communication, W. P. Luckett reported that examination of a number of *T. javanica* in the Hubrecht Collection generally showed 4–6 developing corpora lutea and associated blastocysts, but later stages had only one implanted embryo in each uterine horn. However, preliminary data on 14 *T. longipes* and 14 *T. tana* suggest that two corpora lutea and two embryos are the rule. Luckett also studied the ovaries of 10 *T. minor*, 4 *T. picta*, 2 *T. gracilis*, 2 *T. chinensis*, 3 *Urogale everetti*, and 1 *Dendrogale murina*. A pregnant *Dendrogale* contained 22 accessory corpora lutea. The phylogenetic relationship of the tupaiids to other groups is still controversial, but their fetal membrane characters indicate a closer affinity to "soricoid insectivores and carnivores" than to primates (Luckett, 1969). There is little now known about their ovarian morphology that gives any clear indication of their affinity to other groups.

Superfamily Lemuroidea

Lemuridae and Indridae. W. C. O. Hill and Davies (1954) described the gross anatomy of the ovaries of the broad-nosed gentle lemur (*Hapalemur griseus*) and the sportive lemur (*Lepilemur rificaudatus*). Both have partial bursae with large openings. One *Hapalemur* ovary was slightly lobed.

Petter-Rousseaux (1962) discussed briefly the histology of the ovaries of the mouse lemur (*Microcebus murinus*), dwarf lemur (*Cheirogaleus major*), lemur (*Lemur fulvus*), the weasel lemur (*Lepilemur mustelinus*), and the woolly lemur (*Avahi laniger*). She reported all of them to have great development of "tissu interstitiel" (by which we assume she meant interstitial gland tissue) compared with *Macaca mulatta*; that cortical egg cords (and presumably oogenesis) persist in adults of all species, except those native to Malagasy (Petter-Rousseaux and Bourlière, 1965); and that the corpus luteum persists into late pregnancy in *Avahi*, and at least until parturition in *Lepilemur mustelinus*. No other publications on the structure of the ovary of lemurs could be found.

Infraorder LORISIFORMES

*Lorisidae. Because of the inadequacy of our slow loris (*Nycticebus coucang*) collection, we are not certain of the size of the mature follicle, nor of the formation and time of regression of the corpus luteum. The most unique feature of the ovary is the activity of the germinal epithelium (Duke, 1967; Kumar, 1968). Its tubular invaginations into the cortex (cortical tubes) often contain oogonia. Nests of oogonia just beneath the tunica albuginea are common. There is no obvious interstitial gland tissue. Although the theca interna is well developed, it appears to atrophy during atresia, with resulting hyalinization of atretic follicles. The corpus luteum had regressed to a large fibrous body in a female with 3–4-mm CR embryos.

Petter-Rousseaux (1962) found the ovaries of the slender loris (*Loris tardigradus*) and the bush babies (*Galago senegalensis* and *G. demidovii*) to be essentially like those of the Lemuroidea, including an abundance of interstitial gland tissue. She saw cortical cords in adult ovaries of both genera. W. P. Luckett (personal communication) reported that the ovaries of several *Galago demidovii* he has examined are similar to those of *G. senegalensis*, including the presence of cortical cords. A very rudimentary type of regional lymph node associated with the ovary of a *Galago* is shown in Figure 6.84.

Infraorder TARSIIFORMES

*Tarsiidae. This monogeneric family is generally believed to be more closely related to the Anthropoidea than to any other group of living mammals, and the morphology of both its fetal membranes and its ovary are compatible with this view. The tarsiers (*Tarsius*) have no ovarian bursa, and the relationships of the ovary to the oviduct and mesosalpinx are similar to those of man. W. C. O. Hill (1953) made a similar statement, but presented a figure of the female tract which shows the ovary in a position unknown in any other mammal. The germinal epithelium is of the flattened mesothelial type. The interstitial gland cells are small (7.5 μ diam.) with a low ratio of cytoplasm to nucleus. The rete usually extends well into the central part of the medulla.

Suborder ANTHROPOIDEA

Superfamily Ceboidea

*Cebidae. In his Figure 2, D, of the ovary of a spider monkey (*Ateles geoffroyi*), Dempsey (1939) showed a narrow tubal membrane, apparently better developed than that in our Figure 1.13 of a cacajao (*Cacajao* sp.). The ovaries of *Ateles* and the mantled howler monkey (*Alouatta palliata*) are unusually large relative to the body size. This is especially true during early pregnancy when their interstitial gland development is maximal. Ovaries which we have seen of laboratory-maintained squirrel monkeys (*Saimiri* sp.) also have large amounts of interstitial gland. Although these interstitial gland cells have been described as indistinguishable from the cells of the corpora lutea, the facts that, instead of degenerating, they remain in a less-differentiated state during the nonpregnant period and that they seem to be derived from the theca interna cells of atretic follicles combine to make it highly unlikely that they are actually luteal gland cells. Nonglandular medullary cords are common in all of these genera.

*Callithricidae. During pregnancy, the ovary of Geoffroy's marmoset (*Saguinus geoffroyi*) has large masses of glandular tissue developed from atretic follicles. These so closely resemble corpora lutea that after the limb bud stage Wislocki (1939) was unable to distinguish between them and the primary corpora. They may be true accessory corpora lutea, but more information is needed to be certain of this. *Saguinus* normally ovulates one egg from each ovary. The uterus is simplex, and the two conceptuses fuse and develop in a single allantochorion with anastomotic fetal vessels, yet there are no freemartin effects in case the two are of opposite sex. Most of the large rete is in the mesovarium.

Superfamily Cercopithecoidea

*Cercopithecidae. The rhesus monkey (*Macaca mulatta*) and man are the only primates whose ovarian morphology is reasonably well known. The ovary of *M. mulatta* is a rather typical mammalian ovary. Its most unusual feature is the occasional presence of so-called aberrant corpora lutea (cf. Chap. 6, pp. 219–20). Juveniles have numerous polyovular follicles, presumably of medullary cord origin (see Fig. 7.3). Betteridge, Kelly, and Marston (1970) published color photographs showing the great variability in gross appearance of *M. mulatta* ovaries at ovulation. Conaway (1969) described the unusually well developed gonadal adrenals of the patas monkey (*Erythrocebus patas*). Frommolt (1934) gave a brief description of the ovaries of 34 short-tailed macaques (*Macaca brevicauda*). Their bursa is similar to that of *M. mulatta*. He also described the formation of interstitial gland tissue from degenerating follicles. For an excellent atlas and account of the development of the ovary of *M. mulatta*, see Wagenen and Simpson (1965). Zuckerman and Parkes (1932), Culiner (1946), and Gillman and Gilbert (1946) published on the reproductive cycle of the chacma baboon (*Papio ursinus*), and included some information on ovarian morphology.

Superfamily Hominoidea

*Pongidae. One of our gibbons (*Hylobates lar*) carried a 9.5-mm CR fetus; the others were not pregnant. Except for the complete absence of a bursa, and a somewhat greater amount of thecal gland and thecal type interstitial gland tissue, their ovary probably differs little from that of man and the rhesus monkey. However, much more information is needed, especially from animals during estrus, pregnancy, and lactation. According to the information given by Saglik (1938) and Dempsey (1940), the ovaries of all of the great apes (*Pongo*, *Pan*, and *Gorilla*) are essentially like those of man.

*Hominidae. See Chapter 5.

Summary of the order PRIMATES

Information on the ovaries of Tupaiidae is too scanty to indicate clearly their affinity to any other group. Their most characteristic features are a large rete, a relatively thick thecal gland, small amounts of both thecal and gonadal adrenal type interstitial gland tissue, and medullary cords, often in the form of "testis cords" such as those of some insectivores and rodents.

The ovaries of Lemuroidea and Lorisiformes are much alike. In some species of both groups, there is strong evidence of oogenesis in the adult. However, so far as now known, species native to Malagasy do not have this feature. Some authors have reported little or no interstitial gland tissue, others have indicated that it is plentiful. At present, too little is known about the ovaries of lemurids and lorisoids to indicate special affinity to any other mammalian group.

The ovaries of *Tarsius* somewhat resemble those of man, but the assumption that this indicates a direct phylogenetic relationship is unwarranted until more is known about the biology of reproduction in *Tarsius*.

Ovaries of the few members of the suborder Anthropoidea that have been studied resemble one another as closely as do those of members of other comparable taxonomic groups. Tentatively, it appears that the ovaries of New World anthropoids differ from those of the Old World principally in having more, and more highly differentiated, interstitial gland tissue.

With the interest in primates over the years, and especially with the more recent emphasis on their reproductive physiology, it is amazing that so little is known about the morphology of their ovaries. Experimental work on the rhesus monkey is backed by a fair amount of anatomical information, but that on the baboon and the squirrel monkey and other cebids is not. In fact, the lack of such information on Primates in general is apparent in the chapters on the internal reproductive organs by Eckstein (1958) and by Rosenblum (1968).

Order EDENTATA

Suborder XENARTHRA

Infraorder PILOSA

Superfamilies Myrmecophagoidea and Bradypodoidea

*Myrmecophagidae and *Bradypodidae. About all that can be said, in view of the paucity of material, is that much more investigation of their ovarian morphology is needed. The gross appearance of the ovary and the rudimentary tubal membrane of the tamandua (*Tamandua tetradactyla*) is shown in Figure 1.14.

Infraorder CINGULATA

Superfamily Dasypodoidea

*Dasypodidae. The ovary of the nine-banded armadillo (*Dasypus novemcinctus*) is unusual in at least two respects: 1) it is embedded in the mesovarium and has an ovulation pit similar to the ovary of the horse (*Equus*) (see Fig. 1.30); 2) relatively large well-differentiated accessory cortical adrenals are commonly present at the ovarian hilus or close by in the mesovarium (see Fig. 6.76). Nonglandular medullary cords are always present. From the ovulation pit, radiating columns of oocytes and follicles of increasing size extend into the ovary. These should be investigated further, because it is possible that they result from a thin superficial oogenic zone. Careful search has revealed no adneural gland tissue in the hilus or closely adjacent mesovarium. A. C. Enders (1966) says that the corpus luteum becomes fully developed during delayed implantation and undergoes no striking change at implantation, although there may be some increase in size. It regresses somewhat during the last trimester of gestation. The gross appearance of the ovary and tubal membrane is shown in Figure 1.15. Figure 6.103 shows the ultrastructure of an armadillo luteal cell in mid-gestation.

Summary of the order EDENTATA

From what is known of the paleontology, comparative anatomy, fetal membranes, and reproductive biology of the three living families of Edentata, it seems likely that the Myrmecophagidae (South American anteaters) and Bradypodidae (sloths) are fairly closely related to one another, but that they have little relation to the Dasypodidae (armadillos). Too little is known of the ovaries of the Pilosa (anteaters and sloths) to demonstrate whether or not their morphology also supports this concept.

Order PHOLIDOTA

*Manidae. Only two pairs of ovaries of the pangolin (*Manis temminckii*) were sectioned. Female 1 was in late pregnancy with one fetus (total length 110 mm; CR 50 mm) in the right horn of the uterus. The right ovary had three corpora lutea (see Fig. 6.11). Two of these appeared to be of the same age. One probably developed from the follicle which gave origin to the conceptus. The larger one was approximately 7.5 mm in diameter, the smaller was completely everted and its diameter was estimated to be equivalent to about 5.0 mm. A much smaller, highly vacuolated corpus of irregular shape was wedged between the other two and was no doubt older and degenerating. There was also a corpus albicans (see Fig. 6.10) surrounded by a peculiar double halo of interstitial gland cells and with numerous macrophages laden with pigment resembling hemosiderin in its fibrous core. The left ovary contained an apparently normal follicle 3.0 mm in diameter (see Figs. 6.9 and 6.42), as well as a degenerating corpus luteum in which most of the luteal cells were intact but shrunken and relatively achromatic. There was no pigment in this corpus.

Female 2 was in early pregnancy with one fetus (total length 32 mm) in the left horn of the uterus. The left ovary contained two corpora lutea, one 3.8 mm in diameter, the other 7.0 mm. Both had many large cells completely or almost completely filled with a single vacuole, but the larger corpus had many more of these and its cells averaged noticeably larger than those of the small corpus. It is probable that the smaller corpus was the primary corpus of the current pregnancy. There was a pigmented corpus albicans in the left ovary and a pigmented and nonpigmented one in the right ovary. All of these had a halo of interstitial gland cells. The significance of the extra corpora lutea is not clear, but at least one in each animal probably persisted from a previous pregnancy.

Gonadal adrenal type interstitial gland tissue was very plentiful at the epoophoron region (see Fig. 6.11), and large amounts of exactly similar tissue extended into the mucosa of the infundibulum and its fimbriae. A large mass of this tissue resembling an accessory adrenal, but without the typical cordlike arrangement of the cells, occurred in one mesovarium of both females (see Fig. 6.77).

In both animals, thecal type interstitial gland tissue was plentiful in the cortex, but when mature its cells were indistinguishable from those of the cords of glandular tissue that occupied most of the cortex and medulla. In fact, all of the mature interstitial gland cells, wherever they were, resembled one another closely, hence the question marks in the table. It seems likely that when mature all pangolin interstitial tissue most closely resembles the gonadal adrenal type. Careful comparison and study of this possibility in *Manis* should be made — and also in leporids and talpids, where a somewhat comparable situation seems to exist.

Order LAGOMORPHA

*Ochotonidae. There is only one living genus of pika (*Ochotona*). Pika ovaries differ from those of leporids in being somewhat lobed and fissured, and in having little and poorly differentiated interstitial gland tissue. A ripe follicle is shown in Figure 6.25. It is similar to that of leporids. Pikas usually have small gonadal adrenal nodules in their mesovaria (see Fig. 6.73). No adneural gland cells could be found. We know of no published work on the ovarian morphology of this family, except that of Duke (1952).

*Leporidae. The ovaries of all the leporids (rabbits and hares) we examined are much alike. All have only partial bursae (see Fig. 1.19). Their interstitial gland tissue, believed to be largely of the stromal type, is unusually plentiful, and its cells are large and highly differentiated, rather closely resembling their luteal cells. The eastern cottontail (*Sylvilagus floridanus*) and the New England cottontail (*S. transitionalis*) have the least of this tissue, and more of it appears to be of the thecal type (see Fig. 6.62). The snowshoe hare and white-tailed jack rabbit (*Lepus americanus* and *L. townsendii*) have the most (see Figs. 6.61, 6.68, and 6.70), and the domestic rabbit (*Oryctolagus cuniculus*) is intermediate. Ovaries of the brush rabbit (*Sylvilagus bachmani*) seem to have relatively more interstitial gland tissue than *S. floridanus*. Bloch and Strauss (1958) described the ovary of the common European hare (*L. europaeus*). They mentioned false corpora lutea ("Pseudogelbkörper") which fuse to form a single "corpus luteum," filling the ovary. This, they wrote, was renewed from time to time by "corpora lutea vera." From their description and figures, it is obvious that their "false corpora lutea" are corpora atretica composed of thecal type interstitial gland cells. Also, they failed to notice that the luteal cells of hare corpora lutea degenerate within about two months and never mingle with the interstitial tissue. On the basis of data on known breeding supplied by Lloyd Keith, we are able to state that corpora lutea of the snowshoe hare (*L. americanus*) persist well into a second pregnancy (a total life of about 70 days), but not into a third pregnancy. Very small encapsulated masses of gonadal adrenal tissue are usually present in one or both mesovaria of the domestic rabbit. They are larger and have more cells than those of *Ochotona* (see Fig. 6.73), but their relative size in the two species is about the same.

For the voluminous literature on the morphology of the domestic rabbit ovary, consult Brambell (1956) and Harrison (1962). Duke (1947) described the fibrous connective tissue of the rabbit ovary during development. Pansky and Mossman (1953) showed that the caudal portion of the ovary has an appreciably greater regenerative capacity than the cephalic portion. Guraya and Greenwald (1964*a*) made histochemical studies of the interstitial gland, and Lipner and Cross (1968) felt that they had demonstrated the pseudostratified nature of the follicular epithelium of mature rabbit follicles, and they also gave some evidence that the epithelium may be the source of proteolytic enzymes necessary for follicular rupture at ovulation. Allen (1904) described the development of the domestic rabbit ovary from the 13-day embryo to pregnant and lactating adults.

Summary of the order LAGOMORPHA

The relative scarcity of interstitial gland tissue in *Ochotona* is remarkable in view of its abundance in all the other lagomorphs studied. This is almost certainly correlated with some outstanding difference in the reproductive biology of these two groups. Otherwise their ovaries are much alike. They all have only a partial bursa, the ripe follicles are small with a trabeculated and relatively centrally located cumulus (see Fig. 6.25), and all have well-developed gonadal adrenal tissue which is probably always present.

Order RODENTIA

Suborder SCIUROMORPHA

Superfamily Aplodontoidea

*Aplodontidae. The sewellels (*Aplodontia rufa*) are generally believed to be the most primitive of the sciuromorphs. Their male reproductive tract is much like that of sciurids, but can be interpreted as being somewhat more primitive (Pfeiffer, 1956). The same can be said of their placentation (Harvey, 1959). Pfeiffer (1958) reported that *Aplodontia* has no ovarian bursa, and he did not mention a tubal membrane. If there is more than a rudimentary tubal membrane, we would consider a partial bursa to be present. The juvenile ovary has a very sharp boundary between the cortex and the medulla, and the medulla is filled with medullary follicles (see Figs. 1.2 and 6.50). A study of the origin of the cortex in the fetus would be worthwhile, as one would expect a clearer distinction between it and the medullary portion than has been seen in other species. The epoophoron ductules are of unusually wide caliber and join the rete within the hilus. The rete consists of narrow, widely separated tubules which ramify through the medulla and intergrade with the medullary tubes and cords.

Superfamily Sciuroidea

*Sciuridae. The ovaries of the five genera of squirrels examined are essentially alike, except for differences in amount of gonadal adrenal tissue. To be certain of the complete absence of this tissue, the whole mesovarium must be examined systematically. This was done as carefully for the red squirrel (*Tamiasciurus hudsonicus*) as for the others, yet no gonadal adrenal tissue was found. Its apparent absence could be due to oversight on the part of the observer; however, excepting the male tract of the striped palm squirrels (*Funambulus palmarum* and *F. pennanti*) (Prasad, 1957), the gross anatomy of the male and female tracts of the Tamiasciurini differs greatly from that of the rest of the Sciurinae and Petauristinae so far studied (Mossman, Lawlah, and Bradley, 1932; Mossman, 1940). Thus, the absence of gonadal adrenal tissue may well be another peculiarity of the Tamiasciurini.

All attempts to find adneural gland cells in *Sciurus*, *Tamiasciurus*, *Spermophilus*, and *Tamias* have failed. A few ovaries of the woodchuck (*Marmota monax*), black-tailed prairie dog (*Cynomys ludovicianus*), and southern flying squirrel (*Glaucomys volans*) have been examined and found to be much like those of other Sciuridae, excepting *Tamiasciurus* and *Funambulus*. Other publications on the female reproductive tract of sciurids include those by the following: Völker (1905), Rasmussen (1918), Stockard (1937), Foreman (1962), and Seth and Prasad (1969). Various features of squirrel ovaries are shown in Figures 1.16–1.18, 1.33; 2.2–2.6; 4.1–4.23; 6.1, 6.18, 6.20, 6.63, 6.71, 6.78, and 6.102.

Superfamily Geomyoidea

*Geomyidae and *Heteromyidae. The great thickness and glandular differentiation of the thecal gland is the most remarkable thing about the ovaries of these families (see Figs. 6.36, 6.37, 6.39, 6.40, and 6.45). It was his investigation of the ovary of the plains pocket gopher (*Geomys bursarius*) that convinced Mossman (1937) that the theca interna of a ripe mammalian follicle is an endocrine gland and that it is probably an important source of estrogens at estrus (see also Corner, 1938). The few ovaries examined of the smooth-toothed pocket gophers (*Thomomys* sp.), pocket mice (*Perognathus* sp.) (Duke, 1957), kangaroo mice (*Microdipodops* sp.), and kangaroo rat (*Dipodomys* sp.) (Duke, 1940) are much like those of *Geomys*. No ripe follicles were found, but the theca interna of earlier follicles is unusually thick. The oviducts of all of the geomyoids examined take the form of multiple short, sharp kinks and spirals, the whole bound into an almost straight, narrow, tapered spindle by a sheath of longitudinal muscle continuous with that of the uterus (see Fig. 1.31).

Superfamily Castoroidea

*Castoridae. The ovary of the beaver (*Castor canadensis*) is most like that of the springhare (*Pedetes capensis*). The presence of large and small types of luteal cells, somewhat analogous to those of Artiodactyla, was unexpected. No publication, except Provost's (1962), on the ovarian morphology of the beaver could be found.

Superfamily Anomaluroidea

*Pedetidae. The ovary of the springhare (*Pedetes capensis*) most closely resembles that of *Castor*. The presence of two luteal cell types, one large and one small, was unexpected. The small ones are most numerous in the peripheral half of the corpus. The hilus of the ovary contains a complex net of smooth muscle trabeculae. The tunica albuginea is unusually thick and fibrous (see Fig. 6.15). Other features of the ovary are shown in Figures 6.12 and 6.13.

Summary of the suborder SCIUROMORPHA

Based on the present information on the morphology of their ovaries, the Sciuromorpha may be divided into three rather distinct groups: 1) Aplodontidae and Sciuridae; 2) Geomyidae and Heteromyidae; 3) Castoridae and Pedetidae. The first and third groups have more resemblance to one another than the second does to either. The ovaries of sciurids are in most respects rather generalized, i.e., they have the features common to most mammals, and these features are average in degree of differentiaticn and presumably represent a relatively primitive condition (see Chapter 4). Many members of the ground squirrel group are excellent subjects for experimental study, both because of their docility and because of the ease with which the different endocrine tissues of their ovaries can be distinguished from one another. Many heteromyids also thrive under laboratory conditions, and would make especially useful subjects for studies of the thecal gland.

Suborder MYOMORPHA

Superfamily Muroidea

*Cricetidae and *Muridae. Walker (1964) recognized 201 living genera in these two families, and inasmuch as we have reasonably complete information about the ovarian morphology of only three species — the golden hamster (*Mesocricetus auratus*), Norway rat (*Rattus norvegicus*), and house mouse (*Mus musculus*) — it is risky to attempt to make any generalizatons about Muroidea ovaries. However, the meager morphological information on wild species that can be gleaned from the following literature indicates that they are much like the four species recorded in our tables: *Clethrionomys glareolus* (Brambell and Rowlands, 1936); *Ondatra zibethicus* (Miegel, 1953); *Microtus arvalis* (Delost, 1955, 1956); *Microtus agrestis* (Brambell and Hall, 1939; Breed, 1969; Breed and Clarke, 1970); *Microtus californicus* (Greenwald, 1956); *Microtus pennsylvanicus* (Hamilton, 1941). Duke (1944) produced evidence of oogenetic activity in the ovary of an adult Mexican harvest mouse (*Reithrodontomys mexicanus*). Brown and Conaway (1964) showed that the corpora lutea of white-footed mice (*Peromyscus boylii* and *P. leucopus*) persist in a "healthy" condition until late lactation, when the cells develop multiple lipid vacuoles, which after lactation fuse into single large vacuoles. Large corpora with these unilocular cells persist for about two months, that is, from the end of the reproductive season in the fall until the resumption of breeding activity in the spring, at which time they rapidly regress. This long persistence of corpora lutea may be common; Brambell and Rowlands (1936) reported a Eurasian red-backed mouse (*Clethrionomys glareolus*) with tubal ova, taken in May, which had four generations of corpora lutea in her ovaries.

We know of no published complete description of the ovary of *Mesocricetus auratus*, but the following contain some morphological information: Ward, 1946; Rolle and Charipper, 1949; Knigge and Leathem, 1956; Guraya and Greenwald, 1965; Greenwald, 1965; Greenwald and Pepler, 1968.

There is, of course, an enormous amount of literature on the ovary and reproductive phenomena of the laboratory rat, but only a few papers are cited here. Dawson (1951) and Burkl and Kellner (1956) described follicular development in the immature and adult rat, respectively. Dawson and McCabe (1951) and Burkl and Kellner (1954) discussed the interstitial gland tissue. Bassett (1949) demonstrated the importance of mitotic activity in the growth of the corpus luteum of pregnancy, and Pederson (1951) studied corpus luteum histogenesis during estrous cycles. O. Hall (1952) found true accessory corpora lutea to be rare in the wild *Rattus norvegicus*. Bassett (1943) described in detail the vascular pattern in the ovary at all periods of the estrous cycle.

The morphology of the house mouse ovary is almost identical with that of the rat. Some pertinent references follow: Agduhr (1927) and Wimsatt and Waldo (1945) on the ovarian bursa; Brambell (1927, 1928) and Peters (1969) on development; Sobotta (1896) and Deanesly (1930) on development of the corpus luteum; and Deane and Fawcett (1952) on degeneration of pigmented interstitial cells. Peters' observations are important, but some of her interpretations, such as the outward migration of oocytes through the surface epithelium and her reference to medullary vessels as the central "capillary system," are unorthodox and confusing. There is no mention of gonadal adrenal tissue in any muroid.

Superfamily Gliroidea

Gliridae. The only literature we could find on the morphology of the ovary of glirids is that of Pinho (1925) on the garden dormouse (*Eliomys quercinus*). He described numerous large distinct corpora atretica, i.e., thecal type interstitial gland masses, but gave little other anatomical information.

Superfamily Dipodoidea

*Zapodidae. The few ovaries of jumping mice (*Zapus* and *Napaeozapus*) available were so much alike that their features have been tabulated together. They differ from Muroidea chiefly in the absence of lobulation and in the presence of gonadal adrenal tissue, which has not been reported in muroids. The theca interna is thick, and it is thus almost certain that the thecal gland of ripe follicles will be found to be relatively thick. The absence of a tunica albuginea is apparent in Figure 6.4. Oocytes in the rete and epoophoron are shown in Figures 6.21 and 6.22.

Summary of the suborder MYOMORPHA

The scanty information available indicates that the ovaries of cricetids and murids, and probably of glirids, are essentially alike, but that those of the zapodids differ in the absence of lobulation and in the presence of a much thicker theca interna and a larger rete.

Suborder HYSTRICOMORPHA

Superfamily Hystricoidea

*Hystricidae. The ovary of the African porcupine (*Hystrix cristata*) seems to differ from the Asian species (*H. brachyurum*, *Thecurus pumilis*, and *Atherurus macrourus*) primarily in having more accessory corpora lutea. However, there were so few specimens of each available that the apparent differences may be due merely to differences in the periods of the reproductive cycle represented by the individual animals, or even to differences in interpretation by the two of us. Weir (personal communication) found that a *H. cristata* in estrus had numerous large ripe follicles and large vascular corpora lutea, most of them presumably accessory.

Superfamily Erethizontoidea

*Erethizontidae. The North American porcupine (*Erethizon dorsatum*) is remarkable for its numerous accessory corpora lutea, which form from atretic follicles in both ovaries during proestrus, estrus, and early pregnancy but disappear from the ovary of the nonpregnant horn at the time of formation of the preplacenta (see Figs. 6.100 and 6.101). The luteal cells of both the primary and accessory corpora lutea of the ovary of the pregnant side persist about one year, that is, until implantation of the embryo of the next pregnancy. This species also clearly shows the following features: mitotic division of early luteal cells (see Fig. 6.47); accretion of new luteal cells (paraluteal cells) from surrounding stromal cells (see Figs. 6.46 and 6.47); and eventually, in late pregnancy and lactation, transformation of the cells of the accretionary zone into a definite zone of interstitial gland cells (see Fig. 6.93). Other features are shown in Figures 6.2, 6.3, 6.41 and 6.79.

Superfamily Cavioidea

*Caviidae. The rete of the laboratory guinea pig (*Cavia porcellus*) is frequently cystic. A few coarse medullary tubules extend to the center of the ovary. Their epithelium consists of numerous small basophilic cells and less numerous large achromatic ciliated cells. Equal division of oocytes in atretic follicles, sometimes considered to be parthenogenetic cleavage, is also common (Harman and Kirgis, 1938). The correlations between ovarian condition and reproductive behavior have been described by Myers, Young, and Dempsey (1936) and by Young et al. (1938). Mitotic proliferation and oogenesis have been studied by Schmidt and Hoffman (1941) and by Schmidt (1942). Bujard (1947) described the development of the postnatal guinea pig ovary and (1953) the cyclic growth of follicles in the adult ovary during gestation. Sobotta (1906) described the formation of the corpus luteum, and Rowlands (1956) discussed the corpus luteum and follicular growth in relation to reproductive physiology and aging. The development and history of the thecal gland (see Fig. 6.43) was described by Stafford, Collins, and Mossman (1942) and that of the interstitial gland tissue by Stafford and Mossman (1945).

Rowlands and Weir (personal communication) reported that the ovary of a wild guinea pig (*Galea musteloides*) has a very deep groovelike hilus and that the corpora lutea protrude only on the hilar surface. Their captive animals developed an extreme vacuolation (lipophanerosis) of the interstitial gland cells and eventually of the luteal cells. Accessory corpora lutea are rare, and the primary corpora reach maximal diameter at about the 30th day of their 52-day gestation period. They stated that the ovary of one of the wild cavies (*Cavia aperea*) is very similar to that of *C. porcellus*. Breeding in three species of wild guinea pigs was described by Rood and Weir (1970).

Hydrochoeridae. The one available ovary of a nonpregnant parous capybara (*Hydrochoerus* sp.) had no corpora lutea. It resembled the ovary of *Cavia*, except for unusually large numbers of secondary and small vesicular follicles which were occasionally polyovular and in which Call and Exner bodies were numerous. Small groups of gonadal adrenal cells were present near the epoophoron.

*Dasyproctidae. The agouti (*Dasyprocta aguti*) and acushi (*Myoprocta pratti*) ovaries closely resemble those of the porcupine (*Erethizon dorsatum*). Some individual *Dasyprocta* and *Myoprocta* have well-developed medullary cords (see Fig. 7.2) resembling cryptorchid testis cords. Whether these are in any way correlated with the reproductive condition is unknown. Interstitial gland tissue is better developed in *Dasyprocta* than in *Myoprocta* (Weir, personal communication).

Superfamily Chinchilloidea

*Chinchillidae. In its adaptation for the ovulation of hundreds of ova at one estrus, the ovary of the plains viscacha (*Lagostomus maximus*) is the most aberrant mammalian ovary known (see Figs. 7.6–7.10). Its physiology may well match its anatomical peculiarities, for there is a possibility that it ovulates small numbers of eggs throughout pregnancy. The only vertebrate presently known to do this is the porbeagle shark (*Lamna nasus*) (Shann, 1923).

The mountain viscacha (*Lagidium peruanum*) is also unusual in that at breeding age only one ovary, the right, produces ripe follicles, although the left becomes functional if the right is removed. This is possibly also true of *L. boxi* (Weir, 1971*c*).

Weir (personal communication) has stated that in the chinchilla (*Chinchilla laniger*) "accessory corpora lutea are formed throughout pregnancy and in the later stages may grow as large as or exceed the size of the primary corpora lutea. There is some indication of rejuvenation of corpora lutea at parturition. The interstitial tissue of this species is prominent, and large (600–800 μm) follicles are present at all times."

Superfamily Octodontoidea

*Capromyidae. In spite of the easy availability and economic importance of the nutria (*Myocastor coypus*), its ovarian morphology is inadequately known, and nothing has been published on the ovaries of any of the 27 other living genera in this superfamily. Weir (personal communication) considers its ovary intermediate between that of *Cavia porcellus* and *Chinchilla laniger* in that it has a few accessory corpora lutea, which are smaller than the primary corpora.

*Octodontidae. Weir (personal communication) says of the degu (*Octodon degus*): "The first ovulation does not occur until 14–21 months of age! — Accessory corpora lutea are formed at the end of pregnancy and the primary ones regress rapidly after parturition." She reports that *Octodon* ovaries may be slightly lobed, and that the ovary of the chozchori (*Octodontomys gliroides*) is, for a hystricomorph, very uninteresting.

*Ctenomyidae. Weir (personal communication) says that the data on the tucu-tuco (*Ctenomys torquatus*) may not be entirely representative because the colony on which it is based exhibited symptoms of diabetes. However, this is likely to affect only the follicles and may account for the rather large amount of follicular atresia found. The ovary may be slightly lobed.

*Echimyidae. Weir (personal communication) states that the ovaries of the spiny rat (*Proechimys* sp.) are like those of *Dasyprocta*, but have numerous very small (200 μ) accessory corpora lutea; that the thecae of the follicles are very distinct; that the ovary is filled with thecal type interstitial gland tissue; and that remnants of the zonae of otherwise completely atretic follicles are numerous.

Superfamily Bathyergoidea

*Bathyergidae. The thecal type interstitial gland cells of the corpora atretica of the common mole-rat (*Cryptomys natalensis*) closely resemble true luteal cells, but it is clear that they are derived from the theca interna, not from follicular epithelium. Simpson (1945) classed this family as "Hystricomorpha incertae sedis"; however, their fetal membranes are clearly like other hystricomorphs (Luckett, 1968).

Summary of the suborder HYSTRICOMORPHA

In spite of the few data (4 of 16 living families and 38 of 56 genera [Walker, 1964] are totally unknown), it seems fairly clear that the ovarian morphology of Old and New World hystricomorphs is very similar. It is also clear, from *Lagostomus* and *Lagidium*, that because of the very unusual features present in some genera of this group, further studies of the biology of reproduction of hystricomorphs should be rewarding.

Summary of the order RODENTIA

Rodents are notable for the extremely wide range in the nature of their reproductive phenomena. *Erethizon* ovulates only one egg once each year; most cricetids and murids ovulate up to 10 or 15 eggs at each of several cycles during each breeding season; *Lagostomus* ovulates hundreds at each estrus, but bears only two young at a time; *Geomys* has huge thecal glands; and many hystricomorphs form large numbers of accessory corpora lutea, which in *Erethizon* disappear from the ovary of the nonpregnant side in early pregnancy, but persist along with the primary corpus luteum for over a year on the pregnant side. Little or nothing is known about the ovaries and reproductive biology of scores of rodent genera, among which are such interesting groups as the Anomaluridae, Pedetidae, Gliroidea, Dipodoidea, Ctenodactylidae, and several families of hystricomorphs. Obviously, there are rich materials here for both anatomical and physiological investigation.

Order CETACEA

Suborder ODONTOCETI

Superfamily Delphinoidea

*Monodontidae and *Delphinidae. The ovaries of the white whale (*Delphinapterus leucas*) and pilot whale (*Globicephala melaena*) are alike. The tubal membrane and infundibulum are voluminous, thus the partial bursa they form is much larger than the ovary and has a very wide peritoneal orifice. The mucosa of the infundibulum has delicate folds, but no macroscopically visible fimbriae. The corpora lutea contain two distinct types of luteal gland cells which are uniformly distributed throughout each corpus (see Fig. 6.99). The luteal cells of *Delphinapterus* persist at least to late lactation. The corpora albicantia are yellow to bright orange in color and are densely fibrous. All of the above features are strikingly similar to those of artiodactyls. James Brooks (personal communication) believed that the numerous corpora albicantia result from "sympathetic" ovulations, i.e., they are derived from either secondary or more likely accessory corpora lutea. Since accessory corpora lutea typically have about the same life span as primary corpora lutea, and since we saw only one case of a small corpus luteum accompanying a primary corpus, we believe that most of the corpora albicantia represent corpora lutea of previous pregnancies. If this is true, some of the female white whales studied would be up to forty years of age. It is probable that some stromal type interstitial gland cells occur, but the evidence was inconclusive. Comrie and Adam (1937) said that ovaries of the false killer whale (*Pseudorca crassidens*) contained up to 12 corpora lutea each. However, since they showed only very low-power photomicrographs, it is possible that they were counting pigmented corpora albicantia. They saw one everted corpus luteum. According to the descriptions by Pycraft (1932) of the common dolphin (*Delphinus delphis*) and by Rankin (1961) of Gervais' beaked whale (*Mesoplodon gervaisi*, family Ziphiidae), their bursae are like those of *Delphinapterus* and *Globicephala*.

Suborder MYSTICETI

Balaenopteridae. Lennep (1950) described two types of luteal cells in the finback whale (*Balaenoptera physalus*) and the blue whale (*Sibbaldus musculus*). Corpora lutea of *Sibbaldus* weighed up to 7500 g; those of *Balaenoptera*, 2800 g. Chittleborough (1954) gave 10.5 cm as the diameter of a ripe follicle of the humpback whale (*Megaptera novaeangliae*) and reported that the corpus luteum persists throughout pregnancy. Dempsey and Wislocki (1941) gave the diameter of the follicles of *Megaptera* as 1–5 cm and that of corpora lutea as 3–6 cm. They stated that involution is slow and that "many luteal bodies may be found in the ovaries of mature animals."

Balaenidae. Meek (1918) reported that the funnel of the bow-head whale (*Balaena mysticetus*) is directed forward and that the ovary has numerous furrows, but he made no mention of an ovarian bursa.

Summary of the order CETACEA

The literature on the histology of the ovaries of cetaceans is very meager. All that is known indicates that the ovaries of the two suborders, Odontoceti and Mysticeti, are alike. From the standpoint of phylogenetic relationships, their ovaries, as well as the nature of their bursae and the histology of their corpora lutea and albicantia, are very similar to those of artiodactyls. The fetal membranes of these two orders are also very similar.

Order CARNIVORA

Suborder FISSIPEDA

Superfamily Canoidea

*Canidae. The walls of large vesicular follicles of the dog (*Canis familiaris*) frequently have inwardly projecting folds. Medullary cords and masses of cordlike interstitial gland tissue crowd the medulla of dog ovaries, especially in young animals. Jonckheere (1930) thought that the primary medullary cords of the embryo disappear at about 45 days postpartum and that the medulla is reconstituted by migration inward of anovular cortical follicles which become secondary medullary cords. It is more likely that medullary follicles are formed in the juvenile medullary cords, that these degenerate leaving the masses of thecal type interstitial tissue which characterize the medulla of at least young adults, but that some medullary cords remain throughout life. However, there is also a possibility that some of the interstitial tissue is derived directly from medullary cord epithelium, as in many mustelids. Tsukaguchi and Okamoto (1928) believed that the medullary cords gave rise to follicles, stromal cells, and interstitial gland cells. Evans and Cole (1931) reported corpus luteum "remnants" at the beginning of a following cycle, but did not indicate whether intact luteal cells were still present. Barton (1945) described numerous cortical cords and "egg nests," a feature we confirm, especially in juveniles. The ultrastructure of a luteal cell is shown in Figure 6.106. Raps (1948) described the postnatal development of the dog ovary. The single pair of ovaries of the gray wolf (*Canis lupus*) available to us is essentially like the ovaries of the dog. Except for somewhat more interstitial gland tissue, the ovaries of the red fox (*Vulpes fulva*) and gray fox (*Urocyon* sp.) also closely resemble those of the dog.

*Ursidae. The female reproductive tracts of the black bear (*Ursus americanus*) and the grizzly bear (*U. horribilis*) are very similar. The ovarian bursa has a very small orifice through which a few fimbriae protrude. The oviduct has numerous sharp uniform kinks as it circles in the wall of the bursa (see Fig. 1.22). The mesovarium is a short thick pedicle. Epithelial crypts, cords, and tubules penetrating the tunica albuginea are frequent (see Fig. 6.7). Only relatively small vesicular follicles were seen in the specimens available. Because the theca interna of these was of medium thickness, it is probable that the thecal gland is also of average thickness. Corpora atretica of large size are characteristic (see Fig. 6.65). The luteal cells of the black bear in both mid- and late pregnancy contain numerous large crystals, sometimes as long as the luteal cell is broad, and regressing luteal cells contain distinct vacuoles often at least half the diameter of the cell. The corpora lutea of both species certainly persist into lactation and possibly into the second summer following birth of the cubs. (Breeding occurs only every other year.) In both species, numerous small elongate groups of interstitial gland cells are common in the medulla; these are probably medullary cord type interstitial gland cells.

*Procyonidae. Too few specimens were available to characterize with certainty the ovaries of any of the genera, but the raccoon (*Procyon lotor*), ringtail (*Bassariscus* sp.), coati (*Nasua* sp.), kinkajou (*Potos flavus*), bushy-tailed olingo (*Bassaricyon gabbii*), and lesser panda (*Ailurus fulgens*) all have complete ovarian bursae with a porelike orifice through which one or a few short fimbriae may protrude. Poorly fixed ovaries of five kinkajous showed the following features: many cortical cords in juveniles, but none in adults; a thick fibrous tunica albuginea; a long narrow rete occupying the central medulla; follicles with a large antrum and a thick theca interna; and much thecal and probably medullary cord type interstitial gland tissue. One kinkajou with a limb bud stage embryo and its accompanying corpus luteum had a fibrous corpus albicans in which intact luteal cells were still present. One, probably juvenile, lesser panda had ovaries with a medulla which was dark brown in the fresh state. Sections of them showed no pigment, but the medulla was packed with irregular strands of interstitial gland tissue many cells in thickness. Probably this was of the medullary cord type. A characteristic feature of raccoon ovaries is the presence of many cortical tubules penetrating the tunica albuginea and often spreading out beneath it to form small subsurface crypts (see Fig. 6.6) similar to those described by Harrison-Matthews and Harrison (1949) and Harrison (1950) in the seal. Some procyonid genera breed readily in captivity, and so would seem to be ideal subjects for experimental study of the function of their particular type of medullary interstitial gland. The only literature concerning the histology of procyonid ovaries appears to be that of Sinha, Seal, and Doe (1971*a*) on the fine structure of the corpus luteum of the raccoon (see Fig. 6.105), but much is known of their reproductive biology (Asdell, 1964).

*Mustelidae. The oviducts of mustelids take a characteristic uniform zigzag course around the periphery of the ovarian bursa (see Fig. 1.23). The thecal gland cells of the nearly ripe follicle of the long-tailed weasel (*Mustela frenata*) are well differentiated but form only a thin interrupted zone around the follicle (see Fig. 6.38). All of the genera examined have large amounts of interstitial gland tissue in the medulla, but not at all periods of the cycle (e.g., the European ferret (*Mustela putorius*), see Fig. 6.14). In the mink (*Mustela vison*), it is clearly of medullary cord origin (see Figs. 6.86–6.88). In the others, the relation to medullary cord epithelium is less easy to demonstrate, but, since the location, pattern, and cyclic changes are similar, it is tentatively assumed to be the same (see Figs. 6.89 and 6.90). Thecal, stromal, and medullary cord types are all well developed in juvenile mink and in the badger (*Taxidea taxus*). In all the genera examined, all three of these types are maximally differentiated during proestrus and estrus, and during preimplantation pregnancy, including the whole of the delayed implantation period, if this occurs. They are relatively involuted from midgestation through lactation. Luteal cells are fully differentiated during mid- and late pregnancy. In species with delayed implantation, luteal cells start to differentiate after ovulation (see Fig. 6.66), but soon regress and remain cytologically small and apparently inactive during the whole delay period (see Fig. 6.89) (R. K. Enders and Enders, 1963). At the time of implantation, they suddenly redifferentiate and then behave during the rest of gestation as they do in other mammals (see Fig. 6.90). It is apparent, then, that in mustelids the degree of differentiation of interstitial gland tissue alternates with that of luteal tissue. Because of this, mustelids with long delayed implantation periods are ideal subjects for experimental investigation of interstitial and luteal gland function, especially in relation to the mechanisms of implantation (Canivec, Short, and Bonnin-Laffargue, 1966; Canivec and Bonnin-Laffargue, 1963, 1967; Canivec, 1966, 1968; Fevold and Wright, 1969; Mead and Eik-Nes, 1969).

The badger (*Taxidea taxus*) shows an extreme degree of cyclic fluctuation in amount and character of both the medullary cords and medullary interstitial gland tissue. This should be investigated more thoroughly than we have been able to do with the material available to us. Also in *Taxidea*, during late gestation we have seen massive degeneration of thecal and stromal type interstitial gland tissue along the cortico-medullary border. Massive interstitial tissue degeneraton is unknown to us among other mammals, although periodic dedifferentiation and involution are normal. *Taxidea* also has both medullary cords and long epithelial tubes which may be extensions of the rete.

The ferret (*M. putorius*) was the subject of a classic paper by Robinson (1918) on the formation and rupture of ovarian follicles. This animal has an ideal ovary for study of the follicle (Mainland, 1932) because of the diagrammatic nature and arrangement of the epithelium, particularly that of the cumulus, which is like that of the striped skunk (*Mephitis mephitis*) (see Figs. 6.23 and 6.24).

The ovary of the juvenile striped skunk has large numbers of polyovular follicles which appear to be of cortical origin instead of the more common medullary cord origin. Mead (1968*a*, 1968*b*) has shown that the western form of the spotted skunk (*Spilogale gracilis*) breeds in autumn and has a delayed implantation period of 180–200 days, whereas the eastern form (*S. putorius*) breeds in spring and has a gestation period estimated to be 50–65 days.

Sea otter (*Enhydra lutris*) corpora lutea have unusually large fluid-filled intercellular spaces which are postulated to drain directly into lymphatic capillaries.

Verts (1967) described reproduction in the striped skunk and provided some histological data on its ovary. Other pertinent papers on mustelids have been written by Neal and Harrison (1958) and by Deanesly (1967).

Superfamily Feloidea

*Viverridae. Despite the wide distribution of this group, its numerous genera, and its great ecological importance, the only publication containing any information on the anatomy of their ovaries seems to be that of O. P. Pearson and Baldwin (1953) on the mongoose (*Herpestes auropunctatus*). The ovarian bursa is incomplete with a wide opening in the genet (*Genetta*

genetta), linsang (*Prionodon* sp.), African palm civet (*Nandinia binotata*), dwarf mongoose (*Helogale parvula*), white-tailed mongoose (*Ichneumia albicauda*), and yellow mongoose (*Paracynictis selousi*). In the suricate (*Suricata suricatta*) and banded mongoose (*Mungos mungo*), it is complete with a large porelike orifice similar to that of mustelids. Oogonia were seen in the surface epithelium of a sexually mature palm civet (*Paradoxurus* sp.), and invaginations of surface epithelium into the cortex were common in the oriental civet (*Viverra tangalunga*), small-toothed palm civet (*Arctogalidia trivirgata*), *Paradoxurus* sp., and masked palm civet (*Paguma larvata*). One *Genetta genetta* with embryos in the early gill slit stage had contemporary corpora lutea, together with much smaller old ones which still contained intact luteal cells. A similar situation was seen in *Arctogalidia* and *Paradoxurus*.

*Hyaenidae. In the aardwolf (*Proteles cristatus*), the ovarian bursa is very large and loose, and its broad orifice faces the uterus. The infundibulum opens within the bursa and bears short fimbriae, none of which project beyond the bursa in the preserved specimens we examined. Medullary tubes connect with the unusually large, branched, and dilated rete which extends throughout the medulla. The numerous cords of interstitial gland tissue in the medulla closely resemble those of mustelids and are probably of the medullary cord type. One *Proteles* with 30-mm CR embryos had two sets of corpora lutea, one with large cells, the other with smaller vacuolated often spindle-shaped or radiate cells. Probably the latter were corpora of a previous pregnancy, but in many ways they resembled mustelid corpora during the delayed implantation period; one must therefore await more data on the breeding biology of this species before drawing final conclusions about these corpora. Matthews (1939) indicated that the ovary and bursa of the spotted hyena (*Crocuta crocuta*) are similar to those of *Proteles*. Although this excellent paper emphasizes other features, we were able to glean from it the information in the table, with the following qualifications: the tunica albuginea is thin but distinctly fibrous; the ripe follicle diameter is at least 10.0 mm; and the corpus luteum persists at least somewhat beyond the lactation period.

*Felidae. The ovarian bursae of all the felids examined are alike (see Fig. 1.18); these include the domestic cat (*Felis catus*), serval (*F. serval*), puma (*F. concolor*), leopard (*F. pardus*), lion (*F. leo*), and tiger (*F. tigris*). Samples from a slide collection of over 200 ovaries of the bobcat (*Lynx rufus*) and 14 pumas showed that the ovaries of these animals are so similar to those of the domestic cat that it was unnecessary to tabulate them separately. However, in the bobcat small but variable-sized glandular type luteal cells often occur either singly or in small groups scattered among the typical large luteal cells (Duke, 1949), yet these do not constitute a distinct separate category of luteal cells as they do in the hoofed mammals and whales. Binuclear luteal cells and luteal cells with one or more large basophilic bodies in their cytoplasm are also common. The corpora lutea of both bobcat and puma persist for at least one year. Kingsbury's papers (1913, 1914, 1939) on the domestic cat are classics of the literature on ovarian morphology, and should be consulted along with those of Dawson (1941, 1946) and Dawson and Friedgood (1940) for data on this species. Typical structures in a cat ovary are shown in Figure 6.67.

Suborder PINNIPEDIA

*Otariidae. Rand (1955) described reproduction in the southern fur seal (*Arctocephalus pusillus*), but gave very few data on the morphology of the ovary. Our observations indicate that its ovary is much like that of the northern fur seal (*Callorhinus ursinus*) described by R. K. Enders, Pearson, and Pearson (1946) and by A. K. Pearson and Enders (1951).

Phocidae. Too little has been published on the ovarian morphology of any single species of this family to warrant inclusion in the tables, and we have only one pair of ovaries from a harbor seal (*Phoca vitulina*). It had a fetus weighing 4.7 k. Its corpus luteum has a diameter of about 17 mm, and there is only one type of luteal cell. Its atretic follicles are surrounded by embryonic type theca interna cells. Laws (1956) mentioned a "thin" bursa which "opens dorso-medially into the peritoneal cavity" in the South Atlantic elephant seal (*Mirounga leonina*). A ripe follicle was 15 mm in diameter. He wrote of "luteinized" theca interna cells (thecal type interstitial gland) of atretic follicles, but does not mention the theca interna (thecal gland) of ripe follicles. However, in the Weddell seal (*Leptonychotes weddelli*), Mansfield (1958) described a theca interna consisting of several layers of oval or polyhedral cells, which in a ripe follicle resemble luteal cells. Fisher (1954) and Amoroso et al. (1965) reported that the corpus luteum of *Phoca vitulina* and the gray seal (*Halichoerus grypus*) persists well into the lactation period. All who have examined fetal and newborn seal ovaries report an enormous development of fetal type interstitial gland tissue (Harrison, Matthews, and Roberts, 1952; Amoroso et al., 1965; Harrison, 1960). Harrison-Matthews and Harrison (1949) described subsurface crypts and oogenesis in five genera of phocids and one otariid. Laws (1956) reported that the subsurface crypts of the elephant seal are best developed in late pregnancy, but disappear before parturition. Mansfield (1958) wrote that in the Weddell seal they are best developed during the last few weeks of gestation and the first half of the lactation period. Harrison-Matthews and Harrison (1949) discussed the cytology of the luteal cells in relation to the delayed implantation period in certain species, and indicated that there are changes in the luteal cells somewhat comparable to those in other species with delay periods. Fisher (1954) believed that the delay period in the harbor seal in the area of Nova Scotia extends from late June or early July until September.

Summary of the order CARNIVORA

In gross features of the ovary and its bursa, the ursids, procyonids, mustelids, and pinnipeds are much alike. The viverrids, hyaenids, and felids also resemble one another in these same features. In microscopic features, there is much more variability, at least as shown by the relatively inadequate data available. Mustelid ovaries have an unusual amount of medullary cord type interstitial gland issue. This and thecal and stromal types are maximally differentiated near estrus and in early pregnancy, and especially during the delayed implantation period, if this phenomenon occurs. Mustelids also differ from the others in the rapid degeneration of luteal cells after parturition. Yet, compared with most other large orders (Insectivora, Chiroptera, Rodentia), carnivore ovarian morphology is unusually similar throughout the order. No striking difference exists even between the two suborders, Fissipeda and Pinnipedia.

Order TUBULIDENTATA

*Orycteropodidae. All but one of the ovarian bursae of our aardvark (*Orycteropus afer*) specimens had been more or less mutilated, but that one was voluminous, had a very thin tubal membrane, and had no visible orifice. Horst (1949) described his specimen as having one ovary completely bare and the other completely covered. It is very unlikely that such asymmetry is normal. The ovaries are very irregular in shape, and are split into lobes of various sizes and shapes by deep fissures (see Fig. 6.48). Also, the corpora lutea are everted and protrude, sometimes even having a definite stalk. The corpora albicantia are densely fibrous but are surrounded by a thick zone of cells, presumably macrophages, laden with yellow pigment. Because the theca interna of vesicular follicles is very thick (see Fig. 6.48), the ripe follicles must have unusually large thecal glands, probably comparable with those of *Geomys bursarius*. The epoophoron tubules are very long and of large caliber, but no typical rete could be found in the specimens available. Gonadal adrenal tissue is plentiful, frequently lying directly against the epithelium of the walls of the epoophoron tubules. Similar cells are associated with nerves of the mesovarium, but they resemble adrenal cortex cells rather than Leydig cells, and are thus probably not true adneural interstitial gland cells. Although the aardvark bears only one young, gravid females so far reported have several corpora lutea of the same age.

Order PROBOSCIDEA

Suborder ELEPHANTOIDEA

*Elephantidae. The most characteristic feature of the ovaries of both the African elephant (*Loxodonta africana*) and Asiastic elephant (*Elephas* *maximus*) is the presence at the same time of numerous corpora lutea, which differ in degree of differentiation. Usually, several show rupture stigmata and several do not. Perry (1953) believed that many of these arise from a second ovulation during midpregnancy, and he compared this with the similar condition reported in the horse. The evidence for secondary ovulation is better for the elephant than for the mare (see p. 217), but the problem is by no means settled for either. We have indicated in our tables the presence of accessory corpora lutea, that is, corpora formed at or near ovulation by the lutealization of the follicular epithelium of atretic follicles. We would consider primary corpora lutea to be the corpora derived from the follicles which gave rise to the developing embryos, together with all those of the same cytological and histological character having stigmata. Those without stigmata and of the same age as the primary corpora, we would consider to be accessory corpora. Any with stigmata, but younger than the primary ones, we would class as secondary corpora, that is, corpora originating during the course of a pregnancy. These are the ones which we feel have not yet been proven to exist in either the horse or elephant. Any corpora appreciably older histologically than the primary corpora must simply be regarded as corpora of previous ovulations or of previous accessory corpus formation. Another puzzle in regard to elephant corpora lutea is posed by the work of Short and Buss (1965), who were unable to demonstrate progesterone (measured by 20β-hydroxypregn-4-en-3-one) in elephant corpora lutea which, by their age and histological characteristics, should have been active. Our specimen of a juvenile Asiatic elephant had a grossly visible accessory adrenal in the mesovarium (see p. 190). No further microscopic study of the epoophoron area of this specimen was made.

Order HYRACOIDEA

*Procaviidae. Hyrax ovaries, like those of elephants, are lobed and fissured and their corpora lutea protrude and are often stalked. Also, as with elephants, their corpora offer problems of interpretation (see p. 209). The cortex is unusual in its clear delineation from the medulla, and the medulla is packed with interstitial gland tissue believed to be of the gonadal adrenal type, although its cells differ somewhat from the gonadal adrenal cells associated with the epoophoron (see Figs. 6.8 and 6.72). So far as we can detect, there is little difference in the ovarian morphology of the three genera. Our material shows no clear indication that the luteal cells are derived from the theca interna instead of the follicular epithelium, nor that the luteal cells persist for a very long time or for life, as P. N. O'Donoghue (1963) believed.

Order SIRENIA

Suborder TRICHECHIFORMES

Dugongidae and Trichechidae. We have been unable to find any publication on the histology of the ovaries of any sea cow. W. C. O. Hill (1945) reported that the ovaries of the dugong (*Dugong dugon*) are "very large, softish organs of ovoid form. They measure 66 mm between their poles, 34 mm across and 17 mm thick" (p. 168). Further observations of Hill follow: "Each lies in a peritoneal ovarian pouch which is itself hidden beneath the peritoneum of the dorsal abdominal wall below the kidneys and laterad to the ureters. . . . The surface is finely wrinkled but otherwise smooth. . . . The tubes are not contained within the broad ligament, there being no mesosalpinx. They commence [on] each side in a small fimbriated, flattened funnel near the caudal pole of the ovary, embouching into the ovarian pouch. Proceeding thence beneath the peritoneum, forming a cord-like convoluted tube 42 mm long, they open into the apex of the corresponding uterine cornu at the point where the broad ligament arises from the pelvic parietes" (pp. 168–169). We have one somewhat mutilated ovary of a South American manatee (*Trichechus inunguis*). It is very broad and is flattened against what is presumably an equally broad short mesovarium; in fact, the hilar region appears to be almost as extensive as the ovary itself. The free surface is very irregular and fissured and shows no gross evidence of corpora lutea or follicles. Two microscopic sections taken at random through the cortex and medulla, which together were less than 1 cm thick, contained many follicles in late atresia and much thecal type interstitial gland tissue.

Order PERISSODACTYLA

Suborder HIPPOMORPHA

Superfamily Equoidea

*Equidae. The very deep embedment of the ovary in the mesovarium and the ovulation pit (see Figs. 1.28–1.30) are features that the donkeys, horses, and zebras apparently share with no other group, except the armadillos (Dasypodidae). Groups of primordial and primary follicles occur in rows radiating toward the depths of the ovary from the ovulation pit, as they also do in the nine-banded armadillo (*Dasypus novemcinctus*). Of the domestic mares (*Equus caballus*) examined by Ono et al. (1969), 59% had gonadal adrenal tissue "at the junctional portion of the ovarian medulla and mesovarium, on the side of the plica suspensoria ovarii in single or plural number" (p. 59). This is the approximate position of the epoophoron. These adrenocortical cell nodules ranged from the size of a pinhead to 2.5

cm in diameter and were "yellowish-white or orange-yellow" in color. Hammond and Wodzicki (1941) described a layer of interstitial cells "just under the capsule" which are "brownish-orange to brownish-yellow." Yellowish nodules in this position have also been described to us by L. F. Warszawsky (personal communication). Sections in his collection indicate that these nodules are usually composed of gonadal adrenal tissue (see Figs. 6.74 and 6.75), although thecal type interstitial gland masses also occurred nearby, and it is possible that these also contain yellow pigment. From gross examination, there is an obvious possibility of confusing these yellowish masses with corpora lutea. This contingency is important, because there is still some uncertainty as to the persistence of the primary corpus luteum during pregnancy in the mare and its possible replacement by secondary corpora (see p. 217); see also Kupfer, 1928; Cole, Howell, and Hart, 1931; Amoroso et al., 1948).

In Burchell's zebra (*Equus burchelli*), our microscopic sections of two pairs of ovaries show that each mare had a single prominent cytologically well differentiated corpus luteum, presumably the primary one. Since one of these animals carried a midterm 255-mm CR fetus, the other a near-term fetus, zebra primary corpora probably persist at least to term. There was no evidence of secondary corpora lutea in these zebras, nor in mares of the donkey (*Equus asinus*) and horse studied by Kupfer (1928). Yet Uyttenbroeck and Schueren-Lodeweyckx (1969) showed a photograph of a gross section of the ovary of a mare in the 22d week of pregnancy in which there are several large hollow corpora lutea. These can best be explained as either secondary or accessory corpora. Perhaps the zebra and the domestic horse differ in this respect; but why did Kupfer fail to detect this phenomenon in his mares?

Mare corpora lutea contain two distinct luteal gland cell types (see Fig. 6.94) similar to those of Artiodactyla and Cetacea. Harrison (1946) described the development of the follicles and corpora lutea of the mare and showed a well-developed thecal gland (see also Fig. 6.44). The development of the horse ovary from the 56-mm CR fetal state to a two-year postpartum juvenile has been described by Petten (1933). Petten (1933) and Davies, Dempsey, and Wislocki (1957) have called attention to the unusually large ovaries of the fetal and infantile horse and to the massive amounts of brown-colored fetal type interstitial gland tissue which crams their medullae (see Figs. 6.51 and 6.52).

Suborder CERATOMORPHA

Superfamily Tapiroidea

Tapiridae. Owen (1868) mentioned that the tapir "ovaria are small subcompressed bodies in a widely open peritoneal pouch" (p. 694). Schaüder (1929) reported that the Brazilian tapir (*Tapirus terrestris*) has very small ovaries (2.8 × 1.8 × 1.1 cm) compared with those of the horse, that they are long ovoid in shape, have no ovulation fossa, and lie in a bursa which has a 3–4-cm slitlike orifice.

Superfamily Rhinocerotoidea

Rhinocerotidae. Owen (1850) described the ovaries of an immature great Indian rhinoceros (*Rhinoceros unicornis*) as oblong with a smooth surface and lying in a "very large peritoneal sac" with an opening 3 inches wide. There was one vesicular follicle 1 inch in diameter.

Summary of the order PERISSODACTYLA

The presence of large amounts of gonadal adrenal tissue and the possibility of the formation of a second set of corpora lutea during gestation should make the ovaries of Equidae subjects for profitable new investigations. The total lack of significant histological information on the ovaries of tapirs and rhinoceroses is regrettable.

Order ARTIODACTYLA

Suborder SUIFORMES

Infraorder SUINA

Superfamily Suoidea

*Suidae and Tayassuidae. The partial ovarian bursa of the hog (*Sus scrofa*) is voluminous and similar to that of other artiodactyls and the elephants (see Fig. 1.27). Its thecal type interstitial gland tissue is poorly differentiated and inconspicuous, but the cells are actually numerous. Watzka and Eschler (1933) described "Leydigschen Zwischenzellen" as occuring regularly in the ovarian hilus of young swine, usually in or adjacent to nerves. In adults, these were easily found only during pregnancy. Watzka and Eschler published no photomicrographs, but their drawings and descriptions, showing some in close relation to epoophoron tubules, suggested that these cells may consist of both gonadal adrenal and adneural types of interstitial gland cells. According to Wislocki (1931), Käppeli (1908) found the ovaries of the wild European swine (*Sus scrofa*) to be smaller and contain fewer follicles and corpora lutea than in domestic swine. The average number of young of wild European swine is four. Clough (1969) studied the wart hog (*Phacochoerus aethiopicus*) and reported that the ovary is partially enclosed in a large bursa; ripe follicles are 7–8 mm in diameter; the corpora lutea regress rapidly after parturition; accessory corpora lutea and "luteinized follicles" are absent.

Wislocki (1931) described the ovaries of a single pregnant (two 48-mm CR fetuses) collared peccary (*Tayassu tajacu*). It had one recent and one older corpus luteum in each ovary. The younger corpora, like those of the hog (see Figs. 6.95 and 6.109), had two distinct types of luteal cells, which are very clearly shown in his photomicrograph. The older corpora were highly pigmented. If they were corpora of a previous pregnancy, then they are evidence that peccary corpora lutea persist at least into the middle period of a following gestation.

The prenatal development of the pig ovary was described by Allen (1904) and the postnatal development by Casida (1935).

Infraorder ANCODONTA

Superfamily Anthracotherioidea

Hippopotamidae. Laws and Clough (1966) reported that the ovaries of the hippopotamus (*Hippopotamus amphibius*) are partly enclosed in a membranous ovarian bursa and that gross examination during pregnancy reveals hollow "luteinized follicles" and solid "accessory corpora lutea." Since there was no microscopic examination, the true nature of the accessory corpora is uncertain. The corpora lutea regress rapidly in early lactation. Clough (1970) described medullary cords having the form of "testis cords" in the ovary of a mature *H. amphibius.* Uyttenbroeck and Schueren-Lodeweyckx (1969) published several photographs of grossly sectioned hippopotamus ovaries from the collection of Laws and Clough, but these and the notes concerning them provide little pertinent morphological information.

Suborder TYLOPODA

Camelidae. An alpaca (*Lama pacos*) ovary with a corpus luteum, from an animal in late pregnancy, is ovoid in shape, has a thin indistinct tunica albuginea, thin theca interna, a little poorly differentiated thecal type interstitial gland tissue, two types of glandular luteal cells similar to those of other hoofed animals, and heavily pigmented corpora albicantia like those of the domestic cow.

By gross examination, McIntosh (1930) noted up to 10 corpora lutea in a single ovary of the llama (*Lama* sp.). Bezrukov (1968) described the ovaries of 8 Bactrian camels (*Camelus bactrianus*) of the Astrakhan breed. They had a distinct tunica albuginea. Theca interna development was average, and ovulatory follicles were 25–30 mm in diameter. Both the large follicles and corpora lutea projected well beyond the ovarian surface, and there were 40, or even more, of each to each pair of ovaries. The luteal cells appeared functional into late gestation. A publication by George and Fahmy (1966) on *Camelus dromedarius* was not available to us.

Suborder RUMINANTIA

Infraorder TRAGULINA

Superfamily Traguloidea

*Tragulidae. The ovary of the Asiatic mouse deer (*Tragulus javanicus*) is typical of that of other ruminants, except that the mesovarium is very short, and in older animals, especially during pregnancy, it and the proper ligament are so expanded that the ovary is directly attached over a broad hilar area to the uterine wall. Figure 6.96 shows the two types of glandular luteal cells, which are characteristic of all artiodactyls studied. The luteal cells persist at least into late gestation.

Infraorder PECORA

Superfamily Cervoidea

*Cervidae. The ovaries of the white-tailed deer (*Odocoileus virginianus*) are typical of other ruminants. Call and Exner bodies are very common in the follicular epithelium of this species. The rete is of medium size, complex, and has a high epithelium, and a wide lumen. The two types of glandular luteal cells are shown in Figures 6.97, 6.107, and 6.108.

Two sets of moose (*Alces alces*) ovaries, from specimens collected in autumn, have thick fibrous tunicae albugineae, well-developed thecal gland tissue, scarce thecal type interstitial gland cells, and the follicular epithelium of large vesicular follicles is folded. Most of the folds contain a core of theca interna. This folding is exaggerated in the newly formed corpus luteum. Two generations of corpora lutea and a corpus albicans occur in one ovary, indicating that the corpora lutea probably persist for over a year. One of the few cases indicating that two ova may be ovulated from a single follicle was reported for the moose by Pimlott (1959).

Eight pairs of wapiti (*Cervus canadensis*) ovaries are essentially like those of *Odocoileus*. Their corpora lutea show two distinct types of glandular cell. Morrison (1960) reported a ripe follicle 11.0 mm in diameter in *C. canadensis*. He also described "secondary" and "accessory" corpora lutea in 60% of pregnant wapitis, and found that corpora lutea persist into the lactation period. Halazon and Buechner (1956) reported one postconceptional secondary corpus luteum with a stigma in a wapiti. Douglas (1966) reported accessory corpora lutea in 37% of pregnant red deer (*C. elaphus*). Short and Hay (1966) discussed delayed implantation in the roe deer (*Capreolus capreolus*). The corpora lutea were apparently well developed during the delay period, contrary to their condition at that time in mustelids. Their Figure 8 clearly shows two types of luteal cells.

Superfamily Giraffoidea

Giraffidae. Kellas, Lennep, and Amoroso (1958) provided brief notes on the ovaries of some fetal and prepuberal giraffes (*Giraffa camelopardalis*). They reported large follicles and corpora lutea in full-term fetuses. Their illustrations are entirely inadequate to demonstrate whether these so-called corpora lutea are true luteal glands or thecal type interstitial bodies (corpora atretica), which would be much more likely. They described medullary follicles, but did not mention medullary cords, although they showed what we would consider a medullary cord containing an ovum. We can find no other information on the ovaries of this family.

Superfamily Bovoidea

*Antilocapridae. The remarkable feature of pronghorn (*Antilocapra americana*) reproduction is the fact that these animals ovulate several more eggs than can be implanted. This is known for certain in no other bovoid, but may occur in the Camelidae where many corpora lutea have been reported in one set of ovaries. Figure 6.98 shows the two types of glandular luteal cells of *Antilocapra.*

*Bovidae. As indicated by the tables, the five species of bovids that have been studied carefully have ovaries which are very much alike. The ovaries and bursae of the following other bovids are also like these in gross features: greater kudu (*Tragelaphus capensis*), common eland (*Taurotragus oryx*), African buffalo (*Syncerus caffer*), gray duiker (*Sylvicapra grimmia*), reedbuck (*Redunca arundinum*), sable antelope (*Hippotragus niger*), white-bearded wildebeest (*Connochaetes taurinus*), impala (*Aepyceros melampus*), and springbuck (*Antidorcas marsupialis*).

The development of the ovarian follicles of domestic cattle (*Bos taurus*) has been described by Asdell (1960) and by Marion, Gier, and Choudary (1968). Histological and histochemical features of the bovine ovary during the estrous cycle were studied by Moss, Wren, and Sykes (1954). Erickson (1966) described the characteristics of cow ovaries from birth to the age of 20 years. Between 15 and 20 years of age, their follicles were very scarce or absent. Figure 6.57 of a corpus atreticum of a bison (*Bison bison*) shows the relatively sparse and small thecal type interstitial gland cells characteristic of Artiodactyla.

Morrison (1971) has described in detail and classified the stages in development and degeneration of luteal cells and the corpora lutea in the kob antelope (*Kobus kob*). His studies have apparently led him to believe that the small glandular luteal cells characteristic of artiodactyl corpora lutea are developmental stages of the typical large luteal cells. We question this. No doubt, as in other mammals, there are small differentiating luteal cells (paraluteal cells) at the periphery of young corpora of the kob antelope that do differentiate into typical large luteal cells. However, degenerating corpora, such as shown in Morrison's Figures 9, 10, and 11, also have numerous small glandular cells throughout and show little evidence of transitional forms between them and the large luteal cells. The origin of these small luteal cells of cetaceans, perissodactyls, and artiodactyls is uncertain, but they obviously constitute a separate type of luteal cell which exists side by side with the large type throughout the active life of the corpora of these groups.

The impala (Kayanja, 1969) has medium thick thecae internae, small sparse thecal type interstitial gland cells, and two types of luteal cells. These features are typical of other bovids.

Thwaites and Edey (1970) studied corpora lutea of the ewe (*Ovis aries*) up to the 40th day of gestation, at which time there was no luteal cell regression. They described five types of luteal cells. All of these were probably variations of the large cell type. Figure 6.110 shows the ultrastructure of the two types of luteal cells characteristic of artiodactyls.

Summary of the order ARTIODACTYLA

The data available indicate that the ovaries of Suidae, Camelidae, and Antilocapridae resemble one another but differ from other artiodactyls in being polyovulatory. By and large, however, artiodactyl ovaries are surprisingly alike. They are characterized by large partial ovarian bursae, thin thecal glands, poorly differentiated thecal type interstitial gland cells, and a large and small type of luteal cell, both of which persist throughout the life of the corpus. The latter characteristic they share with Cetacea and Perissodactyla. So far, they are known to have but one type of interstitial gland tissue.

Glossary

Age, menstruation (MA). The age of a conceptus calculated from the start of the last menses.

Age, ovulation (OA). The age of a conceptus calculated from the time the egg was ovulated.

Altricial. Born relatively undeveloped, e.g., marsupials, rats and mice, carnivores. The opposite of precocial.

Amygdaloidal. Shaped like an almond.

Anestrus. The relatively long period of inactivity of the female reproductive organs between breeding periods.

Antrum, follicular. The liquid-filled cavity within a vesicular ovarian follicle.

Area, bare. An area of the surface of a visceral organ which is not covered by peritoneum. The opposite of peritoneal area.

Atresia. As applied to an ovarian follicle, this means degeneration.

Bursa, ovarian. The membranous sac partially to completely isolating an ovary from the general peritoneal cavity. It is composed of the dorsal mesentery of the oviduct (mesosalpinx), the tubal membrane (comparable with an incomplete ventral mesentery of the oviduct), and part or all of the oviduct to which these two mesenteries are attached. Synonyms: bursa ovarii; ovarial bursa; ovarian capsule (a misuse of the term "capsule"); ovarian hood; ovarian pouch; ovarian sheath; periovarian sac; tentorium ovarii.

Call–Exner body. An isolated bit of follicular fluid, usually 10–30 μ in diameter, surrounded by a rosette of follicular cells. Usually located next to the antrum.

Capsule. The connective tissue coat of an organ directly enclosing and supporting the parenchyma of the organ. Examples: tunica albuginea of testis and ovary; renal capsule.

Cell, adneural. Gland cells in or around nerves at or near the ovarian hilus.

Cell, Berger. See *Cell, adneural.*

Cell, hilus. See *Cell, adneural*; and *Gland, interstitial, of ovary, gonadal adrenal type.* Apparently, so-called hilus cell tumors may be composed of either adneural or gonadal adrenal cells.

Cell, paraluteal. A differentiating luteal cell at the periphery of a corpus luteum, especially during its growth period. These are believed to be stromal cells in the process of differentiating into luteal gland cells.

Cell, theca luteal (lutein). A cell of the corpus luteum, said to be derived directly from a thecal gland cell. We have not used the term, for we believe that this does not happen and that these cells are paraluteal cells.

Cleavage. The first few divisions of a developing egg.

Cord, cortical. A cord of ovarian epithelium located in the ovarian cortex, often containing oogonia or oocytes and serving as one source of follicular epithelium. Sometimes called a secondary sex cord.

Cord, medullary. A cord of ovarian epithelium located in the ovarian medulla, in the juvenile often containing oogonia or oocytes and giving rise to follicular epithelium, which surrounds an oocyte to form a medullary ovarian follicle. Homologous to a testis tubule. Sometimes called a primary sex cord.

Cord, primary sex. A cord of epithelium of the embryonic gonad that bears primordial germ cells and gives rise to the seminiferous tubules of the testis and to the medullary cords of the ovary.

Cord, secondary sex. A cord of epithelium usually bearing oogonia or oocytes formed in the cortex of the fetal ovary. The same as cortical cord.

Cord, "testis." A medullary cord persisting in adult ovaries and having the appearance of an immature seminiferous tubule or of the seminiferous tubule of a cryptorchid testis.

Corona radiata. The layer of follicular epithelial cells nearest the egg; these cells are thus attached to the outer surface of the zona pellucida.

Corpus albicans. The fibrous remains of a degenerated corpus luteum. In some species, including man, atretic follicles leave smaller, but otherwise identical, fibrous scars.

Corpus atreticum. The later stage of a degenerating secondary or vesicular follicle which has an outer thick zone of theca interna cells more or less differentiated into thecal type interstitial gland tissue.

Corpus luteum. A body of luteal gland tissue.

Aberrant. An atypical corpus luteum found in rhesus monkey ovaries. These are probably primary corpora lutea of a preceding cycle that did not fully degenerate and that have been stimulated to partial redifferentiation by the hormones of the present cycle.

Accessory. A corpus luteum resulting from lutealization of an atretic follicle. This should not be confused with a corpus atreticum, which has no true luteal cells but only thecal type interstitial gland cells. However, accessory corpora lutea often have a zone of thecal type interstitial gland tissue at their periphery.

Primary. A corpus luteum derived from a follicle that ovulated at the last estrus.

Secondary. A corpus luteum derived from a follicle that ovulated after pregnancy was underway. Said to occur in the mare.

Corpus luteum atreticum. This is a confusing term which should be abandoned. It is never used in the literal sense of a degenerating corpus luteum. Instead, it is incorrectly applied to corpora atretica, which are masses of thecal type interstitial gland tissue resulting from atresia of a follicle.

Cortex, ovarian. The outer zone of an adult ovary derived from the secondary proliferation of the fetal ovary. In the adult ovary, it contains the stroma, follicles, corpora lutea, some types of interstitial gland tissue, and the degeneration products of all of these.

Cumulus oophorus. The hillock of follicular epithelium with its contained oocyte which projects into the antrum of a vesicular follicle.

Cycle, estrous. The interval from the beginning of one estrus to the beginning of the next estrus, provided pregnancy or pseudopregnancy has not intervened.

Cycle, menstrual. The interval from the beginning of one menstrual flow to the beginning of the next, provided pregnancy or pseudopregnancy has not occurred.

Cycle, ovarian. The changes occurring in an ovary in the interval from the beginning of one ovulation period to the beginning of the next. These changes differ, depending upon whether an estrous cycle, pseudopregnancy, or pregnancy intervenes.

Cycle, pregnancy. The interval from the beginning of one pregnancy to the time of the next ovulation that could result in a new pregnancy.

Cycle, pseudopregnancy. The period from one estrus to the next, during which a pseudopregnant period has intervened.

Diestrus. The period of the estrous cycle between metestrus and proestrus, during which recuperative changes occur in the uterine and vaginal mucosae. Applicable to polyestrous animals.

Ductuli aberrantes ovarii. Vestigial mesonephric tubules remaining in the adult mesosalpinx. Homologs of the ductuli aberrantes testis.

Ductuli efferentes ovarii. Small tubules connecting the ovarian rete with the ductus epoophorontis. Homologs of the efferent ductules of the testis.

Ductus deferens femininus. Vestigial mesonephric duct located in the broad ligament. It connects the epoophoron with the region of the uterine cervix where it normally ends blindly. Homolog of the ductus deferens of the male.

Ductus epoophorontis. The gonadal end of the ductus deferens femininus, i.e., the portion to which the ductuli efferentes ovarii connect. Homolog of the ductus epididymis of the male.

Egg nest. A cluster of oogonia or of oocytes in direct contact with one another and not individually surrounded by follicular epithelium so far as can be determined by light microscopy.

Ellipsoidal. Shaped like a cylinder, tapered and rounded symetrically at each end; bluntly fusiform.

Embryo. A developing individual before hatching or birth. More specifically, in mammalian embryology, before its external appearance makes it readily identifiable as belonging to a particular order. The first two months of development of man. Cf. *Fetus.*

Embryonic. Pertaining to the early developmental period of anything. Commonly used to indicate undifferentiated and pluripotential cells and tissues.

Epithelium, follicular. All of the epithelium of an ovarian follicle, including that of the wall and the cumulus oophorous. The same as the granulosa.

Epithelium, germinal. Epithelium capable of giving rise to germ cells. In female mammals, the epithelium of the ovarian surface, cortical cords and crypts, and medullary cords and tubules has been considered to be germinal epithelium.

Epithelium, surface. A general term sometimes applied to the epithelium of the surface of the ovary to avoid the implications of the term "germinal."

Epoophoron. The ductuli efferentes ovarii and the ductus epoophorontis taken together. Homolog of the epididymis of the male.

Estrus. The climactic stage of the reproductive cycle of the female, at which she normally accepts the male in copulation. Ovulation normally occurs during or immediately following estrus.

Fetus. A developing, unhatched or unborn individual after its external appearance makes it readily identifiable as belonging to a particular order. The last seven months of intra-uterine human development. Cf. *Embryo.*

Fimbriae, ovarian. Fringelike projections of the oviduct which, during development, become attached to the ovary and appear to be appendages of it.

Fimbriae, tubal. Fringelike projections on the mucosa of the funnel of the oviduct. They are largely covered by ciliated columnar epithelium.

Fluid, follicular. See *Liquor, follicular.*

Follicle (ovarian). A sphere of ovarian epithelium normally containing an oocyte. The following types are recognized:

Anovular. With no oocyte. Probably actually a segment of a cortical or medullary cord instead of a true follicle.

Atretic. In the process of degeneration.

Mature. Ready for ovulation.

Polyovular. Containing more than one oocyte.

Preovulatory. Same as *Mature.*

Primary. With only simple columnar epithelium.

Primordial. With only simple squamous epithelium.

Secondary. With two or more layers of stratified cuboidal epithelium, but no antrum.

Tertiary. Same as *Vesicular.*

Vesicular. Containing an antrum. This term applies as soon as a space appears. These spaces may be multiple at first, but later coalesce into a single antrum.

Fossa, ovulation. Same as *Ovulation pit.*

Funnel of oviduct. Flared inner (abdominal) end of the uterine tube.

Fusiform. Spindle-shaped.

Germ cell. Any sex cell designed to produce a new individual either by a process of sex cell union (fertilization) or by parthenogenesis.

Primordial. Cells segregated during early development as the mother cells of germ cells only.

Gestation period. The time from fertilization of the egg to birth.

Gland, interstitial, of ovary. Any endocrine secretory cells or tissues in the ovary or mesovarium other than the thecal gland and luteal gland (corpus luteum).

Adneural type. Scattered cells or small groups of cells within or beside nerves close to the hilus of the ovary. They are cytologically similar to testicular interstitial cells, and are known only in Anthropoidea. Tumors (hilus cell tumors) composed of them are masculinizing.

Fetal type. Typically present in the female only in the late fetal and infantile period, and probably homologous anatomically to the interstitial cells of the testis, as they are located mainly in the medulla between the medullary cords, although in the more extreme cases, such as the fetal ovaries of the mare and seal, they also appear in the cortex.

Gonadal adrenal type. Cells which resemble adrenal cortex cells, usually located in small groups near the epoophoron tubules, but often throughout the ovarian medulla. Sometimes they take the form of accessory adrenal glands in the mesovarium. When tumorous (hilus cell tumors), they are masculinizing.

Medullary cord type. Glandular modification of medullary cord epithelium. Common in mustelids.

Rete type. Glandular modificaton of rete epithelium. Known only in one genus of Microchiroptera.

Stromal type. Originates directly from stromal cells, usually from those subjacent to the tunica albuginea. Unusually plentiful in rabbits and hares.

Thecal type. Derived from the theca interna of atretic follicles, often forming clearly defined masses called corpora atretica. This is certainly the most common type of ovarian interstitial gland tissue. It is probably present in all mammals, but its cells are sometimes not well enough differentiated to be easily recognized as glandular.

Gland, luteal. A corpus luteum or glandular cells of the same type, whether or not they are in the form of a definite corpus. Their principal secretion is progesterone (3,20-diketo Δ (4:5)-pregnene).

Gland, thecal. The glandularly differentiated theca interna of a maturing ovarian follicle. In most mammals, it is probably the primary estrogen secretor.

Globose. Spheroidal.

Granulosa. Follicular epithelium.

Gubernaculum. The fibromuscular cord extending from the caudal pole of the gonad through the inguinal canal of the body wall of the embryo and fetus to the subcutaneous tissue in the region of the future scrotum or labium majus. It crosses the female duct at the level of the future tubo-uterine junction. In the female, the portion between the female duct and the ovary becomes the proper ligament of the ovary, and the remainder becomes the round ligament of the uterus. In the male, it becomes the conus testis, which usually disappears completely.

Hilus, ovarian. The region of attachment of the mesovarium to the ovary, or, in a more limited sense, the region of entrance of the main ovarian blood and lymph vessels and nerves. Incorrectly applied to the ovulation pit of the horse ovary.

Hyperplasia. Increase in number of cells by cell division.

Hypertrophy. Increase in size, whether or not involving cell division.

Infundibulum tubae. See *Funnel of oviduct.*

Interstitialization. The formation of interstitial gland cells from any precursor cells.

Juvenile. An individual between infancy and puberty; pertaining to that period.

Length, crown rump (CR). The usual linear measurement of embryos and early fetuses. They are normally strongly flexed ventrally, so that the crown of the head and the rump form the cephalic and caudal limits. The CR length is the straight-line distance between these two points.

Ligament, broad. The mesometrium and mesosalpinx.

Ligament, proper. Fibromuscular thickening of the caudal edge of the mesovarium. It connects the ovary with the region of the tubo-uterine junction.

Ligament of uterus, round. Fibromuscular thickening of the broad ligament, extending from the tubo-uterine junction through the inguinal canal to the labia majora. It forms the boundary between the mesosalpinx and mesometrium.

Liquor, follicular. The liquid contained in vesicular ovarian follicles.

Primary. The first formed.

Secondary. That formed during the preovulatory growth of the follicle.

Tertiary. That formed after ovulation. It forms the gel that fills the residual cavity and plugs the stigma.

Luteal. Pertaining to the corpus luteum, or to a cell or tissue having the same function as the corpus luteum.

Lutealization. The differentiation of a cell into a luteal cell of a corpus luteum, or into a cell that has the same function as a luteal cell.

Lutein. Same as luteal; also the yellow pigment of the corpus luteum.

Luteinization. Same as lutealization. Incorrectly but commonly used to indicate glandular hypertrophy of any cell of the ovary, particularly of the theca interna, whether or not there is any evidence of luteal function.

Maturation. As applied to female germ cells, the process of growth and meiotic cell division by which an oogonium becomes an ovum.

Mature. The fully differentiated condition of a cell, tissue, organ, or individual.

Mediastinum ovarii. The area of the ovarian medulla occupied by the rete and major vessels. It is the distinctly cephalic portion of the hilus, and is the homolog of the mediastinum testis.

Medulla, ovarian. The central portion of the ovary surrounded by cortex, except at the hilus. It contains the major intra-ovarian blood and lymph vessels, nerves, medullary cords and their derivatives, interstitial gland tissue, and the ovarian rete. Homolog of the testis.

Meiosis. The processes of cell division characteristic of maturing germ cells, by which the number of chromosomes (diploid number) in the parent oogonium or spermatogonium is reduced to half (haploid number) in the ovum or spermatozoon.

Membrane, tubal. A peritoneal ridge or membrane along the antimesosalpingian surface of the oviduct. When well developed, it takes part in the formation of the ovarian bursa. It has been called "mesotubarium superior," but this is misleading, as its position is comparable with a ventral mesosalpinx.

Membrane, tubo-uterine. The combined tubal and uterine membranes.

Membrane, uterine. A continuaton of the tubal membrane onto the uterus. It is well developed in the opossum, rudimentary in the armadillo.

Mesenchyme. Connective tissue of the embryo, composed of stellate cells with much intercellular fluid and ground substance and little or no fibrous material; the cells are pluripotential but most are destined to become some form of connective tissue cell (fibrocyte, osteocyte, chondrocyte).

Mesentery. A double layer of peritoneum suspending a visceral organ from the body wall. Specifically, that attached to the small intestine, but applied as a general term to such structures as the mesovarium and mesosalpinx.

Mesometrium. The mesentery of the uterus.

Mesosalpinx. The mesentery of the oviduct.

Mesothelium. The usually simple squamous epithelial lining of the coelom and its derivatives (pleural, pericardial, peritoneal, and scrotal cavities). Hence, it covers the coelomic surfaces of all visceral organs and mesenteries. The surface epithelium of the ovary is the most highly modified mesothelium in mammals.

Mesovarium. The mesentery of the ovary.

Metaplasia. The transformation of one type of adult tissue into another type: simple squamous epithelium to stratified squamous epithelium; fibrous connective tissue to bone; stromal cells to interstitial gland cells; etc.

Mitosis. The usual normal type of cell division in which the daughter cells have the same number of chromosomes as the parent cell.

Monestrous. Having only one estrus during a breeding season.

Moruloid. Shaped like a mulberry.

Neo-oogenesis. Formation of new oogonia after the fetal and infantile periods, which are generally believed to be the only times that oogenesis occurs in mammals.

Nulliparous. Having born no young.

Oblate. Shaped like a sphere which has been flattened by pressure applied at the poles.

Oocyte, primary. A female germ cell which has begun to enlarge and to differentiate beyond the oogonial condition, and which, if normal, will eventually undergo reduction division to become a secondary oocyte and a first polar body, each with only the haploid number of chromosomes.

Oocyte, secondary. The larger daughter cell derived by reduction division of a primary oocyte, hence in mammals an oocyte with the haploid number of chromosomes.

Oogenesis. The formation of female germ cells. This includes differentiation of oogonia from primordial germ cells, mitotic multiplication of oogonia, and the processes of meiosis by which oogonia become oocytes and, finally, ova.

Oogonium. A female germ cell that is still capable of dividing by mitosis to form more oogonia.

Ovoid. Shaped like a hen's egg.

Ovum. In the strict sense, a female germ cell after the final maturation division. Commonly, however, any female germ cell from the oogonial stage to the end of the first cleavage. Human embryologists have traditionally, but incorrectly, applied the term to the product of conception through the first two months of development.

Mature or *ripe.* In the strict scientific sense, the true ovum, i.e., a female germ cell after completion of the final maturation division. In mammals, the spermatozoon usually penetrates the secondary oocyte before the second polar body is given off. Hence, the phase of the mature ovum and the fertilized ovum (zygote) overlap.

Naked. Applied to an oogonium or oocyte that appears to have no follicular epithelium enclosing it.

Parenchyma. The specific functional tissue of an organ, e.g., secretory cells of a gland; secretory cells and germinal epithelium of a gonad.

Parous. Having born one or more young.

Peritoneal area. An area of an abdominal or pelvic visceral organ or of the abdominal or pelvic wall which is covered by peritoneum. The opposite of bare area.

Peritoneal organ. An organ suspended by a mesentery, hence one covered by peritoneum, except where the mesentery is attached. Such an organ is sometimes called a "free" organ because it is movable in contrast with a "fixed," or retroperitoneal, organ.

Peritoneum. The serous membrane lining the coelomic cavity of the abdomen, pelvis, and scrotum. It consists of a surface layer of mesothelium supported by a thin layer of fibro-elastic connective tissue.

Pit, ovulation. The concave area of the antimesovarial surface of the ovary of an equid or armadillo through which all ovulation occurs.

Pluripotential. Said of a cell or tissue that has the capability of differentiating into any of several types of cells or tissues.

Polyestrous. Capable of having two or more consecutive estrous cycles during a breeding season.

Polyovulatory. Ovulating more than one egg at an estrous period.

Pouch, tubo-uterine. Pouch formed by the uterus, broad ligament, and the tubo-uterine membrane in the opossum.

Precocial. Born relatively advanced, e.g., guinea pig, ruminants.

Preplacenta. A mass of trophoblastic tissue containing maternal blood channels before its vascularization by blood vessels of the embryo to form a definitive chorio-allantoic placenta. When relatively small, as in the mouse and rat, it has been called the "Träger."

Proestrus. The period between anestrus or diestrus and estrus during which the female tract shows signs of approaching estrus.

Pseudopregnancy. A state induced in some female mammals by sexual stimulation or infertile copulation during estrus. It resembles a state of pregnancy in behavior of the female, endocrine function, and even in some anatomical features, but no embryos are present.

Puberal or *pubertal.* Pertaining to the period when the sex organs are beginning to function, but before the full function of sexual maturity is achieved.

Reniform. Bean-shaped.

Rete, ovarian. A three-dimensional network of fine, irregularly sized, epithelium-lined spaces and tubules located at the ovarian hilus and often extending far into the ovarian medulla. It is connected by the ductuli efferentes ovarii to the ductus epoophorontis. The homolog of the rete testis.

Retroperitoneal. The condition of an internal organ that has no mesentery and is therefore attached directly to the body wall, hence is only partly covered by peritoneum. Such an organ is also said to be "fixed" in contrast with a "free," or peritoneal, organ which has a mesentery.

Season, breeding. The period of the year during which a species mates.

Season, reproductive. The period of the year during which a species is in some phase of the reproductive process (proestrus through lactation).

Somatopleure. The embryonic layer that makes up the amnion, chorion, and primitive body wall. It consists of ectoderm and mesoderm.

Splanchnopleure. The embryonic layer that makes up the allantois, vascular yolk sac, and the primitive gut. It consists of endoderm and mesoderm.

Stigma, follicular. The point on the surface of an ovary at which a follicle ruptured and would normally have ejected an egg. If rupture is recent, the stigma is usually pink, or even bloody; if older, it may range from a slight indentation to a prominent protruding mass of luteal tissue.

Stroma. The nonfunctional tissue of an organ in contrast with the parenchyma, which is the specific functional tissue. The stroma of the ovarian cortex is a very cellular connective tissue containing a large proportion of pluripotential cells which are the source of follicular epithelium and various ovarian gland cells.

Superovulation. The ovulation at one ovulation period of many more eggs than can be successfully gestated.

Term. The end of a normal gestation period.

Theca externa. The outermost sheath of an ovarian follicle. It consists of elongate cells that, at least in some species, contain contractile filaments. It is often difficult to demonstrate, and is probably absent in some species.

Theca interna. A vascular epithelioid layer adjacent to the basement membrane of the follicular epithelium of secondary and vesicular follicles and rarely of older primary follicles. It becomes the thecal gland of ripening follicles or the thecal type interstitial gland of atretic follicles.

Tube, uterine. The oviduct, the tube that conveys the ova from the peritoneal cavity or ovarian bursa, if one is present, to the uterus.

Tunica albuginea ovarii. The connective tissue capsule of the ovary. It is usually relatively thinner and more cellular than the tunica albuginea testis, and may be absent in very small mammals.

Primary. The capsule of the indifferent gonad, which usually disappears early in ovarian differentiation, but persists in the male as the definitive tunica albuginea testis.

Secondary. The definitive capsule of the ovary.

Zona pellucida. The noncellular, chiefly mucoproteinaceous membrane between the ovarian oocyte and the corona radiata cells. It is secreted by the corona radiata cells and contains ultramicroscopic processes of these during its formation and growth.

Zygote. The fertilized ovum when the egg and sperm nuclei have united as a single nucleus.

Literature cited

Adams, E. C., and A. T. Hertig. 1964. Studies on guinea pig oocytes. I. Electron microscopic observations on the development of cytoplasmic organelles in oocytes of primordial and primary follicles. J. Cell Biol. 21:397–427.

Agduhr, E. 1927. Studies on the structure and development of the bursa ovarica and the tuba uterina in the mouse. Acta Zool. (Stockholm) 8:1–133.

Aichel, O. 1900. Vergleichende Entwicklungsgeschichte und Stammesgeschichte der Nebennierren. Über ein neues normales Organ des Menschen und der Säugetiere. Arch. mikrosk. Anat. Entwicklungsmech. 56:1–80.

Alden, R. H. 1942. The periovarial sac in the albino rat. Anat. Rec. 83:421–433.

Allen, B. M. 1904. The embryonic development of the ovary and testis in mammals. Am. J. Anat. 3:88–153.

Altmann, F. 1927. Untersuchungen über das Ovarium von *Talpa europaea* mit besonderer Berücksichtigung seiner cyclischen Veränderungen. Z. Anat. Entwicklungsgesch. 82:482–569.

Amoroso, E. C. 1956. Discussion, p. 85. *In* G. E. W. Wolstenholme and E. C. P. Millar [eds.] Ciba Found. Colloq. Ageing 2. Little, Brown and Co., Boston.

Amoroso, E. C., and I. W. Rowlands. 1951. Hormonal effects in the pregnant mare and foetal foal. J. Endocrinol. 7:1–lii.

Amoroso, E. C., G. H. Bourne, R. J. Harrison, L. Harrison-Matthews, I. W. Rowlands, and J. C. Sloper. 1965. Reproductive and endocrine organs of foetal, newborn, and adult seals. J. Zool. (London) 147:430–486.

Amoroso, E. C., J. L. Hancock, and I. W. Rowlands. 1948. Ovarian activity in the pregnant mare. Nature (London) 161:355–356.

Andersen, D. H. 1926. Lymphatics and blood-vessels of the ovary of the sow. Contrib. Embryol. Carnegie Inst. Washington 17:107–127.

Anderson, A. C., and M. E. Simpson. 1971. Life span of the corpus luteum in the dog (beagle). Biol. Reprod. 5:88. (Abstr.)

Anderson, L. L. 1966. Pituitary-ovarian-uterine relationships in pigs. J. Reprod. Fert. Suppl. 1:21–32.

Appley, M. B., and K. M. Richter. 1970. Ciliated granulosa cells of the bat, *Myotis grisescens*. Annu. Proc. Electron Microscope Soc. Am. 28:102–103.

Arai, H. 1920. On the post-natal development of the ovary (albino rat), with especial reference to the number of ova. Am. J. Anat. 27:405–462.

Archbald, L. F., R. H. Schultz, M. L. Fahning, H. J. Kurtz, and R. Zemjanis. 1971. Rete ovarii in heifers: a preliminary study. J. Reprod. Fert. 26:413–414.

Asdell, S. A. 1960. Growth of the bovine Graafian follicle. Cornell Vet. 50:3–9.

Asdell, S. A. 1964. Patterns of Mammalian Reproduction. 2d ed. Cornell Univ. Press, Ithaca, N.Y.

Athias, M. 1920. Recherches sur les celluels interstitielles de l'ovaire des cheiroptères. Arch. Biol. 30:89–212.

Athias, M. 1929. Les phénomènes de division de l'oocyte au cours de l'atrésie folliculaire chez les mammifères. Arch. Anat. Microsc. Morphol. Exp. 25:406–425.

Austin, C. R. 1961. The Mammalian Egg. Blackwell, London.

Austin, C. R., and A. Walton. 1960. Fertilization, vol. 1 (pt. 2), pp. 310–416. *In* A. S. Parkes [ed.] Marshall's Physiology of Reproduction. 3d ed. 3 vol. Longmans, Green and Co., London.

Aykroyd, O. E. 1938. The cytoplasmic inclusions in the oogenesis of man. Z. Zellforsch. mikrosk. Anat. 27:691–710.

Bachmann, R. 1949. Ovarialstudien II. Gelbkörper and Lymphgefässe. Z. mikrosk.-anat. Forsch. 55:115–164.

Baer, K. E. von. 1827. De ovi mammalium et hominis genesi. Epistola ad Acad. Imper. Sci. petropolitanam. L. Vossi, Leipzig.

Baker, J. R., and T. F. Bird. 1936. The seasons in a tropical rain forest (New Hebrides). Part 4. Insectivorous bats (Vespertilionidae and Rhinolophidae). J. Linn. Soc. London, Zool. 40:143–161.

Baker, T. G. 1963. Quantitative and cytological study of germ cells in human ovaries. Proc. Roy. Soc., Ser. B 158:417–433.

Baker, T. G., and L. L. Franchi. 1967. The fine structure of oogonia and oocytes in human ovaries. J. Cell Sci. 2:213–224.

Barton, E. P. 1945. The cyclic changes of epithelial cords in the dog ovary. J. Morphol. 77:317–349.

Bassett, D. L. 1943. The changes in the vascular pattern of the ovary of the albino rat during the estrous cycle. Am. J. Anat. 73:251–291.

Bassett, D. L. 1949. The lutein cell population and mitotic activity in the corpus luteum of pregnancy in the albino rat. Anat. Rec. 103:597–610.

Beaumont, H. M., and A. H. Mandl. 1962. A quantitative and cytological study of oogonia and oocytes in the foetal and neonatal rat. Proc. Roy. Soc., Ser. B 155:557–579.

Belt, W. D., and D. C. Pease. 1956. Mitochondrial structure in sites of steroid secretion. J. Biophys. Biochem. Cytol. 2 (Suppl.):369–374.

Belt, W. D., L. F. Cavazos, L. L. Anderson, and R. R. Kraeling. 1970. Fine structure and progesterone levels in the corpus luteum of the pig during pregnancy and after hysterectomy. Biol. Reprod. 2:98–113.

Beneden, E. van. 1880. Contribution à la connaissance de l'ovaire des mammifères. Arch. Biol. 1:475–550.

Ben-or, S. 1963. Morphological and functional development of the ovary of the mouse. I. Morphology and histochemistry of the developing ovary in normal conditions and after FSH treatment. J. Embryol. Exp. Morphol. 11:1–11.

Berger, L. 1922. Sur l'existence de glandes sympathicotropes dans l'ovaire et le testicule humains; leur rapports avec la glande interstitielle du testicule. C. R. Acad. Sci. (Paris) 175:907–909.

Berger, L. 1923. La glande sympathicotrope du hile de l'ovaire; ses homologies avec la glande interstitielle du testicule. Arch. Anat. Histol. Embryol. 2:255–306.

Berrill, N. J., and C. K. Liu. 1948. Germplasm, Weismann, and *Hydrozoa*, Quart. Rev. Biol. 23:124–132.

Betteridge, K. J., W. A. Kelly, and J. H. Marston. 1970. Morphology of the rhesus monkey ovary near the time of ovulation. J. Reprod. Fert. 22:453–459.

Bezrukov, N. I. 1968. On the morphology of ovary in *Camelus bactrianus*. Dokl. Akad. Nauk SSSR, Biol. Sci. Sect. 179:202–205. (Transl.)

Bjersing, L. 1967*a*. On the ultrastructure of follicles and isolated follicular granulosa cells of porcine ovary. Z. Zellforsch. mikrosk. Anat. 82:173–186.

Bjersing, L. 1967*b*. On the ultrastructure of granulosa lutein cells in porcine corpus luteum. With special reference to endoplasmic reticulum and steroid hormone synthesis. Z. Zellforsch. mikrosk. Anat. 82:187–211.

Bjersing, L., M. F. Hay, R. M. Moor, and R. V. Short. 1970*a*. Endocrine activity, histochemistry and ultrastructure of ovine corpora lutea. II. Observations on regression following hysterectomy. Z. Zellforsch. mikrosk. Anat. 111:458–470.

Bjersing, L., M. F. Hay, R. M. Moor, R. V. Short, and H. W. Deane. 1970*b*. Endocrine activity, histochemistry and ultrastructure of ovine corpora lutea. I. Further observations in regression at the end of the oestrous cycle. Z. Zellforsch. mikrosk. Anat. 111:437–457.

Blanchette, E. J. 1966*a*. Ovarian steroid cells. I. Differentiation of the lutein cell from the granulosa follicle cell during the preovulatory state and under the influence of exogenous gonadtrophins (rabbit). J. Cell Biol. 31:501–516.

Blanchette, E. J. 1966*b*. Ovarian steroid cells. II. The lutein cell. J. Cell Biol. 31:517–542.

Blandau, R. J., and R. E. Rumery. 1963. Measurements of intrafollicular pressure in ovulatory and preovulatory follicles of the rat. Fert. Steril. 14:330–341.

Blandau, R. J., B. J. White, and R. E. Rumery, 1963. Observations on the movements of the living primordial germ cells in the mouse. Fert. Steril. 14:482–489.

Bloch, S., and F. Strauss. 1958. Die weiblichen Genitalorgane von *Lepus europaeus* Pallas. Z. Säugetierk. 23:66–80.

Bluntschli, H. 1937. Die Frühentwicklung eines Centetinen (*Hemicentetes semispinosus* Cuv.). Rev. Suisse Zool. 44:271–282.

Bodemer, C. W., and S. Warnick. 1961*a*. Polyovular follicles in the immature hamster ovary. I. Polyovular follicles in the normal intact animal. Fert. Steril. 12:159–169.

Bodemer, C. W., and S. Warnick. 1961*b*. Polyovular follicles in the immature hamster ovary. II. The effects of gonadotropic hormones on polyovular follicles. Fert. Steril. 12:353–363.

Bookhout, C. G. 1945. The development of the guinea pig ovary from sexual differentiation to maturity. J. Morphol. 77:233–263.

Boyd, J. D., and W. J. Hamilton. 1955. The cellular components of the human ovary, pp. 50–78. *In* Kenneth Bowes [ed.] Modern Trends in Obstetrics and Gynaecology. 2d Ser. Butterworth and Co. Ltd., London.

Brambell, F. W. R. 1927. The development and morphology of the gonads of the mouse. 1. The morphogenesis of the indifferent gonad and of the ovary. Proc. Roy. Soc., Ser. B 101:391–408.

Brambell, F. W. R. 1928. The development and morphology of the gonads of the mouse. 3. The growth of the follicles. Proc. Roy. Soc., Ser. B 103:258–272.

Brambell, F. W. R. 1935. Reproduction in the common shrew (*Sorex araneus* Linnaeus). I. The oestrous cycle of the female. Phil. Trans. Roy. Soc. London, Ser. B 225:1–63.

Brambell, F. W. R. 1944. The reproduction of the wild rabbit, *Oryctolagus cuniculus* (L). Proc. Zool. Soc. London 114:1–45.

Brambell, F. W. R. 1956. Ovarian changes, vol. 1 (pt. 1), pp. 397–542. *In* A. S. Parkes [ed.] Marshall's Physiology of Reproduction. 3 vol. Longmans, Green and Co., London.

Brambell, F. W. R., and K. Hall. 1936. Reproduction in the lesser shrew (*Sorex minutus* Linnaeus). Proc. Zool. Soc. London (1936):957–969.

Brambell, F. W. R., and K. Hall. 1939. Reproduction of the field vole, *Microtus agrestis hirtus* Bellamy. Proc. Zool. Soc. London, Ser. A 109:133–138.

Brambell, F. W. R., and I. W. Rowlands. 1936. Reproduction of the bank vole (*Evotomys glareolus*, Schreber). I. The oestrous cycle of the female. Phil. Trans. Roy. Soc. London, Ser. B 226:71–97.

Brambell, F. W. R., A. S. Parkes, and U. Fielding. 1927. Changes in the ovary of the mouse following exposure to x-rays. I. Irradiation at three weeks old. Proc. Roy. Soc., Ser. B 101:29–56.

Breed, W. G. 1969. Oestrus and ovarian histology in the lactating vole (*Microtus agrestis*). J. Reprod. Fert. 18:33–42.

Breed, W. G., and J. R. Clarke. 1970. Ovarian changes during pregnancy and pseudopregnancy in the vole, *Microtus agrestis*. J. Reprod. Fert. 23:447–456.

Breinl, H., C. Andrzejewski, and E. Tonutti. 1967. Zur Feinstruktur der Luteinzellen des Laktationsgelbkörpers der Ratte. Z. mikrosk.-anat. Forsch. 77:442–452.

Broek, A. J. P. van den. 1931. Einige Bemerkungen über den Bau der inneren Geschlechtsorgane der Monotremen. Gegenbaurs morphol. Jahrb. 67:134–156.

Brown, L. N., and C. H. Conaway. 1964. Persistence of corpora lutea at the end of the breeding season in *Peromyscus*. J. Mammal. 45:260–265.

Buechner, H. 1961. Unilateral implantation in the Uganda kob, *Adenota kob thomasi* (P. L. Sclater). Nature (London) 190:738.

Bujard, E. 1947. L'ovaire du jeune cobaye durant la période postnatale. Acta Anat. 4:68–72.

Bujard, E. 1953. L'ovaire du cobaye. I. L'ovaire gravide. Rev. Suisse Zool. 60:615–652.

Bullough, W. S. 1942. Oogenesis and its relation to the oestrous cycle in the adult mouse. J. Endocrinol. 3:141–149.

Burkl, W., and G. Kellner. 1954. Über die Entstehung der Zwischenzellen im Rattenovar und ihre Bedeutung im Rahmen der Oestrogenproduktion. Z. Zellforsch. mikrosk. Anat. 40:361–378.

Burkl, W., and G. Kellner. 1956. Das Wachstum der Follikel und die Reifung der Eizellen in den verschiedenen Zyklusphasen bei der Ratte. Acta Anat. 27:309–323.

Butcher, E. O. 1932. Regeneration in ligated ovaries and transplanted ovarian fragments of the white rat (*Mus norvegicus albinus*). Anat. Rec. 54:87–103.

Butcher, E. O. 1947. The periovarian space and the development of the ovary in the rat. Anat. Rec. 98:547–556.

Butler, H. 1967. The oestrus cycle of the Senegal bushbaby (*Galago senegalensis senegalensis*) in the Sudan. J. Zool. (London) 151:143–162.

Butler, H., and M. B. Juma. 1970. Oogenesis in an adult prosimian. Nature (London) 226:552–553.

Call, E., and S. Exner. 1875. Zur Kenntniss der Graafschen Follikels und des Corpus luteum beim Kaninchen. Sitzungsber. math.-naturwiss. Kl. kaiser. Akad. Wiss. Wien 71:321–328.

Campenhout, E. van, and C. Demuylder. 1946. Contribution à l'étude des cellules sympathicotropes de Berger. Arch. Biol. 57:1–11.

Canivec, R. 1966. A study of progestation in the European badger (*Meles meles* L.) pp. 15–26. *In* I. W. Rowlands [ed.] Comparative Biology of Reproduction in Mammals. Symp. Zool. Soc. London, No. 15. Academic Press, London and New York.

Canivec, R. 1968. Luteal function in the European badger, *Meles meles*, pp. 326–338. *In* A. Jost [ed.] La Physiologie de la Reproduction chez les Mammifères. Editions Centre Nat. Rech. Sci., Paris.

Canivec, R., and M. Bonnin-Laffargue. 1963. Inventory of problems raised by the delayed ova implantation in the European badger (*Meles meles* L.), pp. 115–128. *In* A. C. Enders [ed.] Delayed Implantation. Univ. Chicago Press, Chicago.

Canivec, R., and M. Bonnin-Laffargue. 1967. Luteal asthenia in species with delayed implantation (*Meles meles, Mustela vison*). Eur. Rev. Endocrinol. 4:29–40.

Canivec, R., R. V. Short, and M. Bonnin-Laffargue. 1966. Étude histologique et biochemique du corps jaune du blaireau européen (*Meles meles* L.). Ann. Endocrinol. 27:401–414.

Carlson, R. R., and V. J. DeFeo. 1965. Role of the pelvic nerve *vs.* the abdominal sympathetic nerves in the reproductive function of the female rat. Endocrinology 77:1014–1022.

Casida, L. E. 1935. Prepuberal development of the pig ovary and its relation to stimulation with gonadotropic hormones. Anat. Rec. 61:389–396.

Cavazos, L. F., L. L. Anderson, W. D. Belt, D. M. Hendricks, R. R. Kraeling, and R. M. Melampy. 1969. Fine structure and progesterone levels in the corpus luteum of the pig during the estrous cycle. Biol. Reprod. 1:83–106.

Chester Jones, I., and I. W. Henderson. 1963. The ovary of the 13-lined ground squirrel (*Citellus tridecemlineatus*, Mitchell) after adrenalectomy. J. Endocrinol. 26:265–272.

Child, G., and A. S. Mossman. 1965. Right horn implantation in the common duiker. Science 149:1265–1266.

Chiquoine, A. D. 1954. The identification, origin and migration of the primordial germ cells in the mouse embryo. Anat. Rec. 118:135–146.

Chiquoine, A. D. 1960. The development of the zona pellucida of the mammalian ovum. Am. J. Anat. 106:149–170.

Chittleborough, R. G. 1954. Studies on the ovaries of the humpback whale *Megaptera nodosa* (Bonnaterre), on the Western Australian coast. Aust. J. Mar. Freshwater Res. 5:35–63.

Christian, J. J. 1956. The natural history of a summer aggregation of the big brown bat, *Eptesicus fuscus fuscus*. Am. Midland Natur. 55:66–95.

Claesson, L. 1947. Is there any smooth musculature in the wall of the Graafian follicle? Acta Anat. 3:295–311.

Clark, J. G. 1898. Ursprung, Wachsthum und Ende des Corpus luteum nach Beobachtungen am Ovarium des Schweines und des Menschen. Arch. Anat. Physiol. (1898):95–134.

Clark, J. G. 1899. The origin, growth and fate of the corpus luteum as observed in the ovary of the pig and man. Johns Hopkins Hosp. Rep. 7:181–221.

Clewe, T. H. 1965. Absence of a foramen in the ovarian bursa of the golden hamster. Anat. Rec. 151:446. (Abstr.)

Clough, G. 1969. Some preliminary observations on reproduction in the warthog, *Phacochoerus aetheopicus* Pallas. J. Reprod. Fert. Suppl. 6:323–337.

Clough, G. 1970. A record of "testis cords" in the ovary of a mature hippopotamus (*Hippopotamus amphibius*, Linn.). Anat. Rec. 166:47–50.

Cole, H. H., G. H. Hart, W. R. Lyons, and H. R. Catchpole. 1933. The development and hormonal content of fetal horse gonads. Anat. Rec. 56:275–293.

Cole, H. H., C. E. Howell, and G. H. Hart. 1931. The changes occurring in the ovary of the mare during pregnancy. Anat. Rec. 49:199–209.

Comrie, L. C., and A. B. Adam. 1937. The female reproductive system and corpora lutea of the false killer whale, *Pseudorca crassidens* Owen. Trans. Roy. Soc. Edinburgh 59:521–531.

Conaway, C. H. 1952. Life history of the water shrew (*Sorex palustris navigator*). Am. Midland Natur. 48:219–248.

Conaway, C. H. 1969. Adrenal cortical rests of the ovarian hilus of the patas monkey. Folia Primatol. 2:175–180.

Corner, G. W. 1915. The corpus luteum of pregnancy, as it is in swine. Contrib. Embryol. Carnegie Inst. Washington 222:69–94.

Corner, G. W. 1919. On the origin of the corpus luteum of the sow from both granulosa and theca interna. Am. J. Anat. 26:117–183.

Corner, G. W. 1921. Cyclic changes in the ovaries and uterus of the sow, and their relation to the mechanism of implantation. Contrib. Embryol. Carnegie Inst. Washington 13:117–146.

Corner, G. W. 1938. The sites of formation of estrogenic substances in the animal body. Physiol. Rev. 18:154–172.

Corner, G. W. 1945. Development, organization, and breakdown of the corpus luteum in the rhesus monkey. Contrib. Embryol. Carnegie Inst. Washington 31:117–146.

Corner, G. W., C. W. Bartelmez, and C. G. Hartman. 1936. On normal and aberrant corpora lutea of the rhesus monkey. Am. J. Anat. 59:433–457.

Corner, G. W., Jr. 1956. The histological dating of the human corpus luteum of menstruation. Am. J. Anat. 98:377–401.

Crisp, T. M., A. D. Dessouky, and F. R. Denys. 1970. Fine structure of the human corpus luteum of early pregnancy and during the progestational phase of the menstrual cycle. Am. J. Anat. 127:37–70.

Crombie, P. R., R. D. Burton, and N. Ackland. 1971. The ultrastructure of the corpus luteum of the guinea-pig. Z. Zellforsch. mikrosk. Anat. 115: 473–493.

Crone, M., and H. Peters. 1968. Unusual incorporation of tritiated thymidine in early diplotene oocytes of mice. Exp. Cell Res. 50:664–668.

Culiner, A. 1946. Role of the theca cell in irregularities of the baboon menstrual cycle. S. Afr. J. Med. Sci. 11 (Biol. Suppl.):55–70.

Daudt, W. 1898. Beiträge zur Kenntniss des Urogenitalapparates der Cetaceen. Inaug. Dissertation Zool. Inst., Univ. Jena. G. Fischer, Jena.

Davenport, C. B. 1925. Regeneration of ovaries of mice. J. Exp. Zool. 42:1–12.

Davies, J., and C. D. Broadus. 1968. Studies on the fine structure of ovarian steroid-secreting cells in the rabbit. I. The normal interstitial cells. Am. J. Anat. 123:441–474.

Davies, J., E. W. Dempsey, and G. B. Wislocki. 1957. Histochemical observations on the fetal ovary and testis of the horse. J. Histochem. Cytochem. 5:584–590.

Davis, D. E., and O. Hall. 1950. Polyovuly and anovular follicles in the wild Norway rat. Anat. Rec. 107:187–192.

Dawson, A. B. 1941. The development and morphology of the corpus luteum of the cat. Anat. Rec. 79:155–169.

Dawson, A. B. 1946. The postpartum history of the corpus luteum of the cat. Anat. Rec. 95:29–51.

Dawson, A. B. 1951. Histogenetic interrelationships of oocytes and follicle cells. A possible explanation of the mode of origin of certain polyovular follicles in the immature rat. Anat. Rec. 110:181–197.

Dawson, A. B., and H. B. Friedgood. 1940. The time and sequence of preovulatory changes in the cat ovary after mating or mechanical stimulation of the cervix uteri. Anat. Rec. 76:411–429.

Dawson, A. B., and M. McCabe. 1951. The interstitial tissue of the ovary in infantile and juvenile rats. J. Morphol. 88:543–571.

Deane, H. W. 1952. Histochemical observations on the ovary and oviduct of the albino rat during the estrous cycle. Am. J. Anat. 91:363–413.

Deane, H. W., and D. W. Fawcett. 1952. Pigmented interstitial cells showing "brown degeneration" in the ovaries of old mice. Anat. Rec. 113:239–245.

Deane, H. W., and B. L. Rubin. 1965. Identification and control of cells that synthesize steroid hormones in the adrenal glands, gonads and placentae of various mammalian species. Arch. Anat. Microsc. Morphol. Exp. 54:49–66.

Deane, H. W., M. F. Hay, R. M. Moor, L. E. A. Rowson, and R. V. Short. 1966. The corpus luteum of the sheep: Relationships between morphology and function during the oestrous cycle. Acta Endocrinol. 51:245–263.

Deanesly, R. 1930. Development and vascularisation of the corpus luteum in rabbit and mouse. Proc. Roy. Soc., Ser. B 107:60–76.

Deanesly, R. 1934. The reproductive processes of certain mammals. VI. The reproductive cycle of the female hedgehog. Phil. Trans. Roy. Soc. London, Ser. B 223:239–276.

Deanesly, R. 1966. Observations on reproduction in the mole, *Talpa europaea*, pp. 387–402. *In* I. W. Rowlands [ed.] Comparative Biology of Reproduction in Mammals. Symp. Zool. Soc. London, No. 15. Academic Press, London and New York.

Deanesly, R. 1967. Experimental observations on the ferret corpus luteum of pregnancy. J. Reprod. Fert. 13:183–185.

Deanesly, R. 1970. Oogenesis and the development of the ovarian interstitial tissue in the ferret. J. Anat. 107:165–178.

Deanesly, R., and A. S. Parkes. 1933. The reproductive processes of certain mammals. Part IV. The oestrous cycle in the grey squirrel (*Sciurus carolinensis*). Phil. Trans. Roy. Soc. London, Ser. B 222:47–78.

Delost, P. 1955. Étude de la biologie sexuelle du campagnol des champs (*Microtus arvalis* P.). Arch. Anat. Microsc. Morphol. Exp. 44:150–190.

Delost, P. 1956. Développement sexuel normal du campagnol des champs (*Microtus arvalis* P.), de la naissance à l'âge adulte. Arch. Anat. Microsc. Morphol. Exp. 45:11–47.

Dempsey, E. W. 1939. The reproductive cycle of New World monkeys. Am. J. Anat. 64:381–405.

Dempsey, E. W. 1940. The structure of the reproductive tract in the female gibbon. Am. J. Anat. 67:229–253.

Dempsey, E. W., and G. B. Wislocki. 1941. The structure of the ovary of the humpback whale (*Megaptera nodosa*). Anat. Rec. 80:243–257.

Diczfalusy, E. 1962. Endocrinology of the foetus. Acta Obstet. Gynecol. Scand. 41 (Suppl. 1):45–85.

Diczfalusy, E. 1964. Endocrine functions of the human fetoplacental unit. Fed. Proc. Symp., Fed. Am. Soc. Exp. Biol. 23:791–798.

Diczfalusy, E., R. Pion, and J. Schwers. 1965. Steroid biogenesis and metabolism in the human foeto-placental unit at midpregnancy. Arch. Anat. Microsc. Morphol. Exp. 54:67–84.

Donaldson, L., and W. Hansel. 1965. Histological study of bovine corpora lutea. J. Dairy Sci. 48:905–909.

Douglas, M. J. W. 1966. Occurrence of accessory corpora lutea in red deer, *Cervus elaphus*. J. Mammalogy 47:152–153.

Dryden, G. L. 1969. Reproduction in *Suncus murinus*. J. Reprod. Fert. Suppl. 6:377–396.

Dubreuil, G. 1957. Le déterminisme de la glande thécale de l'ovaire. Induction morphogène à partir de la granulosa folliculaire. Acta Anat. 30:269–274.

Duke, K. L. 1940. A preliminary histological study of the ovary of the kangaroo rat, *Dipodomys ordii columbianus*. Great Basin Natur. 1:63–73.

Duke, K. L. 1941. The germ cells of the rabbit ovary from sex differentiation to maturity. J. Morphol. 69:51–81.

Duke, K. L. 1944. Activity of the germinal epithelium in the ovary of a pregnant harvest mouse. Anat. Rec. 89:135–137.

Duke, K. L. 1947. The fibrous connective tissue of the rabbit ovary from sex differentiation to maturity. Anat. Rec. 98:507–526.

Duke, K. L. 1949. Some notes on the histology of the ovary of the bobcat (*Lynx*), with special reference to the corpora lutea. Anat. Rec. 103:111–132.

Duke, K. L. 1952. Ovarian histology of *Ochotona princeps*, the Rocky Mountain pika. Anat. Rec. 112:737–760.

Duke, K. L. 1957. Reproduction in *Perognathus*. J. Mammalogy 38:207–210.

Duke, K. L. 1964. Histological observations on the ovary of the slow loris, *Nycticebus coucang*. Anat. Rec. 148:414.

Duke, K. L. 1966*a*. Histological observations on the ovary of the white-tailed mole, *Parascaptor leucurus*. Anat. Rec. 154:527–532.

Duke, K. L. 1966*b*. Ovogenesis in some prosimian primates. Anat. Rec. 154:500.

Duke, K. L. 1967. Ovogenetic activity of the fetal-type in the ovary of the adult slow loris, *Nycticebus coucang*. Folia Primatol. 7:150–154.

Duke, K. L., and W. P. Luckett. 1965. Histological observations on the ovary of several species of tree shrews (Tupaiidae). Anat. Rec. 151:450. (Abstr.)

Eckstein, P. 1958. Internal reproductive organs, vol. 3, pp. 542–629. *In* H. Hofer, A. H. Schultze, and D. Starck [eds.] Primatologia: Handbook of Primatology. Karger, Basel and New York.

Edwards, R. G. 1970. Are oocytes formed and used sequentially in the mammalian ovary? pp. 103–105. *In* G. W. Harris and R. G. Edwards [organizers] A Discussion on the Determination of Sex. Phil. Trans. Roy. Soc. London, Ser. B 259:3–206.

Eichner, E., and E. R. Bove. 1954. In vivo studies on the lymphatic drainage of the human ovary. Obstet. Gynecol. 3:287–297.

El-Fouly, M. A., B. Cook, M. Nekola, and A. V. Nalbandov. 1970. Role of the ovum in follicular luteinization (rabbit). Endocrinology 87:288–293.

Emery, F. E., and E. L. Schwabe. 1936. The role of the corpora lutea in prolonging the life of adrenalectomized rats. Endocrinology 20:550–555.

Enders, A. C. 1960. A histological study of the cortex of the ovary of the adult armadillo, with special reference to the question of neoformation of oocytes. Anat. Rec. 136:491–499.

Enders, A. C. 1966. The reproductive cycle in the nine-banded armadillo (*Dasypus novemcinctus*), pp. 295–310. *In* I. W. Rowlands [ed.] Comparative Biology of Reproduction in Mammals. Symp. Zool. Soc. London, No. 15. Academic Press, London and New York.

Enders, A. C., and G. D. Buchanan. 1959. The reproductive tract of the female nine-banded armadillo. Tex. Rep. Biol. Med. 17:323–340.

Enders, R. K. 1952. Reproduction in the mink (*Mustela vison*). Proc. Am. Phil. Soc. 96:691–755.

Enders, R. K., and A. C. Enders. 1963. Morphology of the female reproductive tract during delayed implantation in the mink, pp. 129–139. *In* A. C. Enders [ed.] Delayed Implantation. Univ. Chicago Press, Chicago.

Enders, R. K., O. P. Pearson, and A. K. Pearson. 1946. Certain aspects of reproduction in the fur seal. Anat. Rec. 94:213–227.

Engle, E. T. 1927*a*. Polyovular follicles and polynuclear ova in the mouse. Anat. Rec. 35:341–343.

Engle, E. T. 1927*b*. A quantitative study of follicular atresia in the mouse. Am. J. Anat. 39:187–203.

Erickson, B. H. 1966. Development and senescence of the postnatal bovine ovary. J. Anim. Sci. 25:800–805.

Espey, L. L., and H. Lipner. 1963. Measurements of intrafollicular pressures in the rabbit ovary. Am. J. Physiol. 205:1067–1072.

Espey, L. L., C. Slagter, R. Weymouth, and P. Rondell. 1965. A study of the ultrastructure of the rabbit graafian follicle as it approaches rupture. Physiologist 8:161. (Abstr.)

Evans, H. M., and H. H. Cole. 1931. An introduction to the study of the oestrous cycle in the dog. Mem. Univ. Calif. 9:65–118.

Evans, H. M., and O. Swezy. 1931. Ovogenesis and the normal follicular cycle in adult mammalia. Mem. Univ. Calif. 9:119–225.

Everett, N. B. 1943. Observational and experimental evidence relating to the origin and differentiation of the definitive germ cells in mice. J. Exp. Zool. 92:49–91.

Everett, N. B. 1945. The present status of the germ-cell problem in vertebrates. Biol. Rev. (Cambridge) 20:45–55.

Falck, B. 1959. Site of production of oestrogen in the ovary of the rat. Acta Physiol. Scand. 47 (Suppl. 163):1–101.

Fekete, E., G. Wooley, and C. C. Little. 1941. Histological changes following ovariectomy in mice. I. dba high tumor strain. J. Exp. Med. 74:1–8.

Feremutsch, K. 1948. Der praegravide Genitaltrakt und die Praeimplantation. Rev. Suisse Zool. 55:567–622.

Feremutsch, K., and F. Strauss. 1949. Beitrag zum weiblichen Genitalzyklus der madagassischen Centetinen. Rev. Suisse Zool. 56 (Suppl. 1):1–110.

Fevold, H. R., and P. L. Wright. 1969. Steroid metabolism by badger (*Taxidea taxus*) ovarian tissue homogenates (pregnenolone, androstenedione, dehydroepiandrosterone). Gen. Comp. Endocrinol. 13:60–67.

Fink, G., and G. C. Schofield. 1971. Experimental studies on the innervation of the ovary in cats. J. Anat. 109:115–126.

Fischel, A. 1930. Über die Entwicklung der Keimdrüsen des Menschen. Arch. Anat. Entwicklungsgesch. 92:34–72.

Fischer, T. V. 1967. Local regulation of the corpus luteum. Am. J. Anat. 121:425–442.

Fisher, H. D. 1954. Delayed implantation in the harbour seal, *Phoca vitulina* L. Nature (London) 173:879–880.

Flynn, T. T., and J. P. Hill. 1939. The development of the Monotremata. 4. Growth of the ovarian ovum, maturation, fertilization, and early cleavage. Trans. Zool. Soc. London 24:445–622.

Forbes, T. R. 1942. On the fate of the medullary cords of the human ovary. Contrib. Embryol. Carnegie Inst. Washington 30:9–15.

Foreman, D. 1962. The normal reproductive cycle of the female prairie dog and the effects of light. Anat. Rec. 142:391–407.

Forleo, R. 1961. Anatomy of the human ovary during pregnancy. Riv. Ostet. Ginecol. 16:530–560.

Foster, M. A. 1934. The reproductive cycle in the female ground squirrel, *Citellus tridecemlineatus* (Mitchill). Am. J. Anat. 54:487–511.

Fraenkel, L. 1903. Die Function des Corpus luteum. Arch. Gynaekol. 68:438–545.

Fraenkel, L., and F. Cohn. 1901. Experimentelle Untersuchung über den Einfluss des Corpus luteum und die Insertion des Eies. Anat. Anz. 20:294–300.

Franchi, L. L., A. M. Mandl, and S. Zuckerman. 1962. The development of the ovary and the process of oogenesis, vol. 1, pp. 1–88. *In* S. Zuckerman [ed.] The Ovary. 2 vol. Academic Press, London and New York.

Frommolt, G. 1934. Studien an Makakusovarien. Z. Geburtsh. Gynaekol. 107:165–178.

Gaillard, P. J. 1950. Sex cell formation in explants of the foetal human ovarian cortex. Proc. Kon. Ned. Akad. Wetensch., Ser. C 53:3–30, 1300–1316, 1337–1347.

Garde, M. L. 1930. The ovary of *Ornithorhynchus*, with special reference to follicular atresia. J. Anat. 64:422–453.

George, A. N., and M. F. A. Fahmy. 1966. Histological study of the developing ovary of the dromedary (*Camelus dromedarius*). J. Vet. Sci. U. A. R. 3:93–100 [Original not seen.]

Gerall, A. A., and J. L. Dunlap. 1971. Evidence that the ovaries of the neonatal rat secrete active substances. J. Endocrinol. 50:529–530.

Gérard, P. 1919–1920. Contribution à l'étude de l'ovaire des mammifères. L'ovaire de *Galago mossambicus* (Young). Arch. Biol. 30:357–390.

Gérard, P. 1932. Études sur l'ovogénèse et l'ontogénèse chez les Lémuriens du genre *Galago*. Arch. Biol. 43:93–151.

Gérard, P., and M. Herlant. 1953. Sur la persistance des phénomènes d'ovogénèse chez les Lémuriens adultes. Arch. Biol. 64:97–110.

Gerhardt, V. 1905. Studien über den Geschlechtsapparat der weiblichen Säugetiere. 1. Die Ueberbreitung des Eies in die Tuben. Jena. Z. Naturwiss. 39:649–712.

Gillim, S. W., K. Christensen, and C. E. McLennan. 1969. Fine structure of the human menstrual corpus luteum at its stage of maximum secretory activity. Am. J. Anat. 126:409–428.

Gillman, J. 1948. The development of the gonads in man, with a consideration of the role of fetal endocrines and the histogenesis of ovarian tumors. Contrib. Embryol. Carnegie Inst. Washington 32:81–131.

Gillman, J., and C. Gilbert. 1946. The reproduction cycle of the chacma baboon (*Papio ursinus*) with special reference to the problems of menstrual irregularities as assessed by the behaviour of the sex skin. S. Afr. J. Med. Sci. 11:1–69.

Gillman, J., and H. B. Stein. 1941. Human corpus luteum of pregnancy. Surg. Gynecol. Obstet. 72:129–149.

Godet, R. 1949. Recherches d'anatomie, d'embryologie normale et expérimentale sur l'appareil genital de la taupe (*Talpa europaea* L.). Bull. Biol. Fr. Belg. 83:25–111.

Goetz, R. H. 1937. Studien zur Placentation der Centetiden. II. Die Implantation und Frühentwicklung von Hemicentetes semispinosus (Cuvier). Z. Anat. Entwicklungsgesch. 107:274–318.

Goodman, P., J. S. Latta, R. B. Wilson, and B. Kadis. 1968. Fine structure of sow lutein cells. Anat. Rec. 161:77–90.

Govan, A. D. T. 1970. Follicular activity in the human ovary in late pregnancy. J. Endocrinol. 48:235–241.

Grant, R. 1934. Studies in the physiology of reproduction in the ewe. III. Gross changes in the ovary. Trans. Roy. Soc. Edinburgh 58:36–47.

Green, J. A., and M. Maqueo. 1965. Ultrastructure of the human ovary. I. The luteal cell during the menstrual cycle. Am. J. Obstet. Gynecol. 92:946–957.

Green, J. A., and M. Maqueo. 1966. Histopathology and ultrastructure of an ovarian hilar cell tumor. Am. J. Obstet. Gynecol. 96:478–485.

Green, J. A., J. A. Garcilazo, and M. Maqueo. 1967. Ultrastructure of the human ovary. II. The luteal cell at term. Am. J. Obstet. Gynecol. 99:855–863.

Green, J. A., J. A. Garcilazo, and M. Maqueo. 1968. Ultrastructure of the human ovary. III. Canaliculi of the corpus luteum. Am. J. Obstet. Gynecol. 102:57–64.

Greene, R. R., and W. W. Nelson. 1952. Decidual reaction in the ovary. Quart. Bull. Northwest. Univ. Med. Sch. 26:197–200.

Greenwald, G. S. 1956. The reproductive cycle of the field mouse, *Microtus californicus*. J. Mammalogy 37:213–222.

Greenwald, G. S. 1965. Histologic transformation of the ovary of the lactating hamster. Endocrinology 77:641–650.

Greenwald, G. S., and R. D. Pepler. 1968. Prepubertal and pubertal changes in the hamster ovary. Anat. Rec. 161:447–458.

Groat, R. A. 1943. Adrenocortical-like tissue in the ovaries of the adrenalectomized ground squirrel (*Citellus tridecemlineatus*). Endocrinology 32:488–492.

Groat, R. A. 1944. Formation and growth of adrenocortical-like tissue in the ovaries of the adrenalectomized ground squirrel. Anat. Rec. 89:33–41.

Grob, H. S. 1971. Monolayer culture of ovarian follicular elements derived from isolated mouse follicles. Biol. Reprod. 5:207–213.

Gropp, A., and S. Ohno. 1966. Presence of a common embryonic blastema for ovarian and testicular parenchymal (follicular, interstitial, and tubular) cells in cattle, *Bos taurus*. Z. Zellforsch. mikrosk. Anat. 74:505–528.

Gruenwald, P. 1942. The development of the sex cords in the gonads of man and mammals. Am. J. Anat. 70:359–397.

Guieysse, A. 1901. La capsule surrénale du cobaye. Histologie et functionnement. J. Anat. Physiol. 37:312–341.

Guraya, S. S. 1966. Histochemical analysis of the interstitial gland tissue in the human ovary at the end of pregnancy. Am. J. Obstet. Gynecol. 96:907–912.

Guraya, S. S. 1967. Cytochemical study of interstitial cells in the bat ovary. Nature (London) 214:614.

Guraya, S. S., and G. S. Greenwald. 1964*a*. Histochemical studies on the interstitial gland of the rabbit ovary. Am. J. Anat. 114:405–520.

Guraya, S. S., and G. S. Greenwald. 1964*b*. A comparative histochemical study of interstitial tissue and follicular atresia in the mammalian ovary. Anat. Rec. 149:411–434.

Guraya, S. S., and G. S. Greenwald. 1965. A histochemical study of the hamster ovary. Am. J. Anat. 116:257–268.

Guthrie, M. J., and K. R. Jeffers. 1938. A cytological study of the ovaries of the bats, *Myotis lucifugus lucifugus* and *Myotis grisescens*. J. Morphol. 62:523–557.

Hadek, R. 1963. Electron microscope study on primary liquor folliculi secretion in the mouse ovary. J. Ultrastruct. Res. 9:445–458.

Halazon, G. C., and H. K. Buechner. 1956. Postconception ovulation in elk. Trans. N. Am. Wildlife Conf. 21:545–554.

Hall, E. R., and K. R. Kelson. 1959. The Mammals of North America. 2 vol. The Ronald Press Co., New York.

Hall, O. 1952. Accessory corpora lutea in the wild Norway rat. Tex. Rep. Biol. Med. 10:32–38.

Hamilton, W. J. 1941. Reproduction in the field mouse *Microtus pennsylvanicus* (Ord). Mem. Cornell Univ. Agr. Exp. Sta. 237:2–23.

Hamilton, W. J., and J. H. Gould. 1939. The normal oestrus cycle of the ferret: The correlation of the vaginal smear and the histology of the genital tract, with notes on the distribution of glycogen, the incidence of growth, and the reaction to intravitam staining by trypan blue. Trans. Roy. Soc. Edinburgh 60:87–106.

Hammond, J. 1923. Changes in the reproductive organs of the cow during the sexual cycle and pregnancy. Quart. J. Exp. Physiol. 13 (Suppl.):134–136.

Hammond, J., and F. H. A. Marshall. 1925. Reproduction in the Rabbit. Oliver and Boyd, Edinburgh.

Hammond, J., and F. H. A. Marshall. 1930. Oestrus and pseudopregnancy in the ferret. Proc. Roy. Soc., Ser. B 105:607–629.

Hammond, J., and K. Wodzicki. 1941. Anatomical and histological changes during the oestrous cycle in the mare. Proc. Roy. Soc., Ser. B 130:1–23.

Hansen, P. 1957. Die glatte Muskulatur die Mesovariums und seiner Umgebung. Arch. Gynaekol. 188:299–328.

Hansson, A. 1947. The physiology of reproduction in mink (*Mustela vison* Schreb.) with special reference to delayed implantation. Acta Zool. (Stockholm) 28:1–136.

Hargitt, G. T. 1925. The formation of the sex glands and germ cells of mammals. I. The origin of the germ cells in the albino rat. J. Morphol. 40:517–557.

Hargitt, G. T. 1926. The formation of the sex glands and germ cells of mammals. II. The history of the male germ cells in the albino rat. J. Morphol. 42:253–306.

Hargitt, G. T. 1930*a*. The formation of the sex glands and germ cells of mammals. III. The history of the female germ cells in the albino rat to the time of sexual maturity. J. Morphol. 49:277–331.

Hargitt, G. T. 1930*b*. The formation of the sex glands and germ cells of mammals. IV. Continuous origin and degeneration of germ cells in the female albino rat. J. Morphol. 49:333–353.

Hargitt, G. T. 1930*c*. The formation of the sex glands and germ cells of mammals. V. Germ cells in the ovaries of adult, pregnant, and senile albino rats. J. Morphol. 50:453–473.

Harman, M. T., and H. D. Kirgis. 1938. The development and atresia of the Graafian follicle and the division of intraovarian ova in the guinea pig. Am. J. Anat. 63:79–99.

Harrison, R. J. 1946. The early development of the corpus luteum of the mare. J. Anat. 80:160–166.

Harrison, R. J. 1948*a*. The development and fate of the corpus luteum in the vertebrate series. Biol. Rev. (Cambridge) 23:296–331.

Harrison, R. J. 1948*b*. The changes occurring in the ovary of the goat during the estrous cycle and in early pregnancy. J. Anat. 82:21–47.

Harrison, R. J. 1949. Observations on the female reproductive organs of the Ca'ing whale, *Globiocephala melaena* Traill. J. Anat. 83:238–253.

Harrison, R. J. 1950. Observations on the seal ovary. J. Anat. 84:400. (Abstr.)

Harrison, R. J. 1960. Reproduction and reproductive organs in common seals (*Phoca vitulina*) in the Wash, East Anglia. Mammalia 24:372–385.

Harrison, R. J. 1962. The structure of the ovary. C. Mammals, vol. 1, pp. 143–187. *In* S. Zuckerman [ed.] The Ovary. 2 vol. Academic Press, London and New York.

Harrison, R. J., and L. H. Matthews. 1951. Sub-surface crypts in the cortex of the mammalian ovary. Proc. Zool. Soc. London 120:699–712.

Harrison, R. J., L. H. Matthews, and J. M. Roberts. 1952. Reproduction in some Pinnipedia. Trans. Zool. Soc. London 27:437–540.

Harrison-Matthews, L. 1935. The oestrous cycle and intersexuality in the female mole (*Talpa europaea* Linn.). Proc. Zool. Soc. London (1935): 347–383.

Harrison-Matthews, L., and R. J. Harrison. 1949. Subsurface crypts, oogenesis, and the corpus luteum in the ovaries of seals. Nature (London) 164:587.

Harrison-Matthews, L. See also Matthews, L. H.

Hartman, C. G. 1923. The oestrous cycle of the opossum. Am. J. Anat. 32:353–421.

Hartman, C. G. 1926. Polynuclear ova and polyovular follicles in the opossum and other mammals, with special reference to the problem of fecundity. Am. J. Anat. 37:1–51.

Hartman, C. G., and B. League. 1924. Description of a sex-intergrade opossum, with an analysis of the constituents of its gonads. Anat. Rec. 29:283–297.

Harvey, E. B. 1959. Placentation in Aplodontidae. Am. J. Anat. 105:63–89.

Hashimoto, M., T. Kawasaki, Y. Mori, A. Komori, T. Shimoyama, M. Kosaka, and K. Akashi. 1960. Electron microscopic studies on the fine structure of the rabbit ovarian follicles. I and II. J. Jap. Obstet. Gynecol. Soc. 7:228–235, 267–275.

Haterius, H. O. 1928. An experimental study of ovarian regeneration in mice. Physiol. Zool. 1:45–54.

Hayward, J. N., J. Hilliard, and C. H. Sawyer. 1963. Preovulatory and postovulatory progestins in monkey ovary and ovarian vein blood. Proc. Soc. Exp. Biol. Med. 113:256–259.

Heape, W. 1886. The development of the mole (*Talpa europaea*), the ovarian ovum, and segmentation of the ovum. Quart. J. Microsc. Sci. 26:157–174.

Herlant, M. 1953. Étude comparative sur l'activite genitale des Cheiropteres. Ann. Soc. Roy. Zool. Belg. 84:87–116.

Hertig, A. T. 1968. The primary human oocyte: Some observations on the fine structure of Balbiani's vitelline body and the origin of annulate lamellae. Am. J. Anat. 122:107–138.

Hertig, A. T., and E. C. Adams. 1967. Studies on the human oocyte and its follicle. I. Ultrastructural and histochemical observations on the primordial follicle stage. J. Cell Biol. 34:647–675.

Heys, F. 1929. Does regeneration occur after complete ovariectomy in the albino rat? Science 70:289–290.

Hibbard, B. M. 1961. The position of the maternal ovaries in late pregnancy. Brit. J. Radiol. 34:387–388.

Hill, J. P., and J. W. B. Gatenby. 1926. The corpus luteum of the Monotremata. Proc. Zool. Soc. London, 47:715–763.

Hill, R. T. 1949. Adrenal cortical physiology of spleen grafted and denervated ovaries in the mouse. Exp. Med. Surg. 7:86–98.

Hill, W. C. O. 1945. Notes on the dissection of two dugongs. J. Mammalogy 26:153–175.

Hill, W. C. O. 1953. The female reproductive organs of *Tarsius*, with observations on the physiological changes therein. Proc. Zool. Soc. London 123:589–598.

Hill, W. C. O., and D. V. Davies. 1954. The reproductive organs in *Hapalemur* and *Lepilemur*. Proc. Roy. Soc. Edinburgh, Sect. B 70:251–270.

Hillemann, H. H., and F. D. Tibbitts. 1956. Ovarian growth and development in chinchilla. Northwest Sci. 30:115–126.

Hillemann, H. H., A. I. Gaynor, and H. P. Stanley. 1958. The genital systems of nutria. Anat. Rec. 130:515–532.

Hilliard, J., D. Archibald, and C. Sawyer. 1963. Gonadotropic activation of preovulatory synthesis and release of progestin in the rabbit. Endocrinology 72:59–66.

Hisaw, F. L. 1947. The development of the graafian follicle and ovulation. Physiol. Rev. 27:95–119.

Höfliger, H. 1948. Das Ovar des Rindes in den verschiedenen Lebensperioden unter besonderer Berücksichtigung seiner funktionellen Feinstruktur. Acta Anat. 3 (Suppl. 5):1–196.

Honoré, C. 1900. Recherches sur l'ovaire du lapin. I. Note sur les corps de Call & Exner et la formation du liquor folliculi. II. Recherches sur la formation du corps jaune. Arch. Biol. 16:537–562, 563–599.

Horst, C. J. van der. 1942. Some observations on the structure of the genital tract of *Elephantulus*. J. Morphol. 70:403–430.

Horst, C. J. van der. 1944. Remarks on the systematics of *Elephantulus*. J. Mammal. 25:77–82.

Horst, C. J. van der. 1945. The behaviour of the graafian follicle of *Elephantulus* during pregnancy, with special reference to the hormonal regulation of ovarian activity. S. Afr. J. Med. Sci. 10:1–14.

Horst, C. J. van der. 1949. An early stage of placentation in the aardvark. Proc. Zool. Soc. London 119:1–18.

Horst, C. J. van der. 1954. *Elephantulus* going into anoestrus: Menstruation and abortion. Phil. Trans. Roy. Soc. London, Ser. B 238:27–61.

Horst, C. J. van der, and J. Gillman. 1940. Ovulation and corpus luteum formation in *Elephantulus*. S. Afr. J. Med. Sci. 5:73–91.

Horst, C. J. van der, and J. Gillman. 1941. The number of eggs and surviving embryos in *Elephantulus*. Anat. Rec. 80:443–452.

Horst, C. J. van der, and J. Gillman. 1942. The life history of the corpus luteum of menstruation in *Elephantulus*. S. Afr. J. Med. Sci. 7:21–41.

Horst, C. J. van der, and J. Gillman. 1945. The behaviour of the graafian follicle of *Elephantulus* during pregnancy, with specal reference to the hormonal regulation of ovarian activity. S. Afr. J. Med. Sci. 10 (Biol. Suppl.):1–14.

Horst, C. J. van der, and J. Gillman. 1946. The corpus luteum of *Elephantulus* during pregnancy; its form and function. S. Afr. J. Med. Sci. 11 (Biol. Suppl.):87–102.

Ingram, D. L. 1962. Atresia, vol. 1, pp. 247–274. *In* S. Zuckerman [ed.] The Ovary. 2 vol. Academic Press, New York and London.

Ioannou, J. M. 1968. Oogenesis in adult prosimians. J. Embryol. Exp. Morphol. 17:139–145.

Israel, S. L., A. Rubenstone, and D. R. Meranze. 1954. The ovary at term. I. Decidua-like reaction and surface cell proliferation. Obstet. Gynecol. 3:399–407.

Jacobowitz, D., and E. E. Wallach. 1967. Histochemical and chemical studies of the autonomic innervation of the ovary. Endocrinology 81:1132–1139.

Jonckheere, F. 1930. Contribution à l'histogénèse de l'ovaire des mammifères. L'ovaire de *Canis familiaris*. Arch. Biol. 40:357–436.

Jost, A. 1969. The extent of foetal endocrine autonomy, pp. 79–94. *In* G. E. W. Wolstenholme and Maeve O'Connor [eds.] Fetal Autonomy (Ciba Found. Symp.). J. and A. Churchill Ltd., London.

Jost, A., and J. Prepin. 1966. Données sur la migration des cellules germinales primordiales du foetus de veau. Arch. Anat. Microsc. Morphol. Exp. 55:161–186.

Käppeli, I. 1908. Beiträge zur Anatomie und Physiologie der Ovarien von wildlebenden und gezähmten Wiederkäuern und Schweinen. Dissertation, Vet. Anat. Inst., Bern. [Original not seen.]

Kayanja, F. I. B. 1969. The ovary of the impala, *Aepyceros melampus* (Lichtenstein, 1812). J. Reprod. Fert. Suppl. 6:311–318.

Kellas, L. M., E. W. van Lennep, and E. C. Amoroso. 1958. Ovaries of some foetal and prepuberal giraffes [*Giraffa camelopardalis* (Linnaeus)]. Nature (London) 181:487–488.

Kellogg, M. P. 1941. The development of the periovarial sac in the white rat. Anat. Rec. 79:465–477.

Kelly, G. L. 1939. Effect of opening the ovarian bursa on fecundity in the albino rat. Anat. Rec. 73:401–405.

Kennelly, J. J., and R. H. Foote. 1966. Oocytogenesis in rabbits. The role of neogenesis in the formation of the definitive ova and the stability of oocyte DNA measured with tritiated thymidine. Am. J. Anat. 118:573–589.

Keyes, P. L., and A. V. Nalbandov. 1968. Endocrine function of the ovarian interstitial gland of rabbits. Endocrinology 82:799–804.

Kingsbury, B. F. 1913. The morphogenesis of the mammalian ovary (*Felis domestica*). Am. J. Anat. 15:345–387.

Kingsbury, B. F. 1914. The interstitial cells of the mammalian ovary: *Felis domestica*. Am. J. Anat. 16:59–95.

Kingsbury, B. F. 1939. Atresia and the interstitial cells of the ovary. Am. J. Anat. 65:309–331.

Knigge, K. M., and J. H. Leathem. 1956. Growth and atresia of follicles in the ovary of the hamster. Anat. Rec. 124:679–707.

Koering, M. J. 1969. Cyclic changes in ovarian morphology during the menstrual cycle in *Macaca mulatta*. Am. J. Anat. 126:73–101.

Kohn, A. 1921. Der Bauplan der Keimdrüsen. Arch. Entwicklungsmech. Organismen (Wilhelm Roux) 47:95–118.

Kohn, A. 1926. Über den Bau des Embryonalen Pferdeierstockes. Z. Anat. Entwicklungsgesch. 79:367–390.

Kohn, A. 1928. Über "Leydigsche Zwischenzellen" im Hilus des menschlichen Eierstockes. Endokrinologie 1:3–10.

Kraus, F. T., and R. D. Neubecker. 1962. Luteinization of the ovarian theca in infants and children. Am. J. Clin. Pathol. 37:389–397.

Krohn, P. L. 1967. Factors influencing the number of oocytes in the ovary. Arch. Anat. Microsc. Morphol. Exp. 56:151–159.

Kumar, T. C. A. 1968. Oogenesis in lorises; *Loris tardigradus lydekkerianus* and *Nycticebus coucang*. Proc. Roy. Soc., Ser. B 169:167–176.

Kuntz, A. 1919. The innervation of the gonads in the dog. Anat. Rec. 17:203–219.

Kuntz, A. 1929. The Autonomic Nervous System. Lea and Febiger, Philadelphia.

Kupfer, Max. 1928. The sexual cycle of female domesticated mammals. The ovarian changes and the periodicity of oestrum in cattle, sheep, goats, pigs, and horses (observations on animals in Central Europe and South Africa). 13th and 14th Rep. Vet. Res. Un. S. Afr. pt. 2:1200–1270.

Landau, R. 1938. Der ovariale und tubale Abschnitt des Genitaltraktus beim nicht-graviden und beim früh-graviden *Hemicentetes*-Weibchen. Biomorphosis 1:228–264.

Lane, C. W. 1938. Aberrant ovarian follicles in the immature rat. Anat. Rec. 71:243–247.

Lane-Clayton, J. E. 1905. On the origin and life history of the interstitial cells of the ovary in the rabbit. Proc. Roy. Soc., Ser. B 77:32–57.

Laws, R. M. 1956. The elephant seal (*Mirounga leonina* Linn.). III. The physiology of reproduction. Falkland Isl. Dependencies Surv. Sci. Rep. 15:1–66.

Laws, R. M., and G. Clough. 1966. Observations on reproduction in the hippopotamus, *Hippopotamus amphibius* Linn., pp. 117–140. *In* I. W. Rowlands [ed.] Comparative Biology of Reproduction in Mammals. Symp. Zool. Soc. London, No. 15. Academic Press, London and New York.

Layne, J. N. 1954. The biology of the red squirrel, *Tamiasciurus hudsonicus loquax* (Bangs), in central New York. Ecol. Monogr. 24:227–267.

Leach, B. J., and C. H. Conaway. 1963. The origin and fate of polyovular follicles in the striped skunk. J. Mammalogy 44:67–74.

League, B., and C. G. Hartman. 1925. Anovular graafian follicles in mammalian ovaries. Anat. Rec. 30:1–13.

Lennep, E. W. van. 1950. Histology of the corpora lutea in blue and fin whales ovaries. Proc. Kon. Ned. Akad. Wetensch., Ser. C 53:593–599.

Lipner, H. [J.], and N. L. Cross. 1968. Morphology of the membrana granulosa of the ovarian follicle (rabbit). Endocrinology 82:638–641.

Lipner, H. J., and B. A. Maxwell. 1960. Hypothesis concerning the role of follicular contractions in ovulation. Science 131:1737–1738.

Lloyd, R. S., and B. B. Rubenstein. 1941. Multiple ova in the follicles of juvenile monkeys. Endocrinology 29:1008–1014.

Loeb, L. 1905. Über hypertrophische Vorgänge bei der Follikelatresia nebst

Bemerkungen über die Oocyten in den Marksträngen und über Teilungsercheinungen am Ei im Ovarium des Meerschweinchens. Arch. mikrosk. Anat. Entwicklungsmech. 65:728–753.

Loeb, L. 1906. Über die Entwickelung des Corpus luteum beim Meerschweinchen. Anat. Anz. 28:102–106.

Loeb, L. 1912. Ueber chorionepitheliomartige Gebilde im Ovarium des Meerschweinchens und über ihre wahrscheinliche Entstehung aus parthenogenetisch sich entwickelnden Eiern. Z. Krebsforsch. 11:259–282.

Loeb, L. 1930. Parthenogenetic development of eggs in the ovary of the guinea-pig. Proc. Soc. Exp. Biol. Med. 27:413–416.

Loeb, L. 1932. The parthenogenetic development of eggs in the ovary of the guinea-pig. Anat. Rec. 51:373–408.

Long, J. A., and H. M. Evans. 1922. The oestrous cycle in the rat and its associated phenomena. Mem. Univ. Calif. 6:1–148.

Luckett, W. P. 1968. Phylogenetic relationship of the African mole rat, *Bathyergus janetta*, as indicated by the fetal membranes. Am. Zool. 8:806. (Abstr.)

Luckett, W. P. 1969. Evidence for the phylogenetic relationships of tree shrews (family Tupaiidae) based on the placenta and fetal membranes. J. Reprod. Fert. Suppl. 6:419–433.

Lutwak-Mann, C. 1954. Note on the chemical composition of follicular fluid. J. Agr. Sci. 44:477–480.

McIntosh, W. C. 1930. Notes on a female llama. J. Anat. 64:353.

McKay, D. G., A. T. Hertig, E. C. Adams, and S. Danziger. 1953. Histochemical observations on the germ cells of human embryos. Anat. Rec. 117:201–220.

McKay, D. G., J. H. M. Pinkerton, A. T. Hertig, and S. Danziger. 1961. The adult human ovary: A histochemical study. Obstet. Gynecol. 18:13–39.

MacLeod, J. 1880. Contribution à l'étude de la structure de l'ovaire des mammifères. Arch. Biol. 1:241–278.

McNeill, J. 1931. Aberrant ovarian tissue in the round ligament of the uterus. J. Obstet. Gynaecol. Brit. Empire 38:608–613.

Mainland, D. 1928. The pluriovular follicle, with reference to its occurrence in the ferret. J. Anat. 62:139–158.

Mainland, D. 1932. The early development of the ferret: The zona granulosa, zona pellucida and associated structures. J. Anat. 66:586–601.

Mandl, A., and S. Zuckerman. 1951*a*. The effect of destruction of the germinal epithelium on the number of oocytes. J. Endocrinol. 7:104–111.

Mandl, A., and S. Zuckerman. 1951*b*. Changes in ovarian structure following the injection of carbolic acid into the ovarian bursa. J. Endocrinol. 7:227–234.

Mansfield, A. W. 1958. The breeding behaviour and reproductive cycle of the Weddell seal (*Leptonychotes weddelli* Lesson). Falkland Isl. Dependencies Surv. Sci. Rep. 18:1–41.

Marion, G. B., H. T. Gier, and J. B. Choudary. 1968. Micromorphology of the bovine ovarian follicular system. J. Anim. Sci. 27:451–465.

Martínez-Esteve, P. 1942. Observations on the histology of the opossum ovary. Contrib. Embryol. Carnegie Inst. Washington 30:17–26.

Matthews, L. H. 1937. The female sexual cycle in the British horseshoe bats, *Rhinolophus ferrumequinum insulanus* Barrett-Hamilton and *R. hipposideros minutus* Montagu. Trans. Zool. Soc. London 23:224–248.

Matthews, L. H. 1939. Reproduction in the spotted hyaena *Crocuta crocuta* (Erxleben). Phil. Trans. Roy. Soc. London, Ser. B 230:1–78.

Matthews, L. H. 1941. Notes on the genitalia and reproduction of some African bats. Proc. Zool. Soc. London, Ser. B 111:289–346.

Matthews, L. H. See also Harrison-Matthews, L.

Mead, R. A. 1968*a*. Reproduction in eastern forms of the spotted skunk (genus *Spilogale*). J. Zool. (London) 156:119–136.

Mead, R. A. 1968*b*. Reproduction in western forms of the spotted skunk genus *Spilogale*. J. Mammalogy 49:373–390.

Mead, R. A., and K. B. Eik-Nes. 1969. Oestrogen levels in peripheral blood plasma of the spotted skunk. J. Reprod. Fert. 18:351–353.

Meek, A. 1918. The reproductive organs of Cetacea. J. Anat. 52:186–210.

Meester, J. [ed.] Preliminary Identification Manual for African Mammals. 23. Lipotyphla: Chrysochloridae. Smithsonian Inst., Washington, D.C.

Merker, H. J. 1961. Elektronenmikroskopische Untersuchungen über die Bildung der Zona pellucida in den Follikeln des Kaninchenovars. Z. Zellforsch. mikrosk. Anat. 54:677–688.

Merker, H. J., and J. Diaz-Encinas. 1969. Das elektronmikroskopische Bild des Ovars juveniler Ratten und Kaninchen nach Stimulierung mit PMS und HCG. I. Theka und Stroma (Interstitiellen Drüse). Z. Zellforsch. mikrosk. Anat. 94:605–623.

Meyer, R. 1911. Ueber Corpus luteum-Bildung beim Menschen. Arch. Gynaekol. 93:354–404.

Meyer, R. 1924. Lipoide und Ovarialfunktion. Zentralbl. Gynaecologia 48: 1570–1575.

Miegel, B. 1953. Die Biologie und Morphologie der Fortpflanzung der Bisamratte (*Ondatra zibethica* L.). Z. mikrosk.-anat. Forsch. 58:531–598.

Mintz, B. 1957*a*. Germ cell origin and history in the mouse: Genetic and histochemical evidence. Anat. Rec. 127:335–336.

Mintz, B. 1957*b*. Embryological development of primordial germ-cells in the mouse: Influence of a new mutation, Wj. J. Embryol. Exp. Morphol. 5:396–403.

Mintz, B. 1960. Embryological phases of mammalian gametogenesis. J. Cell. Comp. Physiol. 56:31–48.

Mintz, B. 1961. Formation and early development of germ cells, pp. 1–24. *In* Symposium on Germ Cells and Development. Inst. Intern. d'Embriologie. Fondazione A. Baselli, Milan.

Mintz, B., and E. S. Russell. 1957. Gene-induced embryological modifications of primordial germ cells in the mouse. J. Exp. Zool. 134:207–237.

Moore, C. R., and H. Wang. 1947. Ovarian activity in mammals subsequent to chemical injury of cortex. Physiol. Zool. 20:300–321.

Moricard, R. 1958. Fonction méiogène et fonction oestrogène du follicule ovarian des mammifères (Cytologie golgienne, traceurs, microscopie électronique). Ann. Endocrinol. 19:943–967.

Morris, B., and M. D. Sass. 1966. The formation of lymph in the ovary. Proc. Roy. Soc., Ser. B 164:577–591.

Morrison, J. A. 1960. Ovarian characteristics in elk of known breeding history. J. Wildlife Manage. 24:297–307.

Morrison, J. A. 1971. Morphology of corpora lutea in the Uganda kob antelope, *Adenota kob thomasi* (Neumann). J. Reprod. Fert. 26:297–305.

Moss, S., T. R. Wren, and J. F. Sykes. 1954. Some histological and histochemical observations of the bovine ovary during the estrous cycle. Anat. Rec. 120:409–433.

Mossman, A. S., and H. W. Mossman. 1962. Ovulation, implantation, and fetal sex ratio in impala. Science 137:869.

Mossman, H. W. 1937. The thecal gland and its relation to the reproductive cycle: A study of the cyclic changes in the ovary of the pocket gopher, *Geomys bursarius* (Shaw). Am. J. Anat. 61:289–319.

Mossman, H. W. 1938. The homology of the vesicular ovarian follicles of the mamalian ovary with the coelom. Anat. Rec. 70:643–655.

Mossman, H. W. 1940. What is the red squirrel? Trans. Wis. Acad. Sci. Arts Lett. 32:123–134.

Mossman, H. W. 1952. The embryonic nature of the adult ovary, pp. 196–206. *In* E. T. Engle [ed.] Studies on Testis and Ovary, Eggs and Sperm. Charles C Thomas, Springfield, Ill.

Mossman, H. W. 1953. The genital system and the fetal membranes as criteria for mammalian phylogeny and taxonomy. J. Mammalogy 34:289–298.

Mossman, H. W. 1966. The rodent ovary, pp. 455–470. *In* I. W. Rowlands [ed.] Comparative Biology of Reproduction in Mammals. Symp. Zool. Soc. London, No. 15. Academic Press, London and New York.

Mossman, H. W. 1969. A critique of our progress toward understanding the biology of the mammalian ovary, pp. 187–205. *In* M. Diamond [ed.] Reproduction and Sexual Behavior. Indiana Univ. Press, Bloomington.

Mossman, H. W. 1971. Orientation and site of attachment of the blastocyst: A comparative study, pp. 49–57. *In* R. J. Blandau [ed.] The Biology of the Blastocyst. Univ. Chicago Press, Chicago.

Mossman, H. W., and I. Judas. 1949. Accessory corpora lutea, lutein cell origin, and the ovarian cycle in the Canadian porcupine. Am. J. Anat. 85:1–39.

Mossman, H. W., M. J. Koering, and D. Ferry, Jr. 1964. Cyclic changes of interstitial gland tissue of the human ovary. Am. J. Anat. 115:235–256.

Mossman, H. W., J. W. Lawlah, and J. A. Bradley. 1932. The male reproductive tract of the Sciuridae. Am. J. Anat. 51:89–155.

Motta, P. 1965. Sur l'ultrastructure des "corps de Call et Exner" dans l'ovaire du lapin. Z. Zellforsch. mikrosk. Anat. 68:308–319.

Motta, P. 1969. Electron microscope study of the human lutein cell with special reference to its secretory activity. Z. Zellforsch. mikrosk. Anat. 98:233–245.

Motta, P., and Z. Takeva. 1971. Histochemical demonstration of Δ^5-3β-hydroxysteroid dehydrogenase activity in the interstitial tissue of the guinea pig ovary during the estrous cycle and pregnancy. Fert. Steril. 22:378–382.

Motta, P., Z. Takeva, and D. Palermo. 1971. On the presence of cilia in different cells of the mammalian ovary, fetal and adult. Acta Anat. 78:591–603.

Muta, T. 1958. The fine structure of the interstitial cell in the mouse ovary studied with the electron microscope. Kurume Med. J. 5:167–185.

Myers, H. I., W. C. Young, and E. W. Dempsey. 1936. Graafian follicle development throughout the reproductive cycle in the guinea pig, with especial reference to changes during oestrus (sexual receptivity). Anat. Rec. 65:381–401.

Nahm, L. J., and F. F. McKenzie. 1937. The cells of the adrenal cortex of the ewe during the estrual cycle and pregnancy. Bull. Univ. Mo. Agr. Res. 251:1–20.

Nalbandov, A. V. 1964. Reproductive Physiology. W. H. Freeman and Co., San Francisco and London.

Neal, E. G., and R. J. Harrison. 1958. Reproduction in the European badger (*Meles meles* L.). Trans. Zool. Soc. London 29:67–131.

Neilson, D., G. S. Jones, J. D. Woodruff, and B. Goldberg. 1970. The innervation of the ovary. Obstet. Gynecol. Surv. 25:889–904.

Nelson, W. W., and R. R. Greene. 1953. The human ovary in pregnancy. Surg. Gynecol. Obstet. Int. Abstr. Surg. 97:1–22.

Nelson, W. W., and R. R. Greene. 1958. Some observations on the histology of the human ovary during pregnancy. Am. J. Obstet. Gynecol. 76:66–90.

Niklaus, S. 1950. Die Kerngrösse des Follikelepithels während des Sexualzyklus beim Borstenigel. Z. Zellforsch. mikrosk. Anat. 35:240–264.

Nomina Anatomica. 1966. 3d ed. Excerpta Medica Found., New York.

Nomina Anatomica Veterinaria. 1968. Int. Comm. Vet. Anat. Nomencl., Vienna. [Distributed by Dep. Anat., N.Y. State Vet. Coll., Ithaca.]

Nomina Embryologica. 1970. Fed. Am. Soc. Exp. Biol., Bethesda, Md.

Norrevang, A. 1968. Electron microscopic morphology of oogenesis, vol. 23, pp. 113–186. *In* G. H. Bourne and J. F. Danielli [eds.] International Review of Cytology. 23 vol. to date. Academic Press, New York.

Novak, E., and J. D. Woodruff. 1967. Gynecologic and Obstetric Pathology. 6th ed. Saunders, Philadelphia.

O'Donoghue, C. H. 1912. The corpus luteum in the non-pregnant *Dasyurus* and polyovular follicles in *Dasyurus*. Anat. Anz. 41:353–368.

O'Donoghue, C. H. 1916. On the corpora lutea and interstitial tissue of the ovary in the Marsupialia. Quart. J. Microsc. Sci. 61:433–473.

O'Donoghue, P. N. 1963. Reproduction in the female hyrax (*Dendrohyrax arborea ruwenzorii*). Proc. Zool. Soc. London 141:207–237.

Odor, D. L. 1960. Electron microscope studies on ovarian oocytes and unfertilized tubal ova in the rat. J. Biophys. Biochem. Cytol. 7:567–574.

Odor, D. L., and R. J. Blandau. 1969*a*. Ultrastructural studies on fetal and early postnatal mouse ovaries. I. Histogenesis and organogenesis. Am. J. Anat. 124:163–186.

Odor, D. L., and R. J. Blandau. 1969*b*. Ultrastructural studies on fetal and early postnatal mouse ovaries. II. Cytodifferentiation. Am. J. Anat. 125:177–216.

Oehler, I. E. 1951. Beitrag zur Kenntnis des Ovarialepithels und seiner Beziehungen zur Oogenesis. Untersuchungen an fetalen und kindlichen Ovarien. Acta Anat. 12:1–29.

O'Gara, B. W. 1969. Unique aspects of reproduction in the female pronghorn (*Antilocapra americana*). Am. J. Anat. 125:217–232.

Ohno, S., and J. B. Smith. 1964. Role of fetal follicular cells in meiosis of mammalian oöcytes. Cytogenetics (Basel) 3:324–333.

Ono, H., H. Satoh, M. Miyake, and Y. Fujimoto. 1969. Development of "ovarian adrenocortical cell nodule" in the horse. Exp. Rep. Equine Health Lab. (Tokyo) No. 6:59–90.

O'Shea, J. D. 1970. An ultrastructural study of smooth muscle-like cells in the theca externa of ovarian follicles in the rat. Anat. Rec. 167:127–140.

O'Shea, J. D. 1971. Smooth muscle-like cells in the theca externa of ovarian follicles in sheep. J. Reprod. Fert. 24:283–285.

Osvaldo-Decima, L. 1970. Smooth muscle in the ovary of the rat and monkey. J. Ultrastruct. Res. 30:218–237.

Owen, R. 1850. On the anatomy of the Indian rhinoceros. Trans. Zool. Soc. London 4:31–58.

Owen, R. 1868. The anatomy of vertebrates. 3. Mammals, p. 693. 3 vol. Longmans, Green and Co., London.

Owman, C., and N.-O. Sjöberg. 1966. Adrenergic nerves in the female genital tract of the rabbit. With remarks on cholinesterase-containing structures. Z. Zellforsch. mikrosk. Anat. 74:182–197.

Owman, C., E. Rosengren, and N.-O. Sjöberg. 1967. Adrenergic innervation of the human female reproductive organs. A histochemical and chemical investigation. Obstet. Gynecol. 30:763–773.

Ożdżeński, W. 1967. Observations on the origin of primordial germ cells in the mouse. Zool. Pol. 17:367–379.

Pansky, B., and H. W. Mossman. 1953. The regenerative capacity of the rabbit ovary. Anat. Rec. 116:19–40.

Parkes, A. S., U. Fielding, and F. W. R. Brambell. 1927. Ovarian regeneration in the mouse after complete double ovariotomy. Proc. Roy. Soc., Ser. B 101:328–354.

Pearson, A. K., and R. K. Enders. 1951. Further observations on the reproduction of the Alaskan fur seal. Anat. Rec. 111:695–711.

Pearson, O. P. 1944. Reproduction in the shrew (*Blarina brevicauda* Say). Am. J. Anat. 75:39–93.

Pearson, O. P. 1949. Reproduction in a South American rodent, the mountain viscacha. Am. J. Anat. 84:143–173.

Pearson, O. P., and P. H. Baldwin. 1953. Reproduction and age structure of a mongoose population in Hawaii. J. Mammalogy 34:436–447.

Pearson, O. P., and R. K. Enders. 1943. Ovulation, maturation and fertilization in the fox. Anat. Rec. 85:69–83.

Pearson, O. P., M. R. Koford, and A. K. Pearson. 1952. Reproduction in the lump-nosed bat (*Corynorhinus rafinesquei*) in California. J. Mammalogy 33:273–320.

Pederson, E. S. 1951. Histogenesis of lutein tissue of the albino rat. Am. J. Anat. 88:397–428.

Pencharz, R. I. 1929. Experiments concerning ovarian regeneration in the white rat and white mouse. J. Exp. Zool. 54:319–341.

Perry, J. S. 1953. The reproduction of the African elephant, *Loxodonta africana.* Phil. Trans. Roy. Soc. London, Ser. B 237:93–149.

Personen, S. 1942. Über die Beziehung des Gelbkörpers zur Uteruschleimhaut und Ernährung des Embryos. Ann. Zool. Fenn. 9:1–99.

Peters, H. 1969. Development of the mouse ovary from birth to maturity. Acta Endocrinol. 62:98–116.

Peters, H. 1970. Migration of gonocytes into the mammalian gonad and their differentiation. Phil. Trans. Roy. Soc. London, Ser. B 259:91–101.

Peters, H., and E. Levy. 1966. Cell dynamics of the ovarian cycle. J. Reprod. Fert. 11:227–236.

Petten, J. L. 1933. Beitrag zur Kenntnis der Entwicklung des Pferdeovariums. Z. Anat. Entwicklungsgesch. 99:338–383.

Petter-Rousseaux, A. 1962. Recherches sur la biologie de la reproduction des primates inférieurs. Thesis. Fac. Med., Univ. Paris. 87 p.

Petter-Rousseaux, A., and F. Bourlière. 1965. Persistance des phénomènes d'ovogénèse chez l'adulte de *Daubentonia madagascariensis* (Prosimii, Lemuriformes). Folia Primatol. 3:241–244.

Peyre, A. 1953. Note sur la structure histologique de l'ovaire du desman Pyrénées (*Galemys pyrenaica* G.). Bull. Soc. Zool. Fr. 77:441–447.

Pfeiffer, E. W. 1956. The male reproductive tract of a primitive rodent, *Aplodontia rufa*. Anat. Rec. 124:629–637.

Pfeiffer, E. W. 1958. The reproductive cycle of the female mountain beaver. J. Mammalogy 39:223–235.

Pflüger, E. F. W. 1863. Ueber die Eierstöcke der Säugethiere und des Menschen. W. Engelmann, Leipzig.

Pilton, P. E., and G. B. Sharman. 1962. Reproduction in the marsupial *Trichosurus vulpecula*. J. Endocrinol. 25: 119–136.

Pimlott, D. H. 1959. Reproduction and productivity of Newfoundland moose. J. Wildlife Manage. 23:381–401.

Pincus, G., and E. V. Enzmann. 1937. The growth, maturation and atresia of ovarian eggs in the rabbit. J. Morphol. 61:351–383.

Pines, L., and B. Schapiro. 1930. Über die Innervation des Eierstockes. Z. mikrosk.-anat. Forsch. 20:327–372.

Pinho, A. V. de. 1925. Sur une forme particulière de transformation folliculaire caractéristique de l'ovaire du lérot (*Eliomys quercinus* L.): Faux corps jaunes métaplastiques. Anat. Rec. 30:211–220.

Pinkerton, J. H. M., D. C. McKay, E. C. Adams, and A. T. Hertig. 1961. Development of the human ovary — a study using histochemical techniques. Obstet. Gynecol. 18:152–181.

Poirier, P., and A. Charpy. 1912. Lymphatiques de l'ovaire, vol. 2, p. 1199 and vol. 5, p. 370. *In* Traité d'Anatomie Humaine. 3d ed. 5 vol. Masson and Co., Paris.

Polano, O. 1903. Beiträge zur Anatomie der Lymphbahnen in menschlichen Eierstock. Monatsschr. Geburtsh. Gynaekol. 17:281–295, 466–496.

Popoff, N. 1911. Le tissu interstitiel et les corps jaunes de l'ovaire. Arch. Biol. 26:483–556.

Potter, E. L. 1963. The ovary in infancy and childhood, pp. 11–23. *In* H. G. Grady and D. E. Smith [eds.] The Ovary. Int. Acad. Pathol. Monogr. No. 3. Williams and Wilkins, Baltimore.

Prasad, M. R. N. 1957. Male genital tract of the Indian and Ceylonese palm squirrels and its bearing on the systematics of the Sciuridae. Acta Zool. (Stockholm) 38:1–26.

Price, M. 1953. The reproductive cycle of the water shrew, *Neomys fodiens bicolor* Shaw. Proc. Zool. Soc. London 123:599–621.

Priedkalns, J., and A. F. Weber. 1968*a*. Ultrastructural studies of the bovine graafian follicle and corpus luteum. Z. Zellforsch. mikroskop. Anat. 91:554–573.

Priedkalns, J., and A. F. Weber. 1968*b*. Quantitative ultrastructural analysis of the follicular and luteal cells of the bovine ovary. Z. Zellforsch. mikrosk. Anat. 91:574–585.

Provost, E. E. 1962. Morphological characteristics of the beaver ovary. J. Wildlife Manage. 26:272–278.

Pycraft, W. P. 1932. On the genital organs of the common dolphin (*Delphinus delphis*). Proc. Zool. Soc. London (1932):807–811.

Ramaswami, L. S., and T. C. A. Kumar. 1965. Some aspects of reproduction of the female slender loris, *Loris tardigradus lydekkerianus* Cabr. Acta Zool. (Stockholm) 46:257–273.

Ramaswamy, K. R. 1961. Studies on the sex-cycle of the Indian vampire bat, *Megaderma* (*Lyroderma*) *lyra lyra*. Proc. Nat. Inst. Sci. India, Sect. B 27:285–307.

Rand, R. W. 1955. Reproduction in the female Cape fur seal, *Arctocephalus pusillus* (Schreber). Proc. Zool. Soc. London 124:717–740.

Rankin, J. J. 1961. The bursa ovarica of the beaked whale *Mesoplodon gervaisi* Deslongchamps. Anat. Rec. 139:379–385.

Rao, C. R. N. 1927. On the structure of the ovary and the ovarian ovum of *Loris lydekkerianus* Cabr. Quart. J. Microsc. Sci. 71:57–74.

Raps, G. 1948. The development of the dog ovary from birth to six months of age. Am. J. Vet. Res. 9:61–64.

Rasmussen, A. T. 1918. Cyclic changes in the interstitial cells of the ovary and testis of the woodchuck (*Marmota monax*). Endocrinology 2:353–404.

Regaud, C., and A. Lacassagne. 1913. Les follicules anovulaires de l'ovaire chez la lapine adulte. C. R. Ass. Anat. 15:15–27.

Rhodin, J. A. G. 1963. An Atlas of Ultrastructure. W. B. Saunders Co., Philadelphia.

Richardson, G. S. 1967. Ovarian Physiology. Little, Brown and Co., Boston.

Robinette, W. L., and G. F. T. Child. 1964. The Puku. N. Rhodesia Dep. Game Fish., Occ. Pap. 2:84.

Robinson, A. 1918. The formation, rupture and closure of ovarian follicles in ferret and ferret-polecat hybrids and some associated phenomena. Trans. Roy. Soc. Edinburgh 52:302–362.

Rocereto, T., D. Jacobowitz, and E. E. Wallach. 1969. Observations on the spontaneous contracture of the cat ovary *in vitro*. Endocrinology 84:1336–1341.

Rolle, G. K., and H. A. Charipper. 1949. The effects of advancing age upon the histology of the ovary, uterus and vagina of the female golden hamster (*Cricetus auratus*). Anat. Rec. 105:281–297.

Rondell, P. 1964*a*. Follicular fluid electrolytes in ovulation. Proc. Soc. Exp. Biol. Med. 116:336–339.

Rondell, P. 1964*b*. Follicular pressure and distensibility in ovulation. Am. J. Physiol. 207:590–594.

Rondell, P. 1970. Biophysical aspects of ovulation. Biol. Reprod. Suppl. 2:64–89.

Rood, J. D., and B. J. Weir. 1970. Reproduction in female wild guinea-pigs. J. Reprod. Fert. 23:393–409.

Rosenblum, L. A. 1968. Some aspects of female reproductive physiology in the squirrel monkey, pp. 147–169. *In* L. A. Rosenblum and R. W. Cooper [eds.] The Squirrel Monkey. Academic Press, New York and London.

Rouvière, H. 1938. Anatomy of the Lymphatic System. [Transl. from the French by M. J. Tobias.] Edwards Bros., Ann Arbor, Mich.

Rowlands, I. W. 1956. The corpus luteum of the guinea-pig, pp. 69–83. *In* G. E. W. Wolstenholme and E. C. P. Millar [eds.] Ciba Found. Colloq. Ageing 2. Little, Brown and Co., Boston.

Rowlands, I. W., and R. B. Heap. 1966. Histological observations on the ovary and progesterone levels of the coypu, *Myocastor coypus*, pp. 335–352. *In* I. W. Rowlands [ed.] Comparative Biology of Reproduction in Mammals. Symp. Zool. Soc. London, No. 15. Academic Press, London and New York.

Rowlands, I. W., and A. S. Parkes. 1935. The reproductive processes of certain mammals. VIII. Reproduction in foxes (*Vulpes* spp.). Proc. Zool. Soc. London (1935):823–841.

Rowlands, I. W., W. H. Tam, and D. G. Kleiman. 1970. Histological and biochemical studies on the ovary and of progesterone levels in the systemic blood of the green acouchi (*Myoprocta pratti*). J. Reprod. Fert. 22:533–545.

Rubin, B. L., H. W. Deane, and K. Balogh. 1969. Ovarian steroid biosynthesis and $\Delta^5$3-β- and 20α-hydroxysteroid dehydrogenase activity. Trans. N.Y. Acad. Sci., Ser. II 31:787–802.

Rudkin, G. T., and H. A. Griech. 1962. On the persistence of oocyte nuclei from fetus to maturity in the laboratory mouse. J. Cell Biol. 12:169–175.

Ruge, C. 1913. Ueber Ovulation, Corpus luteum and Menstruation. Arch. Gynaekol. 100:20–48.

Rugh, R. 1968. The Mouse. Its Reproduction and Development. Burgess Publ. Co., Minneapolis.

Saglik, S. 1938. Ovaries of gorilla, chimpanzee, orang-utan and gibbon. Contrib. Embryol. Carnegie Inst. Washington 27:179–189.

Sainmont, G. 1906. Recherches relatives à l'organogenèse du testicule et de l'ovaire chez le chat. Arch. Biol. 22:71–162.

Santoro, A. 1965. On the fine structure of the theca interna cells of the rabbit ovary. Boll. Soc. Ital. Biol. Sper. 40:1636–1637.

Sauramo, H. 1954. Histology and function of the ovary from the embryonic period to the fertile age. Acta Obstet. Gynecol. Scand. 33 (Suppl. 2): 3–25.

Savard, K. 1968. The biogenesis of steroids in the human ovary, pp. 10–26. *In* H. C. Mack [ed.] The Ovary. Charles C Thomas, Springfield, Ill.

Schaüder, W. 1929. Zur vergleichenden Anatomie der inneren weiblichen Geschlechtsorgane, embryonalen Anhangsorgane und Plazenta des Pferdes und Tapirs, pp. 273–283. *In* Hermann Baum — Zur feier seines 65 Geburtstages in Verehrung und Dankbarkeit gewidmet von Kollegen, Freunden und Schülern. M. and H. Schaper, Hanover.

Schmidt, I. G. 1942. Mitotic proliferation in the ovary of the normal mature guinea pig treated with colchicine. Am. J. Anat. 71:245–270.

Schmidt, I. G., and F. G. Hoffman. 1941. Proliferation and ovogenesis in the germinal epithelium of the normal mature guinea pig ovary, as shown by the colchicine technique. Am. J. Anat. 68:263–273.

Schrön, O. 1863. Beitrag zur Kenntniss der Anatomie und Physiologie des Eierstocks der Säugethiere. Z. Wiss. Zool. 12:409–426.

Seliger, W. G., A. J. Blair, and H. W. Mossman. 1966. Differentiation of adrenal cortex-like tissue at the hilum of the gonads in response to adrenalectomy. Am. J. Anat. 118:615–629.

Selye, H. 1946. Ovary. Vol. 7, Ovarian Tumors, sect. 4. *In* Encyclopedia of Endocrinology. Vol. 1–4, 7. Richardson, Bond, and Wright, Montreal.

Sergeant, D. E. 1962. The biology of the pilot or pothead whale, *Globicephala melaena* (Traill) in Newfoundland waters. Fish. Res. Board Can. Bull. 132:1–84.

Seth, P., and M. R. N. Prasad. 1967. Effect of bilateral adrenalectomy on the ovary of the five striped Indian palm squirrel, *Funambulus pennanti* (Wroughton). Gen. Comp. Endocrinol. 8:152–162.

Seth, P., and M. R. N. Prasad. 1969. Reproductive cycle of the female five-striped Indian palm squirrel, *Funambulus pennanti* (Wroughton). J. Reprod. Fert. 20:211–222.

Shann, E. W. 1923. The embryonic development of the porbeagle shark, *Lamna cornubica*. Proc. Zool. Soc. London (1923):161–171.

Sherman, H. B. 1937. Breeding habits of the free-tailed bat. J. Mammalogy 18:176–187.

Short, R. V. 1966. Oestrous behaviour, ovulation and the formation of the corpus luteum in the African elephant, *Loxodonta africana*. East Afr. Wildlife J. 4:56–68.

Short, R. V. 1969. Notes on the teeth and ovaries of an African elephant (*Loxodonta africana*) of known age. J. Zool. (London) 158:421–426.

Short, R. V., and I. O. Buss. 1965. Biochemical and histological observations on the corpora lutea of the African elephant, *Loxodonta africana*. J. Reprod. Fert. 9:61–68.

Short, R. V., and M. F. Hay. 1966. Delayed implantation in the roe deer, *Capreolus capreolus*, pp. 173–194. *In* I. W. Rowlands [ed.] Comparative Biology of Reproduction in Mammals. Symp. Zool. Soc. London, No. 15. Academic Press, London and New York.

Simkins, C. S. 1932. Development of the human ovary from birth to sexual maturity. Am. J. Anat. 51:465–505.

Simon, Doris. 1960. Contribution à l'étude de la circulation et du transport des gonocytes primaires dans les blastodermes d'oiseau cultivés in vitro. Arch. Anat. Microsc. Morphol. Exp. 49:93–176.

Simpson, G. G. 1945. The principles of classification and a classification of mammals. Bull. Am. Mus. Natur. Hist. 85:1–350.

Sinha, A. A., C. H. Conaway, and K. W. Kenyon. 1966. Reproduction in the female sea otter. J. Wildlife Manage. 30:121–130.

Sinha, A. A., U. S. Seal, and R. P. Doe. 1971*a*. Fine structure of the corpus luteum of the raccoon during pregnancy. Z. Zellforsch. mikrosk. Anat. 117:35–45.

Sinha, A. A., U. S. Seal, and R. P. Doe. 1971*b*. Ultrastructure of the corpus luteum of the white-tailed deer during pregnancy. Am. J. Anat. 132:189–206.

Sneider, M. E. 1940. Rhythms of ovogenesis before sexual maturity in the rat and cat. Am. J. Anat. 67:471–499.

Snell, G. D. [ed.] 1941. Biology of the Laboratory Mouse. Blakiston, Philadelphia.

Sobotta, J. 1896. Ueber die Bildung des Corpus luteum bei der Maus. Arch. mikrosk. Anat. Entwicklungsmech. 47:261–308.

Sobotta, J. 1906. Über die Bildung des Corpus luteum beim Meerschweinchen. Anat. Hefte 32:89–142.

Sotelo, J. R., and K. R. Porter. 1959. An electron microscope study of the rat ovum. J. Biophys. Biochem. Cytol. 5:327–341.

Soupart, P., and R. W. Noyes. 1964. Sialic acid as a component of the zona pellucida of the mammalian ovum. J. Reprod. Fert. 8:251–253.

Stafford, W. T., and H. W. Mossman. 1945. The ovarian interstitial gland tissue and its relation to the pregnancy cycle in the guinea pig. Anat. Rec. 93:97–107.

Stafford, W. T., C. F. Collins, and H. W. Mossman. 1942. The thecal gland in the guinea pig ovary. Anat. Rec. 83:193–207.

Stearns, M. L. 1940. Studies on the development of connective tissue in transparent chambers in the rabbit's ear. I & II. Am. J. Anat. 66:133–176; 67:55–97.

Stegner, H. E. 1967. Die electronmikroskopische Struktur der Eizelle. Ergeb. Anat. Entwicklungsgesch. 39:1–112.

Stegner, H. E., and H. Wartenberg. 1961. Electronenmikroskopische und histotopochemische Untersuchungen über Struktur und Bildung der Zona pellucida menschlicher Eizellen. Z. Zellforsch. mikrosk. Anat. 53:702–713.

Sternberg, W. H., A. Segaloff, and C. J. Gaskill. 1953. Influence of chorionic gonadotropin on ovarian hilus cells. (Leydig-like cells). J. Clin. Endocrinol. Metab. 13:139–153.

Stockard, A. H. 1937. Studies on the female reproductive system of the prairie dog, *Cynomys leucurus*. 2. Normal cyclic phenomena of the ovarian follicles. Pap. Mich. Acad. Sci. Arts Lett. 22:671–689.

Strassmann, I. O. 1961. The theca cone: The pathmaker of growing human and mammalian follicles. Int. J. Fert. 6:135–142.

Stratz, K. H. 1898. Der geschlechtsreife Säugethiereierstock. M. Nijhoff, The Hague.

Strauss, F. 1938. Die Befruchtung und der Vorgang der Ovulation bei *Ericulus* aus der Familie der Centetiden. Biomorphosis 1:281–312.

Strauss, F. 1939. Die Bildung des Corpus luteum bei Centetiden. Biomorphosis 1:489–544.

Strauss, F. 1966. Weibliche Geschlechtsorgane, vol. 8, pp. 1–96. *In* J.-G. Helmcke, H. v. Lengerken, D. Starck, and H. Wermuth [eds.] Handbuch der Zoologie. 8 vol. to date. Walter de Gruyter and Co., Berlin.

Stricht, O. van der. 1912. Sur le processus de l'excretion des glandes endocrines; le corps jaune et la glande interstitielle de l'ovaire. Arch. Biol. 27:585–721.

Sturgis, S. H. 1949. Rate and significance of atresia of the ovarian follicle of the rhesus monkey. Contrib. Embryol. Carnegie Inst. Washington 33: 67–70.

Tam, W. H. 1970. Function of the accessory corpora lutea in the hystricomorph rodents. J. Endocrinol. 48:liv–lv. (Abstr.)

Tarkowski, A. K. 1957. Studies on reproduction and prenatal mortality of the common shrew (*Sorex araneus* L.). II. Reproduction under natural conditions. Ann. Univ. Mariae Curie-Sklodowska, Sect. C Biol. 10:177–244.

Tarkowski, A. K. 1970. Germ cells in natural and experimental chimeras in mammals. Phil. Trans. Roy. Soc. London, Ser. B 259:107–111.

Teilum, G. 1971. Special Tumors of Ovary and Testis, and Related Extragonadal Lesions. L. B. Lippincott Co., Philadelphia and Toronto.

Thwaites, C. J., and T. N. Edey. 1970. Histology of the corpus luteum of the ewe: Changes during the estrous cycle and early pregnancy, and in response to some experimental treatments. Am. J. Anat. 129:439–447.

Tripp, H. R. H. 1971. Reproduction in elephant-shrews (Macroscelididae) with special reference to ovulation and implantation. J. Reprod. Fert. 26:149–159.

Tsukaguchi, R., and T. Okamoto. 1928. Der Ursprung der interstitiellen Zellen des Ovariums beim Hunde. Folia Anat. Jap. 6:663–686.

Unsicker, K. 1970. Zur Innervation der interstitiellen Drüse im Ovar der Maus (*Mus musculus*, L.). Eine fluoreszenz- und electronenmikroskopische Studie. Z. Zellforsch. mikrosk. Anat. 109:46–54.

Uyttenbroeck, F., and M. Van der Schueren-Lodeweyckx. 1969. Luteinization and corpus luteum formation. A comparative study with findings in animals. Acta zool. pathol. Antverpiensia 48:97–121.

Valdes-Dapena, M. A. 1967. The normal ovary of childhood. Ann. N.Y. Acad. Sci. 142:597–613.

Vanneman, A. S. 1917. The early history of the germ cells in the armadillo, *Tatusia novemcincta*. Am. J. Anat. 22:341–357.

Velloso de Pinho, A. See Pinho, A. V. de.

Vermande-Van Eck, G. J. 1956. Neo-ovogenesis in the adult monkey; consequences of atresia of ovocytes. Anat. Rec. 125:207–224.

Verts, B. J. 1967. The Biology of the Striped Skunk. Univ. Illinois Press, Urbana, Chicago, and London.

Völker, O. 1905. Ueber die Histogenese des Corpus luteum beim Ziesel (*Spermophilus cit.*). Arch. Anat. Physiol. (1905):301–320.

Wagenen, G. van, and M. E. Simpson. 1965. Embryology of the Ovary and Testis (*Homo sapiens* and *Macaca mulatta*). Yale Univ. Press, New Haven.

Walczak, M., and J. Pieńkowska-Mikotajczyk. 1960. A case of rare feminizing tumor of the ovary — interstitioma ovarii. Acta Med. Pol. 1:197–202.

Waldeyer-Hartz, W. 1870. Eierstock und Ei. W. Engelmann, Leipzig.

Walker, E. P. 1964. Mammals of the World. 1st ed. 3 vol. Johns Hopkins Press, Baltimore.

Wallart, J. 1927. Sur le tissu paraganglionnaire de l'ovaire humain. Arch. Anat. Histol. Embryol. 7:3–39.

Wallart, J. 1936. Sur l'innervation de l'epoophoron. Arch. Biol. 47:87–90.

Warbritton, V. 1934. The cytology of the corpora lutea of the ewe. J. Morphol. 56:181–202.

Ward, M. C. 1946. A study of the estrous cycle and the breeding of the golden hamster, *Cricetus auratus*. Anat. Rec. 94:139–161.

Warszawsky, L. F., W. G. Parker, N. L. First, and O. J. Ginther. 1971. Gross changes of internal genitalia during the estrous cycle in the mare. Am. J. Vet. Res. 33:19–26.

Watzka, M. 1940. Mikroskopisch-anatomische Untersuchungen über die Ranzzeit und Tragdauer des Hermelins (*Putorius ermineus*). Z. mikrosk.-anat. Forsch. 48:359–374.

Watzka, M. 1949. Ueber die Beziehung zwischen Corpus luteum und verlängerter Tragzeit. Z. Anat. Entwicklungsgesch. 114:366–374.

Watzka, M. 1957. Weibliche genitalorgane. Das Ovarium, vol. 7, pp. 1–178. *In* W. v. Möllendorf and W. Bargmann [eds.] Handbuch der mikroskopischen Anatomie des Menschen. 7 vol. Springer, Berlin.

Watzka, M., and J. Eschler. 1933. Extraglanduläre Zwischenzellen im Eierstockhilus des Schweines. Z. mikrosk.-anat. Forsch. 34:238–248.

Weakley, B. S. 1968. Comparison of cytoplasmic lamellae and membranous elements in the oocytes of five mammalian species. Z. Zellforsch. mikrosk. Anat. 85:109–123.

Weakley, B. S. 1969*a*. Differentiation of the surface epithelium of the hamster ovary. An electron microscopic study. J. Anat. 105:129–147.

Weakley, B. S. 1969*b*. Initial stages in the formation of cytoplasmic lamellae in the hamster oocyte and identification of associated electron-dense particles. Z. Zellforsch. mikrosk. Anat. 97:438–448.

Weir, B. J. 1967. Aspects of reproduction in some hystricomorph rodents. Ph.D. thesis, Univ. of Cambridge.

Weir, B. J. 1970. Some observations on reproduction in the female green acouchi, *Myoprocta pratti*. J. Reprod. Fert. 24:193–201.

Weir, B. J. 1971*a*. Some observations on reproduction in the female agouti, *Dasyprocta aguti*. J. Reprod. Fert. 24:203–211.

Weir, B. J. 1971*b*. The reproductive organs of the female plains viscacha, *Lagostomus maximus*. J. Reprod. Fert. 25:365–373.

Weir, B. J. 1971*c*. Some notes on reproduction in the Patagonian mountain viscacha, *Lagidium boxi* (Mammalia: Rodentia). J. Zool. (London) 164:463–467.

Weismann, A. 1883. Die Entstehung der Sexualzellen bei den Hydromedusen. Gustav Fischer, Jena.

Weismann, A. 1892. Das Keimplasma. Eine Theorie der Vererbung. Gustav Fischer, Jena.

White, R. F., A. T. Hertig, J. Rock, and E. C. Adams. 1951. Histological and histochemical observations on the corpus luteum of human pregnancy with special reference to corpora lutea associated with early normal and abnormal ova. Contrib. Embryol. Carnegie Inst. Washington. 34:55–74.

Wilcox, D. E., and H. W. Mossman. 1945. The common occurrence of "testis" cords in the ovaries of a shrew (*Sorex vagrans*, Baird). Anat. Rec. 92:183–195.

Wimsatt, W. A. 1944. Growth of the ovarian follicle and ovulation in *Myotis lucifugus lucifugus*. Am. J. Anat. 74:129–174.

Wimsatt, W. A. 1963. Delayed implantation in the Ursidae, with particular reference to the black bear (*Ursus americanus* Pallas), pp. 49–76. *In* A. C. Enders [ed.] Delayed Implantation. Univ. Chicago Press, Chicago.

Wimsatt, W. A. [ed.] 1970. Biology of Bats. Academic Press, New York.

Wimsatt, W. A., and F. C. Kallen. 1957. The unique maturation response of the Graafian follicles of hibernating bats and the question of its significance. Anat. Rec. 129:115–131.

Wimsatt, W. A., and H. F. Parks. 1966. Ultrastructure of the surviving follicle of hibernation and the ovum-follicle cell relationship in the vespertilionid bat, *Myotis lucifugus*, pp. 419–454. *In* I. W. Rowlands [ed.] Comparative Biology of Reproduction in Mammals. Symp. Zool. Soc. London, No. 15. Academic Press, London and New York.

Wimsatt, W. A., and H. Trapido. 1952. Reproduction and the female reproductive cycle in the tropical American vampire bat, *Desmodus rotundus murinus*. Am. J. Anat. 91:415–445.

Wimsatt, W. A., and C. M. Waldo. 1945. The normal occurrence of a peritoneal opening in the bursa ovarii of the mouse. Anat. Rec. 93:47–57.

Winiwarter, H. [de]. 1901. Recherches sur l'ovogenèse et l'organogenèse de l'ovaire des mammifères (lapin et homme). Arch. Biol. 17:33–199.

Winiwarter, H. de. 1942. Y a-t-il néoformation d'ovules dans l'ovaire des mammifères adultes? Arch. Biol. 53:259–280.

Winiwarter, H. [de], and G. Sainmont. 1908–1909. Nouvelles recherches sur l'ovogenèse et l'organogenèse de l'ovaire des mammifères (chat). Arch. Biol. 24:1–142, 165–276, 373–431, 628–650.

Wislocki, G. B. 1930. On a series of placental stages of a platyrrhine monkey (*Ateles geoffroyi*) with some remarks upon age, sex and breeding period in platyrrhines. Contrib. Embryol. Carnegie Inst. Washington 133:173–192.

Wislocki, G. B. 1931. Notes on the female reproductive tract (ovaries, uterus and placenta) of the collared peccary (*Pecari angulatus bangsi* Goldman). J. Mammalogy 12:143–149.

Wislocki, G. B. 1932. On the female reproductive tract of the gorilla with a comparison of that of other primates. Contrib. Embryol. Carnegie Inst. Washington 23:165–204.

Wislocki, G. B. 1939. Observations on twinning in marmosets. Am. J. Anat. 64:445–483.

Wislocki, G. B., and E. W. Dempsey. 1939. Remarks on the lymphatics of the reproductive tract of the female rhesus monkey (*Macaca mulatta*). Anat. Rec. 75:341–363.

Wislocki, G. B., and O. P. van der Westhuysen. 1940. The placentation of *Procavia capensis*, with a discussion of the placental affinities of the Hyracoidea. Contrib. Embryol. Carnegie Inst. Washington 28:67–88.

Witschi, E. 1948. Migration of the germ cells of human embryos from the yolk sac to the primitive gonadal folds. Contrib. Embryol. Carnegie Inst. Washington 32:67–97.

Witschi, E. 1963. Embryology of the ovary, pp. 1–10. *In* H. G. Grady and D. E. Smith [eds.] The Ovary. Int. Acad. Pathol. Monogr. No. 3. Williams and Wilkins Co., Baltimore.

Wotton, R. M., and P. A. Village. 1951. The transfer function of certain cells in the wall of the Graafian follicle as revealed by their reaction to previously stained fat in the cat. Anat. Rec. 110:121–127.

Wright, P. L. 1942. Delayed implantation in the long-tailed weasel (*Mustela frenata*), the short-tailed weasel (*M. cicognani*) and the marten (*Martes americana*). Anat. Rec. 83:341–353.

Wright, P. L. 1963. Variations in reproductive cycles in North American mustelids. pp. 77–97. *In* A. C. Enders [ed.] Delayed Implantation. Univ. Chicago Press, Chicago.

Wright, P. L. 1966. Observations on the reproductive cycle of the American badger (*Taxidea taxus*), pp. 27–45. *In* I. W. Rowlands [ed.] Comparative Biology of Reproduction in Mammals. Symp. Zool. Soc. London, No. 15. Academic Press, London and New York.

Young, W. C. 1944. Genital cycles following incomplete oöphorectomy in the chimpanzee. Anat. Rec. 89:457–493.

Young, W. C. 1961. The mammalian ovary, pp. 449–496. *In* W. C. Young [ed.] Sex and Internal Secretions. 3d ed. Vol. 1. Williams and Wilkins, Baltimore.

Young, W. C., E. W. Dempsey, H. I. Myers, and C. W. Hagquist. 1938. The ovarian condition and sexual behavior in the female guinea pig. Am. J. Anat. 63:457–487.

Zachariae, F., and C. E. Jensen. 1958. Histochemical and physicochemical investigations on genuine follicular fluid. Acta Endocrinol. 27:343–355.

Zalesky, M. 1934. A study of the seasonal changes in the adrenal gland of the thirteen-lined ground squirrel (*Citellus tridecemlineatus*), with particular reference to its sexual cycle. Anat. Rec. 60:291–321.

Zamboni, L. 1970. Ultrastructure of mammalian oocytes and ova. Biol. Reprod. Suppl. 2:44–63.

Zamboni, L., D. R. Mishell, Jr., J. H. Bell, and M. Baca. 1966. Fine structure of the human ovum in the pronuclear stage. J. Cell Biol. 30:579–600.

Zietzschmann, O., E. Ackerknecht, and H. Grau. 1943. Ellenberger-Baum Handbuch der vergleichenden Anatomie der Haustiere. Springer, Berlin.

Zuckerkandl, E. 1897. Zur vergleichenden Anatomie der Ovarialtaschen. Anat. Hefte 8:707–799.

Zuckerman, S. 1962. The Ovary. Academic Press, London and New York.

Zuckerman, S., and A. S. Parkes. 1932. The menstrual cycle of the primates. Part V. The cycle of the baboon. Proc. Zool. Soc. London (1932):138–191.

Index

Page numbers directly following the scientific and vernacular names of taxa refer to either the Synoptic Tables (pp. 298–347) or Supplementary Notes (pp. 348–87), or both. Italicized numbers refer to pages on which pertinent figures occur. Latinized anatomical terms found in the *Nomina Anatomica*, *Nomina Embryologica*, and *Nomina Anatomica Veterinaria* are indicated by the appropriate initials *NA*, *NE*, and *NAV*.

DESIGNED BY WILLIAM NICOLL OF EDIT, INC.

COMPOSED BY THE NORTH CENTRAL PUBLISHING CO., ST. PAUL, MINNESOTA

PRINTED BY MERIDEN GRAVURE CO., MERIDEN, CONNECTICUT

BOUND BY GEORGE BANTA, INC., MENASHA, WISCONSIN

TEXT IS SET IN LINOTYPE BODONI BOOK

DISPLAY LINES IN BULMER, HELVETICA

AND HELVETICA BOLD

Library of Congress Cataloging in Publication Data
Mossman, Harland Winfield, 1898–
Comparative morphology of the mammalian ovary.
Bibliography: p. 399–434
1. Ovaries. 2. Anatomy, Comparative. 3. Mammals — Anatomy.
I. Duke, Kenneth Lindsay, 1912– joint author. II. Title.
QL881.M67 599'.04'6 72–143765
ISBN 0-299-05930-8

Richter 284-3232